Cold Regions Hydrology in a Changing Climate

Cold Regions Hydrology in a Changing Climate

Edited by

DAQING YANG
National Hydrology Research Center, Canada

PHILIP MARSH
National Hydrology Research Center, Canada

ALEXANDER GELFAN
Water Problem Institute, Russia

Proceedings of an international symposium, H02, held during IUGG2011, the XXV General Assembly of the International Union of Geodesy and Geophysics, at Melbourne, Australia, July 2011.

IAHS Publication 346
in the IAHS Series of Proceedings and Reports

Published by the International Association of Hydrological Sciences 2011

IAHS Publication 346

ISBN 978-1-907161-21-6

British Library Cataloguing-in-Publication Data.
A catalogue record for this book is available from the British Library.

The papers included in this volume have been reviewed and some were extensively revised by the Editors, in collaboration with the authors, prior to publication.

IAHS is indebted to the employers of the Editors for the invaluable support and services provided that enabled them to carry out their task effectively and efficiently.

Publications in the series of Proceedings and Reports are available from:
IAHS Press, Centre for Ecology and Hydrology, Wallingford, Oxfordshire OX10 8BB, UK
tel.: +44 1491 692442; fax: +44 1491 692448; e-mail: jilly@iahs.demon.co.uk

Printed by Information Press

Preface

The high latitude and lowland cold regions of the globe are experiencing some of the most rapid changes in climate. These regions include many of the most severely ungauged basins on Earth and suffer from sparse meteorological observations. Hydrology of these regions is dominated by snow and ice. Our understanding of the hydrological response to a changing climate over these cold regions is incomplete due to a lack of understanding of the controlling processes, and a paucity of hydrological and meteorological observations. Changes in hydrology related to changing frozen soils, snowfall/rainfall ratio, snow cover, river and lake ice, glacier cover and vegetation are not well known. In addition, our ability to model the effect of these changes on the fluxes of both energy and water between the land surface and the atmosphere, and soil and water bodies needs improvement. For example, a particular issue for modelling is the impracticability of model calibration due to the sparse gauge network and rapid climate change. There is also a lack of knowledge on process emergence with scale change across these regions.

To address the major issues and challenges in cold regions hydrology research and applications, a special symposium (H02) on *Cold Regions Hydrology in a Changing Climate* was organized by the IAHS International Commission on Snow and Ice Hydrology and the Predictions in Ungauged Basins (PUB) initiative at the 2011 IUGG Assembly. The emphasis of this symposium was on snow and ice hydrology, in particular, changes in the characteristics and functioning of rivers, lakes and wetlands in cold regions, and their interactions with changing human activities and ecosystems. This symposium also explored the biological, physical, and social impacts of hydrological and climatic change in the cold regions.

This book presents 28 papers from the symposium. These papers, coming from colleagues in 14 countries around the World, clearly demonstrate the international interest and attention on cold regions hydrology/climate research and applications. They cover a very broad domain, including snow cover, glaciers, permafrost, streamflow, temperature, precipitation, groundwater and ecosystems. They report new research results based on field observations, modelling and remote sensing in a great range of geographical regions from Chile to the Arctic. Collectively, these papers highlight recent progress in cold regions hydrology research and its linkage with climate change at various space and time scales. They also identify gaps and needs for future research.

As the editors and symposium conveners, we truly appreciate the contributions from our colleagues, and the interactions and communications with them regarding their papers and the publication of this book. Peer review of the manuscripts was critical to ensuring the quality of the papers in this session and book. The three editors shared this duty and work with several colleagues, and we are grateful to acknowledge the insightful reviews by Chris Spence, Yinsheng Zhang, Ross Brown, Thian Yew Gan and William Bolton.

Daqing Yang
National Hydrology Research Center, Canada

Philip Marsh
National Hydrology Research Center, Canada

Alexander Gelfan
Water Problem Institute, Russia

Contents

2 Snow Cover, Permafrost and Glaciers

3 Climate

1 Basin Hydrology

Precipitation trends contribute to streamflow regime shifts in northern Canada

CHRISTOPHER SPENCE[1], STEVEN V. KOKELJ[2] & EGHBAL EHSANZADEH[1]

1 *Environment Canada, Saskatoon, Saskatchewan S7N 3H5, Canada*
chris.spence@ec.gc.ca

2 *Indian and Northern Affairs Canada, Yellowknife, Northwest Territories X1A 2R3, Canada*

Abstract Autumn runoff events rivalling the size of the spring freshet peak as well as sustained winter streamflow have become more common in the northwestern Canadian Shield since the mid 1990s. Previous circumpolar and large regional-scale studies have implied these phenomena are due to increased water inputs from thawing permafrost. However, results from an investigation of the precipitation and temperature trends provide an alternate explanation for this region. A shift from a nival to a combined nival/pluvial streamflow regime, particularly in small watersheds, can be attributed to trends in the timing and state of autumn precipitation. Because these trends are subtle, careful consideration of hydrological processes, and the temporal and landscape context in which they operate, is important when attempting to explain the observed shifts in regional streamflow. It is important to correctly explain why streamflow regimes are changing because of close relationships with variations in ground thermal conditions and aquatic chemistry, which are of significance to society. These relationships are discussed.

Key words streamflow; precipitation; trends; shifts; Canadian Shield; permafrost

INTRODUCTION

The last half century in the circumpolar north has been characterized by relatively rapid environmental change (Arctic Council, 2005) including the increase of cold season streamflow in watersheds with subarctic nival streamflow regimes (Smith *et al.*, 2007). While most have speculated about the mechanisms behind this increase (e.g. St. Jacques & Sauchyn, 2009), only a few studies have focused on understanding the process and drivers of seasonal hydrological change (Jones & Reinhart, 2010) which are likely to vary across scales and environments (Rawlins *et al.*, 2009). This study focused on the subarctic Precambrian Shield portion of the Canadian Northwest Territories to evaluate if climatic mechanisms could be responsible for streamflow change in this particular region. Many large regional studies evaluate relatively coarse climate data (i.e. annual or seasonal). However, only subtle changes in precipitation and temperature are needed in cold regions to enact significant hydrological changes. In this study, monthly data are evaluated to determine if: (1) there have been changes to the streamflow regimes in the northwestern portion of the subarctic Canadian Shield, and, if so, (2) whether these changes are due to changes in climatic drivers.

DATA AND METHODS

Streamflow data were obtained for the period of record from the four longest operating hydrometric stations in the region (http://www.wsc.ec.gc.ca/hydat /H2O/index_e.cfm) (Table 1). Monthly precipitation and temperature data that overlapped the geographic extent of the basins gauged by these four hydrometric stations (bounded by 62°N, 112.5°W, 66°N, 115.5°W) were extracted from the CANGRID data set (Louie *et al.*, 2002). This experimental Meteorological Service of Canada data set provides precipitation and temperature values across Canada at a 50-km resolution. There is uncertainty associated with this data set. All the gridded precipitation data products available for northern Canada rely on the same sparse network for calibration, but CANGRID is the only one to address undercatch and homogeneity (Mackay *et al.*, 2003), a key source of precipitation estimation error in northern Canada (Metcalfe *et al.*, 1994). The selected four hydrometric gauges represent a spectrum of catchment areas to permit investigation of spatial scale effects on regime change. Analyses were performed on monthly data averaged from daily

Table 1 Traits and statistics for the selected streams. Standard normal variate (*Z*) for trends significant at 90% are in bold and the change points detected with a probability of greater than 70% are listed. Q_f/Q_a is the fraction of basin yield provided in May and June (June and July for Cameron and Snare rivers) and Q_w/Q_a is the fraction of basin yield during winter (October–March).

Station Name	Baker Creek		Cameron River		Indin River		Snare River	
Station ID	07SB013		07SB010		07SA004		07SA002	
Record period	1972–2009		1976–2009		1977–2009		1985–2009	
Location Lat.:	62°30.8′N		62°29.5′N		64°23.3′N		63°58.4′N	
Location Long.:	114°24.3′W		113°32.2′W		115°1.3′W		115°26.0′W	
Area (km^2)	155		3630		1520		13300	
Lake fraction	0.18		0.18		0.04		0.10	
Statistic	*Z*	Shift	*Z*	Shift	*Z*	Shift	*Z*	Shift
January	**3.0**	1997	**3.0**	n/a	2.0	n/a	0.1	n/a
February	**3.1**	1997	**3.1**	n/a	1.6	n/a	0.3	n/a
March	**3.3**	n/a	**3.1**	n/a	**2.7**	1997	0.7	n/a
April	**2.3**	1997	**2.9**	n/a	**2.4**	n/a	0.9	1997
May	0.8	n/a	**2.0**	n/a	–0.3	n/a	0.1	n/a
Jun	0.5	n/a	1.0	n/a	0.5	n/a	–0.9	1998
July	–0.9	n/a	1.0	n/a	0.3	n/a	–0.8	n/a
August	–0.7	n/a	1.6	n/a	0.5	n/a	–0.8	n/a
September	–0.4	n/a	1.7	n/a	1.5	n/a	–0.8	n/a
October	0.5	n/a	**2.6**	n/a	1.8	n/a	–0.3	1997
November	**1.7**	n/a	**3.0**	n/a	1.9	n/a	–0.1	n/a
December	**2.7**	n/a	**2.8**	n/a	2.0	n/a	0.2	n/a
Before/following 1997	<1997	>1997	<1997	>1997	<1997	>1997	<1997	>1997
Q_f/Q_a	0.76	0.50	0.33	0.27	0.55	0.52	0.30	0.30
Q_w/Q_a	0.07	0.19	0.19	0.21	0.08	0.11	0.17	0.21

extracted values. Gradual changes in the mean (i.e. trends) were analysed using the non-parametric Mann-Kendall test with pre-whitening (Mann, 1945; Kendall, 1975; Yue *et al.*, 2002) often used with hydrological data because it is thought to be more suitable for non-normally distributed and censored hydro-meteorological time series. Shifts (i.e. abrupt changes in the mean) were identified using a Bayesian change-point detection model (Seidou & Ouarda, 2007; Ehsanzadeh *et al.*, 2010).

RESULTS

Analysis of monthly precipitation revealed only two significant trends at the 90% confidence level. Precipitation in September is increasing (Z score = 2.0) but decreasing in October (Z score = –1.8). There were upward shifts in precipitation in January, February and April, all in 1997. Temperature trends were limited to January and December (Table 2), which were both positive. The trend in December temperature was interpreted as an upward shift in 1997 by the Bayesian change-point detection model. Upward shifts in temperature were also detected in 1995 (February) and 1991 (March). The overall upward temperature trends and wetter conditions in the autumn have also been recently documented by Zhang *et al.* (2011).

The summer storage deficits that accumulate in this landscape occur because of low precipitation to evapotranspiration ratios (Spence & Rouse, 2002) that prevent a basin runoff response until ~100 mm of rainfall has fallen between mid-July and the end of October (Fig. 1(a)). The observed increase in September precipitation was offset by a reduction in October precipitation. This is the period during which the 0°C air temperature threshold is crossed so that precipitation in September tends to fall as rain, whereas in October it accumulates as snow (Fig. 1(b)). Any increase in rain is not due to an increase in autumn temperatures because no statistically significant trend or shift was found. A trend towards earlier autumn precipitation has resulted in an increase in the frequency with which the rainfall threshold of ~100 mm has occurred

Table 2 Traits and statistics for the subset of the CANGRID domain. Standard normal variate (*Z*) for trends significant at 90% are in bold and the change points detected with a probability of greater than 70% are listed.

	Precipitation		Temperature	
Statistic	*Z*	Shift	*Z*	Shift
January	0.9	1997	**1.7**	n/a
February	0.2	1997	1.0	1995
March	1.1	n/a	0.6	1991
April	0.2	1997	1.0	n.a
May	–0.7	n/a	–1.2	n/a
Jun	0.6	n/a	0.1	n/a
July	0.7	n/a	1.0	n/a
August	0.8	n/a	0.5	n/a
September	**2.5**	n/a	1.0	n/a
October	**–1.8**	n/a	1.1	n/a
November	1.5	n/a	1.2	n/a
December	0.8	n/a	**1.9**	1997

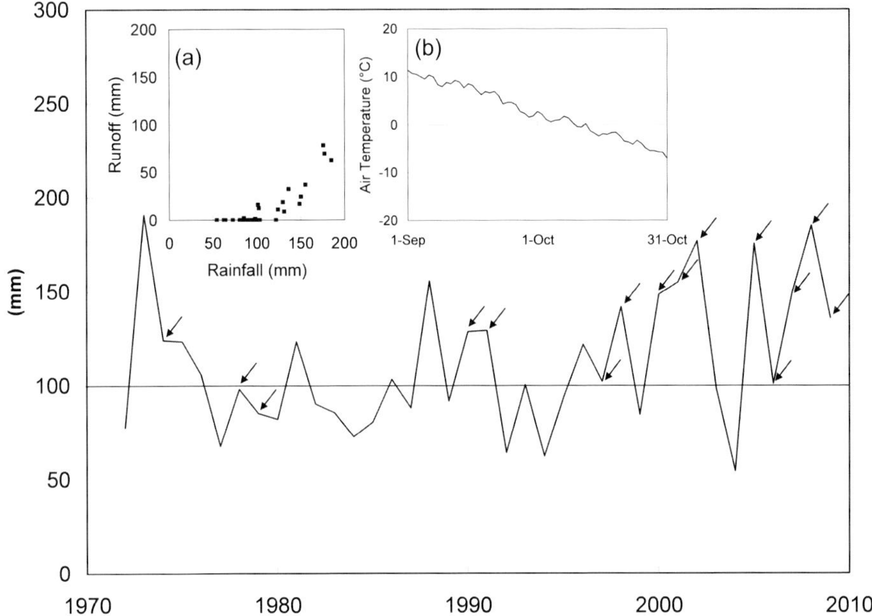

Fig. 1 Seasonal rainfall (15 July–31 October) at Yellowknife. The increased frequency with which autumn runoff was generated is demonstrated by the increase in the number of arrows denoting those years with autumn runoff events since 1997. The black horizontal line is the threshold evident in inset (a) at which runoff tends to be generated. Insets illustrate: (a) the threshold mediated nature of runoff response to seasonal rainfall, and (b) mean daily air temperatures at Yellowknife in September and October for the period of record under study.

since 1997 (Fig. 1). Threshold driven responses, such as runoff generation, can be highly influenced by small changes in forcing variables near the threshold (Zehe *et al.*, 2005), and over longer periods this can account for significant trends and shifts in streamflow (McClelland *et al.*, 2004).

Discharge trends in cold season months were common to all streams except the Snare, and they were all of similar magnitude. The combined data from the four rivers yielded 148 record-years of hydrometric data, but shifts were only detected in 1997 and 1998 (Table 1). Dividing the period of record into pre- and post-1997 may provide insight into changes in the streamflow regime of the region. An increase in the fraction of discharge in the cold season (Q_w/Q_a) was consistent among all four streams since 1997 (Fig. 2). At the smallest scale (Baker Creek), yield distribution during the spring freshet and winter before (after) 1997 was 76% (50%) and 7% (19%), respectively (Table 1). Annual hydrographs from Baker Creek do not display the classic nival streamflow regimes that they once did, as there are now frequent secondary peaks that rival the freshet in magnitude, with recessional limbs extending well into the winter season. Mean monthly Baker Creek streamflow (± two standard deviations) in January through April since 1997 is outside the range documented before 1997. Early winter streamflow becomes attenuated by large lakes in the larger basins, and the yield distribution does not change at these scales. However, secondary autumn peaks are relatively common so that the smooth and consistent recessions between annual snowmelt runoff events in basins of >1000 km^2 are now being represented by uneven recession curves (Fig. 2). As with late winter streamflow in Baker Creek, post-1997 mean October streamflow (± two standard deviations) in the larger rivers is outside the range observed before 1997.

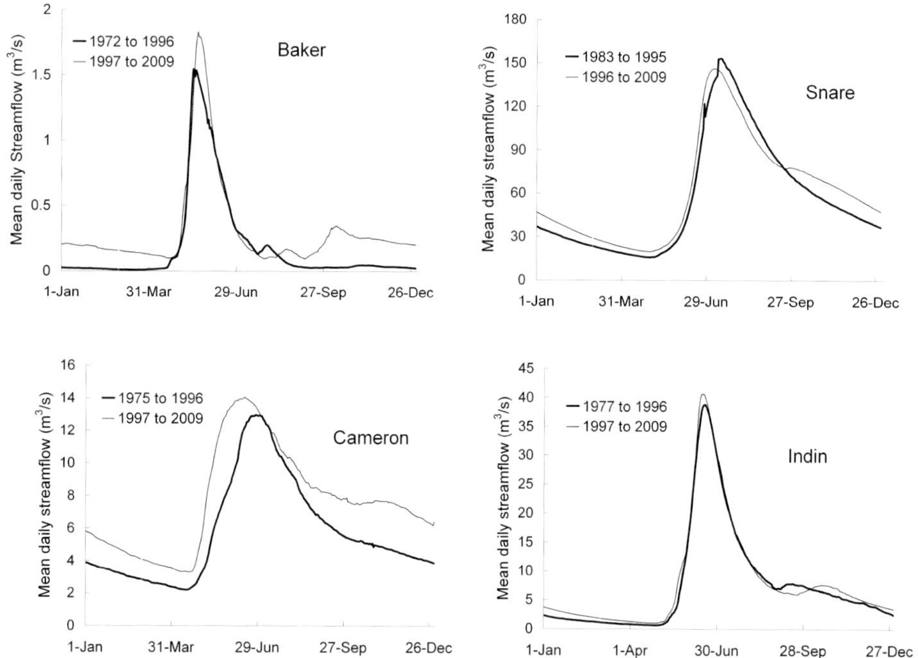

Fig. 2 Average annual hydrographs at each gauge before and since 1997.

DISCUSSION

The results indicate a shift from a nival to a combined nival/pluvial streamflow regime at the small basin scale in this region. This does not manifest in the larger basins of the subarctic Canadian Shield. Watersheds of different sizes may exhibit different streamflow regimes within the same physioclimatic region because of the relative dominance of different hydrological processes with changes in scale. Even though the lake fraction in the Cameron, Baker and Snare watersheds is similar (Table 1), the larger Cameron and Snare watersheds each contain at least two lakes larger than the entire Baker Creek watershed. The presence of large lakes creates a situation where streamflow attenuation is such a significant process that the streamflow regime becomes prolacustrine. This regime is perhaps more resilient to change than the nival regime common to smaller watersheds that are influenced more by hillslope processes.

The changes in the precipitation regime documented here are subtle, but changes near 0°C can have dramatic hydrological effects (Quinton & Carey, 2008). Observed streamflow changes between small and large watersheds were different because of different predominant hydrological processes acting at different scales. Furthermore, small changes in atmospheric circulation patterns that are known to influence northern Canadian streamflow, such as the specific location of cyclones south of the Aleutian Islands (Dery *et al.*, 2009), are not always resolved by many currently available circulation indices (Overland & Wang, 2005). These factors, and perhaps others, mean the relationships between northern Canadian streamflow and circulation indices are complex (Woo & Thorne, 2008), and that caution should be exercised when predicting any future changes to northern streamflow regimes in specific locales.

The impacts of increasing autumn rainfall extend beyond shifts in streamflow regimes. Increasing early winter soil moisture with heavier autumn rains, in conjunction with observed trends towards a deeper thawed layer (Burn & Kokelj, 2009) can further prolong active layer freezeback (Osterkamp & Romanovsky, 1999). The precipitation trends documented in this study in conjunction with warming permafrost are likely to have significant hydrogeochemical consequences. Thawing of near-surface permafrost can make previously sequestered soluble materials available for transport by late season runoff (Kokelj & Burn, 2003). The rise in late summer rainfall and early winter runoff, in correspondence with the increased duration of active layer freezeback, suggests that the winter solute flux from terrestrial to aquatic environments may be amplified. It is likely that the net effects of these changes would be an increase in stream nutrient and solute loads (Frey & McClelland, 2009). Through multidisciplinary efforts, small basin hydrological research is needed to test the hypothetical drivers of changing hydrological and geochemical flux in streams across a range of northern environments.

CONCLUDING REMARKS

A trend towards more autumn rainfall in the northwestern subarctic Canadian Shield has been sufficient to permit a more frequent exceedence of runoff generation thresholds, causing late season peaks in discharge and higher winter baseflows. The emergence of a new nival/pluvial regime in small subarctic Canadian Shield streams is causing an increase in the relative winter discharge yields and it is most clearly manifested in small watersheds. Streamflow responses to climate change differ across the diverse circumpolar north because each landscape filters precipitation differently. The results presented here also imply that basins of different sizes filter precipitation differently. Therefore, assumptions that do not account for the distinct physical processes acting within landscapes and across scales will result in poor analysis of the regional impacts of climate change. Because of the integrated nature of relationships between autumn precipitation, streamflow, ground thermal regimes and aquatic chemistry, it will be important to consider these physical processes if sound information is to be generated for present and future environmental conditions in cold regions.

REFERENCES

Arctic Council (2005) *Arctic Climate Impact Assessment*. Cambridge Univ. Press, New York, USA.

Burn, C. R. & Kokelj, S. V. (2009) The environment and permafrost of the Mackenzie Delta area. *Permafrost and Periglacial Processes* **20**, 83–105.

Dery, S. J., Hernández-Henriquez, M. A., Burford, J. E. & Wood, E. F. (2009) Observational evidence of an intensifying hydrological cycle in northern Canada. *Geophys. Res. Lett.* **36** (L13402), doi:10.1029/2009GL038852.

Ehsanzadeh, E., Ouarda, T. B. M. J., & Saley, H. M. (2010) A simultaneous analysis of gradual and abrupt changes in Canadian low streamflows. *Hydrol. Processes* doi:10.1002/hyp.7861.

Frey, K. E. & McClelland, J. W. (2009) Impacts of permafrost degradation on arctic river biogeochemistry. *Hydrol. Processes* **23**, 169–182.

Jones, J. B. & Reinhart, A. J. (2010) The long term response of stream flow to climatic warming in headwater streams in interior Alaska. *Can. J. For. Res.* **40**, 1210–1218.

Kendall, M. G. (1975) *Rank Correlation Methods*. Griffin, London, UK.

Kokelj, S. V. & Burn, C. R. (2003) Ground ice and soluble cations in near-surface permfrost, Inuvik, Northwest Territories, Canada. *Permafrost and Periglacial Processes* **14**, 275–289.

Louie, P. Y. T., Hogg, W. D., Mackay, M. D., Zhang, X. & Hopkinson R. F. (2002) The water balance climatology of the Mackenzie Basin with reference to the 1994/95 water year. *Atmosphere Ocean* **40**, 159–180.

Mackay, M. D., Seglenieks, F., Verseghy, D., Soulis, E. D., Snelgrove, K. R., Walker, A. & Szeto, K. (2003) Modeling Mackenzie Basin surface water balance during CAGES with the Canadian Regional Climate Model. *J. Hydromet.* **4**, 748–766.

Mann, H. B. (1945) Nonparametric tests against trend. *Econometrica* **13**, 245–259.

McClelland, J. W., Holmes, R. M. & Peterson, B. J. (2004) Increasing river discharge in the Eurasian Arctic: consideration of dams, permafrost thaw, and fires as potential agents of change. *J. Geophys Res.* **109** (D18102), doi: 10.1029/2004JD004583.

Metcalfe, J. R., Ishida, S. & Goodison, B. E. (1994) A corrected precipitation archive for the Northwest Territories. In: *Mackenzie Basin Impact Study, Interim Report #2* (Proc. 6th Biennial Meeting on Northern Climate and mid study workshop of the MBIS (ed. by S. Cohen), 110–117.

Osterkamp, T. E. & Romanovsky, V. E. (1997) Freezing of the active layer on the coastal plain of the Alaskan arctic. *Permafrost and Periglacial Processes* **8**, 23–44.

Overland, J. E. & Wang, M. (2005) The third Arctic climate pattern: 1930s and early 2000s. *Geophys. Res. Lett.* **32** (L23808), doi:10.1029/2005GL024254.

Quinton, W. L. & Carey, S. K. (2008) Towards an energy-based runoff generation theory for tundra landscapes. *Hydrol. Processes* **22**, 4649–4653.

Rawlins, M. A., Ye, H., Yang, D., Shiklomanov, A. & McDonald, K. C. (2009) Divergence in seasonal hydrology across northern Eurasia: emerging trends and water cycle linkages. *J. Geophys Res.* **114** (D18119), doi: 10.1029/2009JD011747.

Seidou, O. & Ouarda, T. B. M. J. (2007) Recursion-based multiple changepoint detection in multiple linear regression and application to river streamflows. *Water Resour. Res.* **43** (W07404), doi: 10.1029/2006WR005021.

Smith, L. C., Pavelsky, T. M., MacDonal, G. M., Shiklomanov, A. I. & Lammers, R. B. (2007) Rising minimum daily flows in northern Eurasian rivers: A growing influence of groundwater in the high-latitude hydrological cycle. *J. Geophys. Res.* **112** (G04S47), doi: 10.1029/2006JG000327.

Spence, C. & Rouse, W. R. (2002) The energy budget of subarctic Canadian Shield terrain and its impact on hillslope hydrology. *J. Hydromet.* **3**, 208–218.

St Jacques, J. M. & Sauchyn, D. J. (2009) Increasing winter baseflow and mean annual streamflow from possible permafrost thawing in the Northwest Territories, Canada. *Geophys. Res. Lett.* **36**, L01401, doi: 10.1029/2008GL035822.

Woo, M. K. & Thorne, R. (2008) Analysis of cold season streamflow response to variability of climate in north-western North America. *Hydrol. Res.* **39**, 257–265.

Yue, S., Pilon, P., Phinney, B. & Cavadias, G. (2002) The influence of autocorrelation on the ability to detect trend in hydrological series. *Hydrol. Processes* **16**, 1807–1829.

Zehe, E., Becker, R., Bárdossy & Plate, E. (2005) Uncertainty of simulated catchment runoff response in the presence of threshold processes: Role of initial soil moisture and precipitation. *J. Hydrol.* **315**, 183–202.

Zhang, X., Brown, R., Vincent, L., Skinner, W., Feng, Y. & Mekis, E. (2011) *Canadian Climate Trends, 1950–2007*. Canadian Biodiversity: Ecosystem Status and Trends 2010 Technical Thematic Report no. 5. Canadian Councils of Resource Ministers. Ottawa, Ontario, Canada.

Streamflow responses and trends between permafrost and glacierized regimes in northwestern Canada

J. RICHARD JANOWICZ

Water Resources Branch, Yukon Department of Environment, PO Box 2703, Whitehorse, Yukon Territory Y1A 2C6, Canada

richard.janowicz@gov.yk.ca

Abstract An assessment of the streamflow response of glacierized basins in southwestern Yukon was carried out to determine if there are apparent trends associated with climate warming. The study area includes portions of the sporadic and discontinuous permafrost zones. Annual mean, maximum and minimum flows, as well as the timing of the maximum and minimum annual discharge, were assessed using the Mann-Kendall test. A slight positive trend in annual mean discharge was generally observed throughout the study region, likely a result of combined precipitation increases and glacier melt contributions. Annual maximum flow trends are more variable with the majority of station records exhibiting a positive trend. Permafrost likely has a significant role in controlling annual peak discharge trends. Basins with little permafrost exhibited positive trends in response to additional meltwater contributions, while basins with significant permafrost exhibited negative trends, likely a result of the degrading permafrost enhancing subsurface flow processes. Positive trends in annual minimum flows were generally obtained, presumably due to greater groundwater contributions to baseflow.

Key words glacierized; discontinuous; sporadic permafrost; Mann-Kendall; trend analysis; streamflow response; Yukon Territory, Canada

INTRODUCTION

Permafrost has a dominant control over hydrological response in the northern regions by producing short pathways to the stream channel, with little interaction with subsurface processes (Hinzman *et al.*, 2005). A thicker active layer enhances infiltration and associated groundwater recharge, which in turn results in greater groundwater contributions to streamflow. Yukon hydrological response follows this principle, and is closely tied to the underlying permafrost (Janowicz, 2008). Similar to all cold regions, Yukon spring streamflow response is characterized by a rapid rise in discharge as a result of snowmelt contributions. Minimum annual discharge occurs in March or April, coinciding in timing to minimum annual groundwater inputs. The southwestern portions of Yukon Territory also have considerable glacier and ice cap coverage. Unlike nival regimes, which experience their peak flow in May or June due to snowmelt inputs, peak flows in glacierized basins are delayed until later in the summer due to supplemental inputs from glacier melt (Janowicz, 2004, 2008). Stahl & Moore (2006) found basins with as little as 2–3% glacierized area to supplement summer flows.

While there is as yet no definitive evidence to prove that climate variability in northern Canada is anthropogenic, air temperature and precipitation in Yukon Territory have fluctuated significantly over the last century. Air temperatures have generally increased throughout Yukon. Summer precipitation has generally increased, while winter precipitation has increased in northern regions and decreased in southern regions (Janowicz, 2010). There is also some evidence suggesting the trends may be associated with teleconnections between large-scale oceanic and atmospheric processes, such as the El Nino-Southern Oscillation (ENSO) and the Pacific Decadal Oscillation (PDO) (Barlow *et al.*, 2001).

Changing climate appears to be resulting in a likewise change in the permafrost and glacier distributions of northern regions, including the Yukon Territory. Increasing air temperatures are resulting in permafrost warming and associated thawing, which in turn results in a thicker active layer. Permafrost degradation is expected to be greatest within the discontinuous and sporadic permafrost zones since these permafrost classes are warmer, and therefore more susceptible to thawing (Hinzman *et al.*, 2005).

On a global scale, glaciers and ice caps have generally declined significantly since the end of the Little Ice Age (Zemp *et al.*, 2007). On a regional basis, glaciers throughout western North America have generally retreated since the Little Ice Age (Zemp *et al.*, 2007; Moore *et al.*, 2009). Considerable work has been carried out in recent years on glacier mass balance studies of northwestern North America. Schiefer *et al.* (2007) estimated the annual change in glacier volume for British Columbia for the 15 year period (1985–1999) to be 22.48 ± 5.53 km^3. The estimated loss varies regionally with the greatest decline observed in the Coast and St Elias Mountains, which have the greatest glacier and ice cap coverage. According to Barrand & Sharp (2010) Yukon glaciers have experienced a surface area loss of 22% in the last 52 years (0.42%/year), equating to 0.78 ± 0.34 m/year water equivalent, which accounts for 1.12 ± 0.49 mm/year of global sea level rise.

There have been a number of studies carried out in northern regions of North America on the impact of climate change on hydrological response (Kite, 1993; Burn, 1994; Loukas & Quick, 1996, 1999; Leith & Whitfield, 1998; Whitfield & Taylor, 1998; Spence, 2002). Dery & Wood (2005) investigated the discharge of 64 arctic or subarctic Canadian rivers from 1964 to 2003. They found a general 10% decline in mean annual discharge to the Arctic and North Atlantic Oceans over that period with potential teleconnection links.

There has been only limited work to date on the impact of climate change specific to Yukon hydrology. Janowicz & Ford (1994) carried out an assessment of the impacts of climate warming on the water supply to the upper Yukon River with their results indicating that annual inflows to the glacierized upper Yukon River would increase by 39% due to increasing temperature and precipitation. Whitfield & Cannon (2000) and Whitfield (2001) assessed recent climatic and hydrological variations for stations in British Columbia and Yukon, finding hydrological response to be characterized with higher annual and peak flows and lower summer and fall flows. Similarly, Janowicz (2001) noted a dramatic change in annual peak flows with increases in glacierized regions of southwestern Yukon, and a progressive decrease moving northward into more dominant permafrost regions. Zhang *et al.* (2001) and Yue *et al.* (2003) found that winter low flows in northern British Columbia and Yukon have increased significantly, while annual mean and peak flows increased in glacierized basins of southern Yukon and northern British Columbia. Walvoord & Striegl (2007) found a general increase in winter discharge within the Yukon River basin. Fleming & Clarke (2003) carried out an assessment of the streamflow response of five glacierized and four non-glacierized basins in northern British Columbia and southwest Yukon in response to climate warming. They found a trend of increasing and decreasing annual flow in glacial and nival regime streams, respectively. Janowicz (2008) carried out an assessment of annual mean, maximum and minimum flows using 43 hydrometric station records distributed between continuous, discontinuous and sporadic permafrost zones in Yukon Territory and adjacent areas of northern British Columbia and western Northwest Territories. Annual mean flows were observed to have slight positive trends within continuous and discontinuous permafrost zones, with variable results within sporadic permafrost regions. Annual peak flows have largely decreased within the continuous permafrost regions, and less so within the discontinuous regions, and results within the sporadic permafrost zones are variable. Winter low flows have experienced significant changes within the continuous and discontinuous permafrost regions over the last three decades, and with variable results within the sporadic permafrost regions.

This paper summarizes the results of a study carried out to assess apparent trends of streamflow characteristics of glacierized basins with Yukon Territory drainage.

SETTING

Yukon Territory, in northwestern Canada, consists of three permafrost regions: continuous, discontinuous and sporadic, while the immediate study area is underlain by sporadic and discontinuous permafrost only (Fig. 1) (Natural Resources Canada, 1995). Not including the Arctic islands, the greatest glacier and ice cap coverage in Canada occurs in the western Cordillera

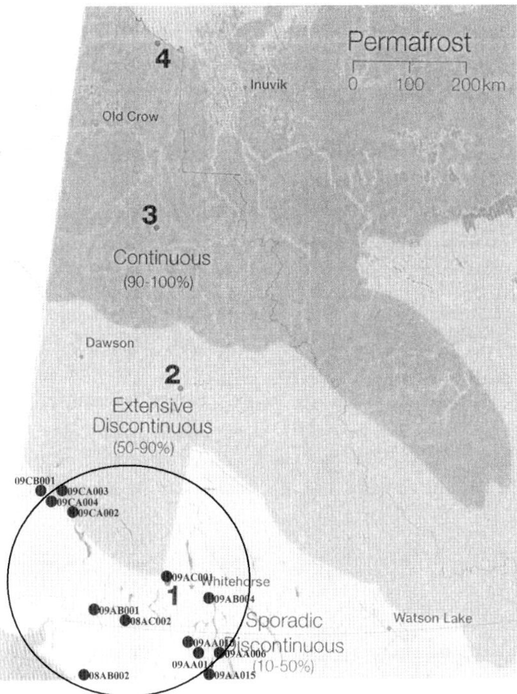

Fig. 1 Location plan, permafrost zones, hydrometric stations (adapted from Smith *et al.*, 2004).

with areas of 22 000 and 4300 km^2 for the Coast and St Elias Mountains, respectively (Moore *et al.*, 2009). The Yukon Territory portions of the Yukon and Alsek River basins have approximately 7400 and 3500 km^2 of glacier and ice cap coverage, respectively, representing 15 and 20% of total basin drainage area.

METHODOLOGY

Data from active hydrometric stations downstream of glacierized basins in the upper Yukon and Alsek River basin in southwest Yukon, northwest British Columbia and southeast Alaska were used in the analyses. Table 1 provides a summary of station and drainage area, glacier area and mean annual temperature.

Table 1 Study drainage basin parameters.

Station name	Drainage area (km^2)	Glacier area (%)	Mean annual temp. (°C)
Atlin River nr Atlin	6 810	12.7	–0.6
Marsh Lake nr Whitehorse	19 400	8.3	–1.5
Takhini River nr Whitehorse	6 930	5.1	–3.0
Duke River nr Mouth	631	9.5	–4.9
Kluane River at Mouth Kluane Lk	4 950	6.0	–5.7
White River at Alaska Hwy	6 240	38.6	–8.7
Tatshenshini River at Dalton Post	1 750	7.0	–2.5
Alsek River at Bates R	16 200	21.0	–3.5
Alsek River nr Yakatat	28 000	31.9	–2.4

Trend analyses were carried out on the time series of annual mean, maximum and minimum discharge, as represented by the 7-day average minimum annual low flow, and timing of minimum and maximum daily discharge, as represented by the Julian day. The Mann-Kendall trend test was used to assess trends in the three hydrograph parameters (Mann, 1945; Kendall, 1975). The standard normal variate value (Z) is associated with a specific level of significance. The significance level provides an indication of the strength of the trend. A significance level of 0.001 (99.9%) indicates a very strong trend, 0.01 (99%) indicates a strong trend, 0.05 (95%) indicates a moderate trend, and 0.1 (90%) indicates a weak trend. A level of significance of greater than 0.1 indicates there is no statistically significant trend.

RESULTS AND DISCUSSION

Table 2 provides a summary of the trend analyses. The assessment of annual mean discharge generally indicates a slight positive trend throughout the study region. Based on temperature and precipitation trends, increases in the southern portions of the study area, within the sporadic permafrost zone, may be a result of combined precipitation increases and glacier melt contributions. The nominal increases in the remainder of the study basin may be a result of glacier melt contributions.

Table 2 Mann-Kendall trend statistics.

Station name	Station number	Area (km^2)	Record period	n (years)	Mann-Kendall Z statistic: QAnnual	QMax	QMin	JDyMx	JDyMin
Atlin	09AA006	6 810	1951–08	57	3.05***	2.66**	2.46***	–0.41	–2.10**
Marsh	09AB004	20 300	1951–09	48		2.28**		0.61	
Takhini	09AC001	6 930	1949–08	58	0.32	–1.88*	1.69*	–1.14	–1.05
Duke	09CA004	631	1981–08	25	0.56	0.42	2.08**	–0.07	0.14
Kluane	09CA002	4 950	1953–08	55	0.19	1.08	4.51****	0.06	0.56
White	09CB001	6 240	1975–08	32	0.43	–1.77*	3.08***	–1.80*	0.6
Tatshenshini	08AC002	1 750	1989–08	19	0.18	1.05	1.19	–1.58	0.23
Alsek –Bate	09AB001	16 200	1975–08	32	1.72*	–0.1	0.3	–2.06*	0
Alsek - Yak	08AB002	28 000	1993–08	16	0.16	0.54	–0.05	–0.95	–0.09

Level of significance: *0.10; **0.05; ***0.01; ****0.001

Annual maximum flow trends are more variable in comparison to the annual mean, with the majority of station records exhibiting a positive trend, two of which are significant, one with no change, and two with significant negative trends. Permafrost likely has a considerable role in controlling annual peak discharge trends, since air temperature has increased relatively uniformly over the study area. The two stations with strong positive trends are in the sporadic permafrost zone, supporting the theory of increased glacier melt contributions; while the two stations with strong negative trends are in the discontinuous permafrost zone. The difference may be due to overlapping signals of permafrost degradation and glacier melt where greater meltwater runoff may be entering the groundwater system rather than running off near the surface, resulting in smaller peak flow events.

Positive trends in annual minimum flows were obtained for all stations within the study region, with the exception of two stations within the southwestern portion which remain unchanged. Increases in air temperature suggest the increase to be largely related to permafrost degradation with greater groundwater contributions to baseflow, though in some cases increased annual and peak flows may also contribute to winter flows. Basins with little permafrost appear to experience minimal increases in winter baseflow. The timing of annual maximum flows was generally observed to advance, while the timing of annual minimum flows is not consistent.

CONCLUSIONS

An assessment of streamflow response of the glacierized drainage basins within the sporadic and discontinuous permafrost zones in southwestern Yukon was carried out to determine if there are apparent trends as a result of recent temperature and precipitation changes. Study results revealed a slight positive trend of mean annual discharge throughout the study region, while trends of annual maximum flow were observed to be more variable, with the majority of station records exhibiting a positive trend and two stations having significant negative trends. Positive trends in annual minimum flows were generally observed for all the stations within the study region,

As permafrost properties change with climate warming, hydrological response in the northern regions would likewise presumably change. Degrading permafrost increases the thickness of the active layer, decreases the overall thickness of the permafrost, and in certain areas eliminates the presence of underlying permafrost entirely. These actions place a greater reliance on the interaction between surface and subsurface processes. Melting glaciers and ice caps are likewise altering hydrological response by increasing runoff and altering its distribution.

Glacierized basins with little permafrost seem to exhibit a classic glacial response trend, while those with significant amounts of permafrost show more classic permafrost responses and trends. Other glacierized basins with variable amounts of permafrost have overlapping signals, having mixed streamflow trends which are difficult to interpret.

Acknowledgements The work carried out by Colin Abbot in performing much of the data manipulation and Miranda Allison in text design is greatly appreciated.

REFERENCES

Barlow, M., Nigam, S. & Berbery, E. H. (2001) ENSO, Pacific decadal variability, and US summertime precipitation, drought and streamflow. *J. Climate* **14**, 2105–2127.

Barrand, N. E. & Sharp, M. J. (2010) Sustained rapid shrinkage of Yukon glaciers since the 1957–1958 International Geophysical Year. *Geophys. Res. Lett.* **37**(7), L07501. DOI: 10.1029/2009GL042030.

Burn, D. H. (1994) Hydrologic effects of climate change in west-central Canada. *J. Hydrol.* **160**, 53–70.

Dery, S. J. & Wood, E. F. (2005) Decreasing river discharge in northern Canada. *Geophys. Res. Lett.* **32**, L10401. doi:10.1029/2005GL022845.

Fleming, S. W. & Clarke, G. K. C. (2003) Glacial control of water resource and related environmental responses to climate warming: empirical analysis using historical streamflow data from northwestern Canada. *Can. Water Resour. J.* **28**(1), 69–85.

Hinzman, L. D., Bettez, N. D., Bolton, W. R., Chapin, F. S., Dyurgerov, M. B., Fastie, C. L., Griffith, B., Hollister, R. D., Hope, A., Huntington, H. P., Jensen, A. M., Jia, G. J., Jorgenson, T., Kane, D. L., Klein, D. R., Kofinas, G., Lynch, A. H., Lloyd, A. H., McGuire, A. D., Nelson, F. E., Nolan, M., Oechel, W. C., Osterkamp, T. E., Racine, C. H., Romanovsky, V. E., Stone, R. S., Stow, D. A., Sturm, M., Tweedie, C. E., Vourlitis, G. L., Walker, M. D., Walker, D. A., Webber, P. J., Welker, J., Winker, K. S. & Yoshikawa, K. (2005) Evidence and implications of recent climate change in northern Alaska and othr Arctic regions. *Climate Change* **72**, 251–298.

Janowicz, J. R. (2001) Impact of recent climatic variability on peak streamflow in northwestern Canada with implications for the design of the proposed Alaska Highway gas pipeline. *Proceedings of the 13th Northern Research Basins International Symposium & Workshop*, 19–24 August 2001, Saariselka, Finland and Murmansk, Russia, 161–169.

Janowicz, J. R. (2004) Yukon overview: watersheds and hydrologic regions. In: *Ecoregions of the Yukon Territory – Biophysical Properties of Yukon Landscapes* (ed. by C. A. S Smith, J. C. Meikle, & C. F. Roots). Agriculture and Agrifood Canada PARC Technical Bulletin 04-0115-18.

Janowicz, J. R. (2008) Apparent recent trends in hydrologic response in permafrost regions of northwest Canada. *Hydrology Research* **39**(4), 267–275.

Janowicz, J. R. (2010) Observed trends in the river ice regimes of northwest Canada. *Hydrology Research* **41**(6), 462–470.

Janowicz, J. R. & Ford, G. (1994) Impact of climate change on water supply in the upper Yukon River. *Proceedings of the 62nd Annual Western Snow Conference*, 18–21 April 1994, Sante Fe, New Mexico.

Kendall, M. G. (1975) *Rank Correlation Methods*. Charles Griffin, London, UK.

Kite, G. W. (1993) Application of a land class hydrological model to climate change. *Water Resour. Res.* **29**(7), 2377–2384.

Leith, R. M. & Whitfield, P. H. (1998) Evidence of climate change effects on the hydrology of streams in south-central B.C. *Can. Water Resour. J.* **23**, 219–230.

Loukas, A. & Quick, M. C. (1996) Effect of climate change on hydrologic regime of two climatically different watersheds. *J. Hydrol. Engng* **1**(2), 77–87.

Loukas, A. & Quick, M. C. (1999) The effect of climate change on floods in British Columbia. *Nordic Hydrology* **30**, 231–256.

Mann, H. B. (1945) Nonparametric tests against trend. *Econometrica* **13**, 245–259.

Moore, R. D., Fleming, S. W., Menounos, B., Wheate, R., Fountain, A., Stahl, K., Holm, K. & Jakob, M. (2009) Glacier change in western North America: influences on hydrology, geomorphic hazards and water quality. *Hydrol. Processes* **23**, 42–61. doi: 10.1002/hyp.7162.

Natural Resources Canada (1995) *National Atlas of Canada* (5th edition). MCR 4177. Geological Survey of Canada, Terrain Sciences Division, Ottawa.

Schiefer, E., Menounos, B. & Wheate, R. (2007) Recent volume loss of British Columbian glaciers, Canada. *Geophys. Res. Lett.* **34,** L16503. doi:10.1029/2007GL030780.

Smith, C. A. S., Meikle, J. C. & Roots, C. F. (2004) Ecoregions of the Yukon Territory – biophysical properties of Yukon landscapes. Agriculture and Agri-food Canada, PARC *Technical Bulletin 04-01*, Summerland, British Columbia.

Spence, C. (2002) Streamflow variability (1965–1998) in five Northwest Territories and Nunavut rivers. *Can. Water Resour. J.* **27**, 135–155.

Stahl, K. & Moore, R. D. (2006). Influence of watershed glacier coverage on summer streamflow in British Columbia, Canada. *Water Resour. Res.* **42**, W06201. doi: 10.1029/2006WR005022.

Walvoord, M. A. & Striegl, R. G. (2007) Increased groundwater to stream discharge from permafrost thawing in the Yukon River basin: potential impacts on lateral export of carbon and nitrogen. *Geophys. Res. Lett.* **34.** L12402.

Whitfield, P. H. (2001) Linked hydrologic and climate variations in British Columbia and Yukon. *Environ. Monitoring & Assess.* **67,** 217–238.

Whitfield, P. H. & Cannon, A. J. (2000) Recent climate moderated shifts in Yukon hydrology. *Proceedings of the AWRA Conference Water Resources in Extreme Environments* (ed. by D. L. Kane) (May 2000, Anchorage, Alaska).

Whitfield, P. H. & Taylor, E. (1998) Apparent recent changes in hydrology and climate of coastal British Columbia. In: *Mountains to Sea: Human Interaction with the Hydrologic Cycle* (Proc. 51st Annual Canadian Water Resource Conference (ed. by Y. Alila), 22–29.

Yue, S., Pilon, P. & Phinney, B. (2003) Canadian streamflow trend detection: impacts of serial and cross-correlation. *Hydrol. Sci. J.* **48,** 51–63.

Zhang, X., Harvey, K. D., Hogg, W. D. & Yuzyk, T. R. (2001) Trends in Canadian streamflow. *Water Resour. Res.* **37**, 987–998.

Zemp, M., Hoezle, M. & Haeberli, W. (2007) Six decades of glacier mass-balance observations: a review of the worldwide monitoring network. *Ann. Glaciology* **50**, 101–111.

Yukon River hydrological and climatic changes, 1977–2006

SHAOQING GE, DAQING YANG & DOUGLAS L. KANE
Water and Environment Research Center, University of Alaska, Fairbanks, Alaska, USA
geshaoqing@gmail.com

Abstract This paper analyses long-term hydrology and climate data over the Yukon River basin. It uses regression analysis to define the relationship between the climate and discharge data over the basin. Discharge at the outlet of the basin shows low runoff in the cold season (November to April), with small variations. Flow is high (28 483–177 000 ft^3/s; 807–5012 m^3/s) with high fluctuations in the warm season (May to October). The discharge in May has a positive trend (177 000 ft^3/s; 5012 m^3/s). The mean annual flow is about 227 912 ft^3/s (6454 m^3/s), with high fluctuations; it has increased by 18 213 ft^3/s (or 8%) during the study period. Basin air temperature from 1977 to 2006 increased by 3.9°F (2.2°C) in June and decreased by 10.5°F (5.8°C) in January. Basin precipitation has negative trend in June (0.6 inch; 15.2 mm) with a confidence over 93%. Regression analysis shows a strong and positive correlation between temperature and discharge in May, and a strong and negative correlation between May temperature and June discharge. Precipitation in August and September has strong and positive correlations with basin discharge in September and October.

Key words cold region hydrology; Arctic climate; Yukon River basin; correlation analysis

1 INTRODUCTION

During the last several decades, climate over the Arctic region has experienced significant changes, such as warmer winters (Serreze *et al.*, 2000; Houghton *et al.*, 2002), increasing winter and autumn precipitation in northern Eurasia (Wang & Cho, 1997), greater winter snow depth and enhanced thawing of permafrost in the Arctic and subarctic Russia (Pavlov, 1994). River runoff in the high latitude regions is a critical source of freshwater to the Arctic Ocean. Freshwater systems in arctic and subarctic regions are undergoing profound changes (IPCC, 2001; ACIA, 2005). Freshwater discharge from the northern rivers plays an important role in regulating the thermohaline circulation of the world's oceans (Aagaard & Carmack, 1989). Studies show that both the amount and the timing of freshwater inflow to the ocean systems affect ocean circulation, salinity, sea ice dynamics, and climate (Aagaard & Carmack, 1989; Macdonald, 2000). Therefore, it is of critical importance to understand and quantify the hydrological regimes and changes over large rivers in the northern regions.

Significant climate change has occurred in the Yukon River Basin. Shulski & Wendler (2007) have detected a slight increase in the average annual temperature over the last 30 years in Alaska. Zhao (2004) detected significant warming trends in the northern river basins, including the Yukon River. Yang *et al.* (2009) identified a clear correspondence of river discharge to seasonal snow cover change for the Yukon basin. Brabets *et al.* (2009) found that annual discharge had remained relatively unchanged during 1944–2005, but a few glacier-fed rivers demonstrate positive trends due to enhanced glacier melt. The Yukon River is very important to the Bering Sea ecosystem because it provides most of the freshwater runoff, sediments, and dissolved solutes in the eastern part of the Bering Sea (Lisitsysn, 1969). Therefore, it is necessary to analyse the hydrological regime and its response to climate change over the Yukon River.

2 BASIN DESCRIPTION, DATA SETS AND METHODS

The Yukon River basin (Fig. 1) is located in northwestern Canada and central Alaska. It is the fourth largest river in the North America, with a drainage area of 331 005 sq. miles (857 299 km^2) and average annual discharge of 226 014 ft^3/s (6400 m^3/s) (Brabets *et al.*, 2002, 2009). The discharge data for 1977–2006 (US Geological Survey, http://www.waterdata.usgs.gov/nwis) are from the Pilot station near the basin outlet (Fig. 1). Temperature and precipitation data are from the Alaska Climate Research Center (ACRC, http://climate.gi.alaska.edu) at University of Alaska

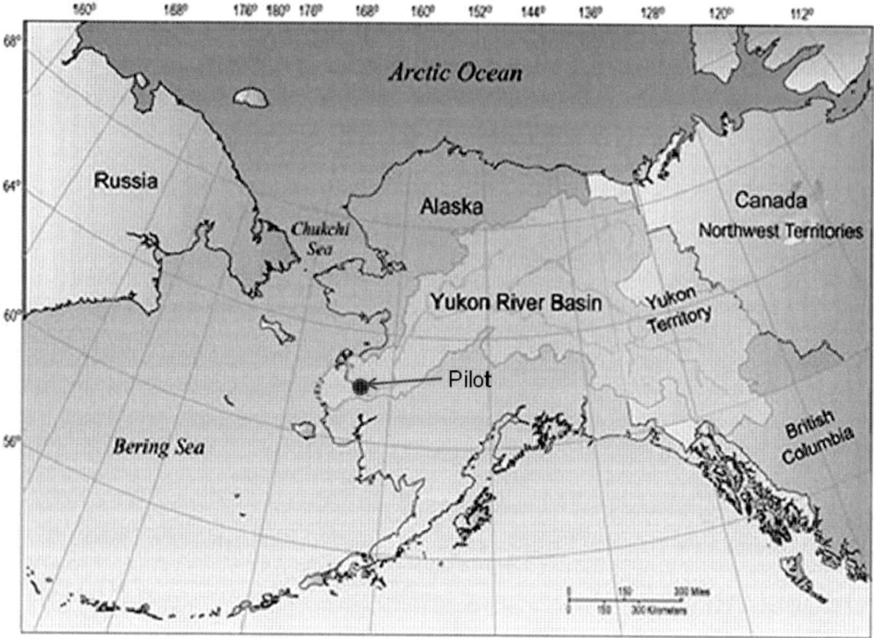

Fig. 1 The Yukon River watershed and location of the Pilot station near the basin outlet (USGS, 2001).

Fairbanks. The monthly temperature and precipitation data from five stations within (and closest to) the Yukon River basin have an overlapping time period (1977–2006) with discharge data.

Monthly mean and standard deviation for monthly discharge, temperature, and precipitation have been calculated to define the hydrological and climatic regimes over the basin. The long-term hydro-climatic changes have been determined by a linear trend analysis. Statistical significance tests are conducted for the trends. Correlation analysis and statistical significance tests are used to assess the associations among climatic and hydrological variables. Based on these analyses, the consistency in flow changes over the basin is examined.

3 DISCHARGE REGIME AND CHANGE

The statistical analysis of monthly flow records at the Pilot station (Fig. 2) shows a low flow period (46 639–125 587 ft^3/s; 1321–3556 m^3/s) during November to April, and a high flow period (252 087–579 747 ft^3/s; 7138–16 417 m^3/s) from May to October, with the maximum flow (579 747 ft^3/s; 16 417 m^3/s) occurring usually in June due to snowmelt floods and ice jams. Trend analyses show an obvious increase by 177 000 ft^3/s (5012 m^3/s) in May, which is significant at 97%; the increase of 4 586 ft^3/s (130 m^3/s) in April is significant at over 85%; the other months have small changes with low confidence levels.

The annual discharge of the Yukon River from 1977 to 2006 is given in Fig. 3. Trend analysis indicates an increase of mean annual flow by 18 213 ft^3/s (516 m^3/s, 8.0%) over 1977–2006. This change is probably caused by the high discharge event in 2005. The peak flow (784 400 ft^3/s; 22 211 m^3/s) in May 2005 has pulled the trend up, as this high flood event caused the higher mean annual flow for this year.

Daily flow data (Fig. 4) show that the flow in the cold season (November to April) is very low and does not change much. This is because the flow in the cold season is dominated by groundwater, which does not change much in winter. But for the warm season (May to October), flow variations are quite large due to rainfall variations. The timing of peak flow at the Pilot Station shifted to an earlier date due to a warmer spring.

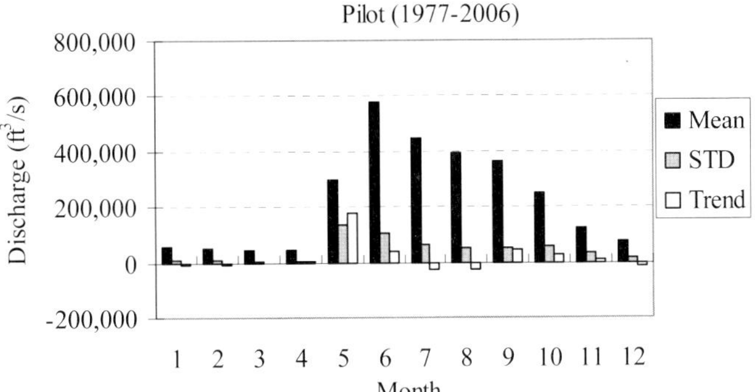

Fig. 2 Mean monthly discharge, standard deviation, and trend at the Pilot Station during 1977–2006.

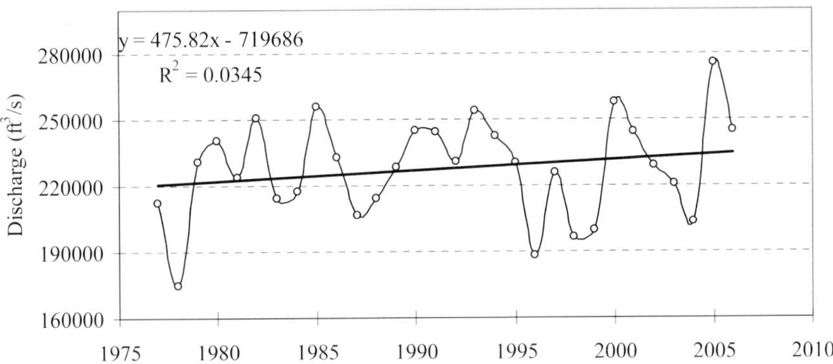

Fig. 3 Yearly discharge record at the Pilot Station during 1977–2006.

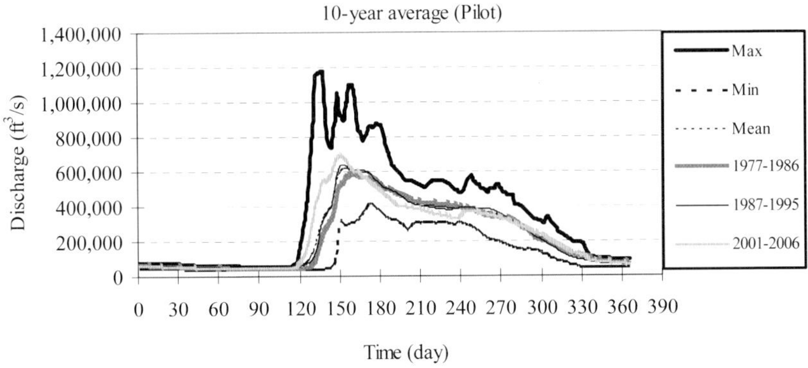

Fig. 4 The 10-year average of daily discharge at the Pilot Station.

4 BASIN TEMPERATURE AND CORRELATION WITH DISCHARGE

Figure 5 shows the monthly mean air temperature over the Yukon River Basin during 1977–2006. The warm season is from May to September (43.9–59.6°F; 6.6–15.3°C) with the maximum temperature in July (59.6°F; 15.3°C); while the cold season is from October to April (–2.3°F to 28.8°F; –19.1°C to –1.8°C). Trend analyses suggest an increase by 3.9°F (2.2°C) in June

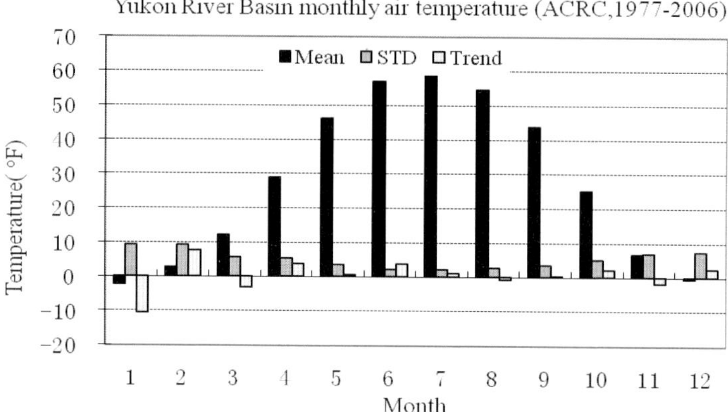

Fig. 5 Mean monthly temperature, standard deviation, and trend over the Yukon River basin (1977–2006).

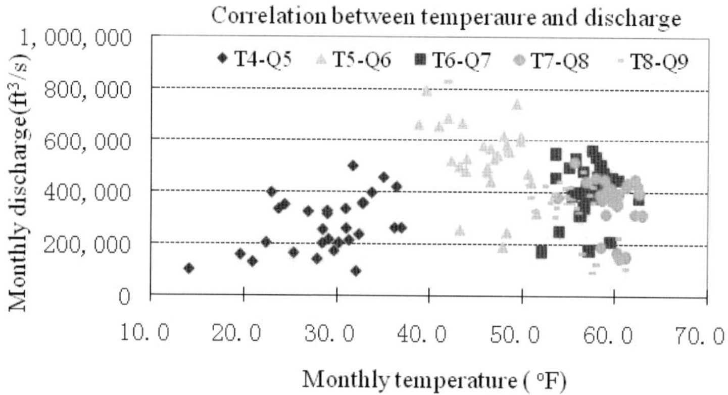

Fig. 6 Correlation between monthly discharge at the Pilot Station and basin mean temperature, with a time lag of one month.

(confidence level at 99.7%); there is a slight decrease of 0.8°F (0.4°C) in August but the confidence level is below 60%; weak positive trends (statistically insignificant) are detected in May (0.6°F; 0.3°C), July (1.0°F; 0.6°C), and September (0.4°F; 0.2°C). During the winter season, the temperature decreased by 10.5°F (5.8°C) in January with a confidence level of 92.3%. It increased in February (7.7°F; 4.3°C), April (3.7°F; 2.1°C), October (2.2°F; 1.2°C), and December (2.6°F; 1.4°C). The positive trends in these months are not statistically significant. The negative trends in March (3.2°F; 1.8°C) and November (1.8°F; 1°C) have confidence levels below 60%. Mean annual temperature during 1977–2006 increased by 0.5°F (0.3°C). Brabets *et al.* (2002) found that the warming rate over the Yukon River during 1949–1996 has been about 0.4°F (0.2°C) per hundred years.

Correlation analyses between basin air temperature and discharge at the Pilot Station (Fig. 6) show a strong positive relation in May. This is probably caused by a stronger and faster snowmelt in warmer springs. Temperature in May has a strong negative correlation with the discharge in June. It is likely that the increased temperature in May causes more snowmelt in May, and subsequently reduces discharge in June.

5 BASIN PRECIPITATION AND CORRELATION WITH DISCHARGE

Figure 7 displays results of statistical analyses for monthly precipitation from 1977 to 2006. It indicates high precipitation period (1.7–2.5 inch; 43.2–63.5 mm) from June to September, with the maximum value in August (2.5 inches; 63.5 mm). The low precipitation period (0.5–1.1 inch; 12.7–27.9 mm) is from October to May. An obvious negative trend is observed in June (0.6 inch; 15.2 mm) with a confidence level over 93%. Precipitation decreased by 0.4 inch (10.2 mm) in December, but with a confidence level lower than 80%. Weak precipitation decreases are detected in March (0.1 inch; 2.5 mm), April (0.1 inch; 2.5 mm), and November (0.1 inch; 2.5 mm). Positive trends are observed in February (0.5 inch; 12.7 mm) with a confidence level greater than 87%, and in January (0.2 inch; 5.1 mm), May (0.3 inch; 7.6 mm), July (0.3 inch; 7.6 mm), August (0.6 inch; 15.2 mm), September (0.3 inch; 7.6 mm), and October (0.1 inch; 2.5 mm), with confidence levels less than 80%. Annual precipitation during 1977–2006 shows an increase by 1.1 inch (27.9 mm).

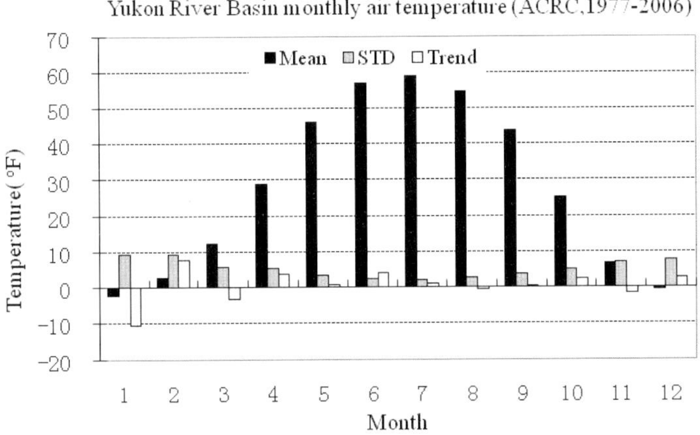

Fig. 7 Mean monthly precipitation, standard deviation, and trend for the Yukon River basin (1977–2006).

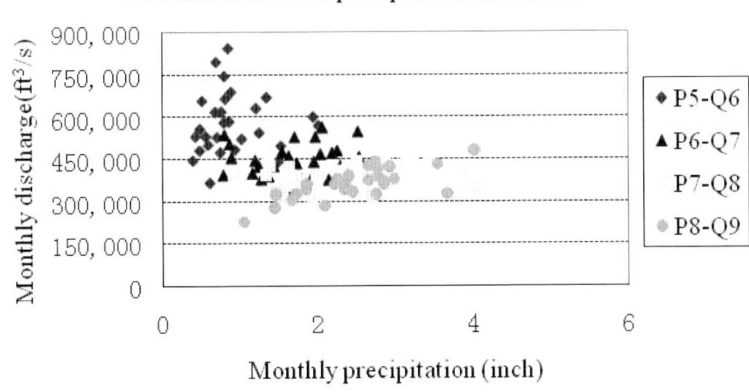

Fig. 8 Correlation between monthly discharge at the Pilot Station and basin mean precipitation, with time lag of one month.

Correlation analyses between basin precipitation and discharge at the Pilot Station (Fig. 8) reveal that the precipitation in August has a strong positive relation with runoff in September. Precipitation in April and May weakly and positively correlates with runoff in May and June, respectively.

6 CONCLUSION

The Yukon River has low flow with small variations in the cold season (November to April) and high flow with large variations in warm season (May to October). Monthly discharge at the Pilot Station increased significantly by 177 000 ft^3/s (5012 m^3/s) or 59% in May and 42 303 ft^3/s (1198 m^3/s) or 7% in June due to a warm spring. Daily discharge analyses show that the peak flow at the Pilot station increased by 67 400 ft^3/s (1909 m^3/s), with timing shifting to an earlier date due to early warming in spring. The annual discharge at the Pilot station has large fluctuations and an increase trend of 18 213 ft^3/s (516 m^3/s, 8.0%) during 1977–2006.

Basin temperature during 1977–2006 strongly increased in winter (December to February) and spring (March to May) by 2.2 to 7.7°F (1.2–4.3°C). The warming is especially significant (3.7°F; 2.1°C) in April, which caused earlier snowmelt over the basin. The basin temperature in April has a strong positive correlation with discharge in May. From 1977 to 2006, the basin precipitation during June–August increased by 0.3 to 0.6 inch (7.6–15.2 mm); and during September to November increased by 0.1 to 0.3 inch (2.5–7.6 mm). Precipitation in August has a strong positive correlation with discharge in September. The total precipitation increase (1.0 inch; 25.4 mm) during May to September contributes to the flow increase in September at the Pilot Station.

Acknowledgements This study was supported by USGS and NSF (Award no. ARC- 0612334).

7 REFERENCES

Aagaard, K. & Carmack, E. C. (1989) The role of sea ice and other fresh water in the arctic circulation. *J. Geophys. Res.* **94**(C10), 14,485– 14,498.

ACIA (2005) *Arctic Climate Impact Assessment.* Cambridge Univ. Press, New York, USA. 1042 pp.

Brabets, T. P. & Walvoord, M. A. (2009) Trends in streamflow in the Yukon River Basin from 1944 to 2005 and the influence of the Pacific Decadal Oscillation. *J. Hydrol.* **371**(1-4), 108–119.

Brabets, T. P., Wang, B. & Meade, R. H. (2000) Environmental and hydrologic overview of the Yukon River Basin, Alaska and Canada. *US Geological Survey Water-Resources Investigations Report 99-4204*, 106 pp.

Houghton, J. T., Ding, Y., Griggs, D. J., Noguer, M., van der Linden, P. J., Dai, X., Maskell, K. &. Johnson, C. A. (2002) *Climate Change 2001: The Scientific Basis.* Cambridge University Press, Cambridge, UK.

IPCC (2001) *Climate Change 2001: Impacts, Adaptation and Vulnerability - Contribution of Working Group II to the Third Assessment Report of the Intergovernmental Panel on Climate Change (IPCC)*, 801–824, Cambridge Univ. Press.

Lisitsysn, A. P. (1969) *Recent Sedimentation in the Bering Sea.* Washington, D.C., National Science Foundation, The Israel Program for Scientific Translations (translated from Russian).

Macdonald, R. W. (2000) Arctic estuaries and ice: A positive-negative estuarine couple. In: *The Freshwater Budget of the Arctic Ocean: Proceedings of the NATO Advanced Research Workshop*, edited by E. L. Lewis *et al.*, 383–407, Kluwer Acad., Norwell, Massachusetts, USA.

Pavlov, A. V. (1994) Current change of climate and permafrost in the Arctic and subarctic of Russia, *Permafrost and Periglacial Processes* **5**, 101–110.

Serreze, M. C., Walsh, J. E., Chapin, E. C., Osterkamp, T., Dyugerov, M., Romanovsky, V., Oechel, W. C., Morison, J., Zhang, T. & Barry, R. G. (2000) Observation evidence of recent change in the northern high-latitude environment. *Climate Change* **46**, 159–207.

Shulski, M. & Wendler, G. (2007) *The Climate of Alaska.* University of Alaska Press, Fairbanks, USA. 143 pp.

Wang, X. L. & Cho, H.-R. (1997) Spatial-temporal structures of trend and oscillatory variabilities of precipitation over Northern Eurasia. *J. Clim.* **10**, 2285–2298.

Yang, D., Zhao, Y., Armstrong, R. & Robinson, D. (2009) Yukon River streamflow response to seasonal snow cover changes, *Hydrol. Processes* **23**, 109–121.

Zhao, Y. (2004) Snow cover runoff assessment in five large northern watersheds, Master Thesis. University of Alaska Fairbanks, Fairbanks, Alaska, USA.

Distinguishing human and climate influences on the Columbia River: changes in the disturbance processes

PRADEEP K. NAIK[1] **& DAVID A. JAY**[2]

1 *Agricultural Engineering and Water Resources, Ministry of Municipality Affairs and Urban Planning, PO Box 31126, Kingdom of Bahrain*
pradeep.naik@water.net.in

2 *Department of Civil and Environmental Engineering, Portland State University, PO Box 751, Portland, Oregon 97207, USA*

Abstract This paper distinguishes human and climate influences on the Columbia River streamflow disturbance regime, examines how this disturbance regime has changed over the last 150 years, and discusses downstream impacts. Flow management and withdrawal have greatly curtailed exceedence of the natural bankfull level of approx. 20 000 $m^3 s^{-1}$. The frequency distribution of Columbia River flow has also changed. Sediment transport is positively correlated with streamflow standard deviation, and has been greatly reduced by flow regulation. Three kinds of spring freshet styles have been identified; there are also four kinds of winter freshets. Flow regulation and regional climate warming have changed freshet styles and reduced their maximum spring intensities. Downstream effects of hydrological alterations include increased salinity intrusion length, loss of shallow water habitat area during the freshet season, increased tides throughout most of the year, and a decrease in area of the Columbia River offshore plume during spring and summer. Although climate changes and variations have played a substantial role in changing the hydrological disturbance regime, their influence is still less than that of human manipulation of the flow cycle.

Key words Columbia River; climate impact; human impact; freshet style; salmonid; overbank flow; disturbance frequency; flow regulation; reservoir manipulation; irrigation depletion

INTRODUCTION

Columbia River Basin (Fig. 1) hydrology has changed due to both human and climate influences (Sherwood *et al.*, 1990; Hamlet & Lettenmaier, 1999; Jay & Naik, 2002; Naik & Jay, 2005). Naik & Jay (2010, 2011) separated anthropogenic and climate impacts on the Columbia River mean flow and sediment transport regimes. We show here that direct human manipulation of river flow through flood control, water withdrawal and hydropower generation, has been the largest single source of disturbance to the physical processes in the system. Still, climate fluctuations and other human activities, such as navigational development, diking and filling, and changes in land use (especially timber harvest), are individually important and interact with river flow manipulation. The hydrological factors discussed here affect not only the river, but also the estuary and the coastal ocean. Moving in a seaward direction, the effects of hydrological alterations on the river are fairly well defined, those on the estuary are becoming clearer, and those on the coastal ocean are quite uncertain.

The Cascade Mountains divide the Columbia River drainage basin into two parts: (1) a Western Sub-basin and (2) an Interior Sub-basin covering ~92% of the total drainage east of the Cascade Range. Understanding the response of the Columbia River Basin to human influences and climate perturbations requires attention to the diverse properties of its sub-basins. There is, for example, wide variability in the percentage flow from the various parts of the basin during the spring freshet and over the water year. The West sub-basin (~8% of basin area, 24% of total flow at the mouth) is very wet. The Canadian part of the Interior sub-basin also has a high runoff production per unit area – Canada accounts for ~50% of the flow at The Dalles (that measures 95% of the flow of the Interior sub-basin), but has only 25% of the total surface area of the Interior sub-basin. The Snake River is relatively dry, with ~40% of the Interior sub-basin surface area, but only 30% of the total flow at The Dalles. In very high flow years, half or more of the streamflow at Dalles is derived from Canada. During the largest known freshet (1894), for example, the peak flow at Grand Coulee was ~20 500 $m^3 s^{-1}$ compared to a maximum flow at The Dalles of 34 800 $m^3 s^{-1}$ and at Beaver of ~39 400 $m^3 s^{-1}$. The Canadian contribution to the spring freshet has

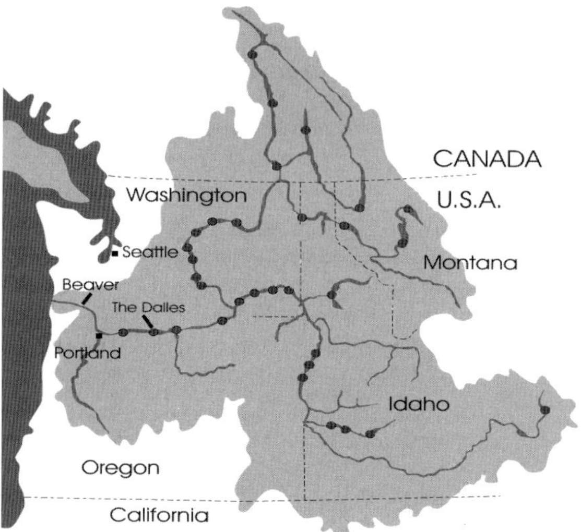

Fig. 1 The Columbia River basin; the Interior basin is east of The Dalles (after US Army Corps of Engineers).

been declining since 1970 due to the large residence time of the dams in Canada. These dams are very effective in changing the seasonality of the flow in the Upper Columbia, which in turn affects streamflow in the entire mainstem (Naik & Jay, 2010).

RESULTS

1 Disturbance frequency The magnitude and frequency of disturbance to the Columbia River system is important to the survival of salmonids. The historical bankfull flow level was approx. 20 000 $m^3 s^{-1}$ for the Columbia mainstem below Vancouver; modern bankfull level has now been set by the standard project flood level of ~24 000 $m^3 s^{-1}$ for the lower river. Some overbank flows did occur in many years before 1900, but now flow regulation (after 1970) and water withdrawal have made overbank flows (above 24 000 $m^3 s^{-1}$) rare, with significant events occurring only five times since 1948. With regard to the incidence of overbank flow now, climate is a secondary factor. Even during cold PDO (Pacific Decadal Oscillation) phases, overbank flow is now rare; it was totally absent during the warm phase of 1977–1995 (Fig. 2).

2 Habitat availability Reduction in overbank flows in the lower Columbia River due to decrease in maximum flow levels, disposal of dredged material, and diking/flood protection measures, have resulted in the reduction of shallow water habitat availability/opportunity during the freshet season, when juvenile salmonid densities are high, by limiting flow out over the historic flood plain and into areas that were previously forested swamp or other types of seasonal wetland.

3 River flow frequency distribution High frequency variations associated with power peaking have been greatly augmented by the dam system, whereas low-frequency flow variations with periods of between ~2 years and 6 months have been substantially suppressed. Also, the diurnal tidal signal in the river has been perturbed by the daily power peaking cycle (Fig. 3).

4 Spring freshet styles The flow cycle is different each year, but three recurring patterns of spring freshets have been identified: (a) a large winter snow pack without exceptional spring rain, (b) a normal winter snow pack followed by a very wet spring, and (c) a large winter snow pack combined with heavy spring rains. The largest known freshet that occurred in 1894 was of type (c), and the second largest freshet that occurred in 1948 was of type (b).

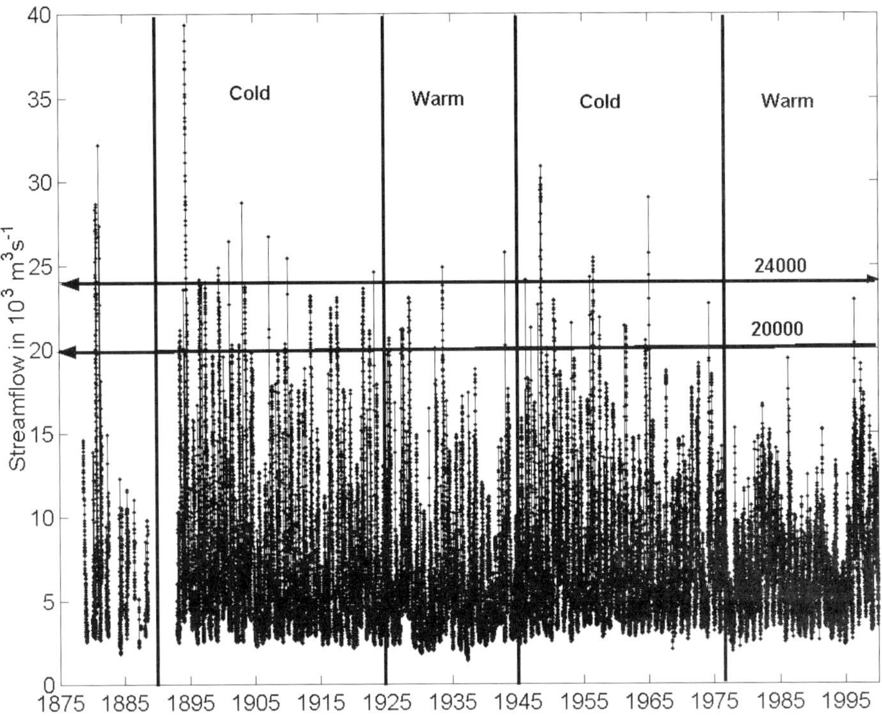

Fig. 2 Overbank flows exceeding 20 000 $m^3 s^{-1}$ and 24 000 $m^3 s^{-1}$ in cold and warm PDO phases in the Columbia River at Beaver.

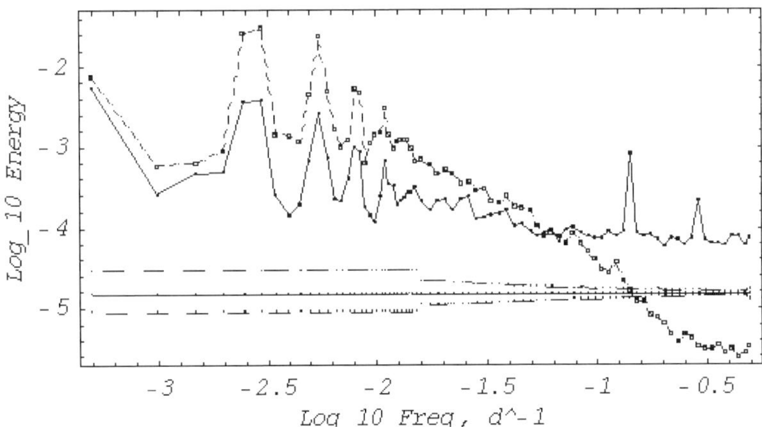

Fig. 3 Power spectra of The Dalles observed daily flow 1878–1910 and 1970–1999. The peaks at 1 year and 6, 4, and 3 months, clearly visible in the 1878–1910 record, have been greatly reduced by flow regulation and irrigation depletion. The power peaking cycle has added energy to the system at frequencies above ~20 days, but especially at 7 and 3.5 days. Also shown are 95% confidence limits.

5 Winter freshet styles There are also four types of winter freshets, based on the source of streamflow: (a) primarily Western sub-basin with extensive snowmelt, (b) Interior and Western sub-basins, (c) primarily Interior sub-basin, and (d) primarily Western sub-basin without extensive

(a)

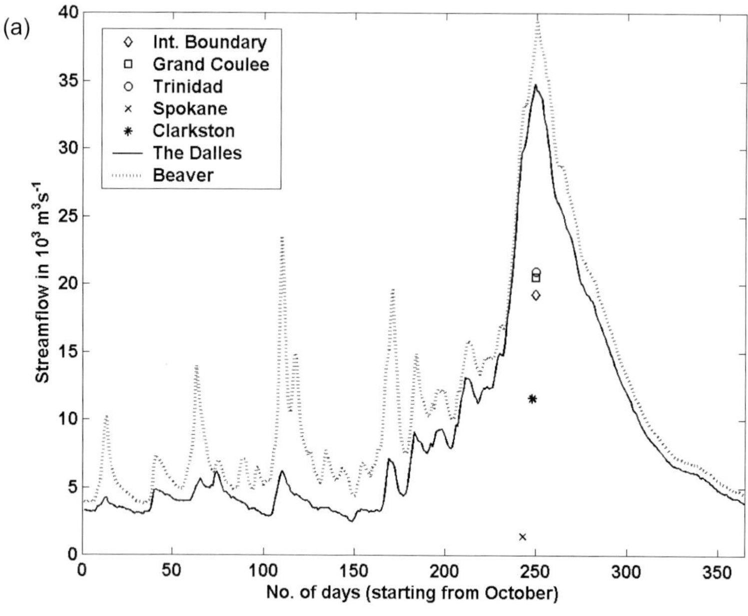

(b)

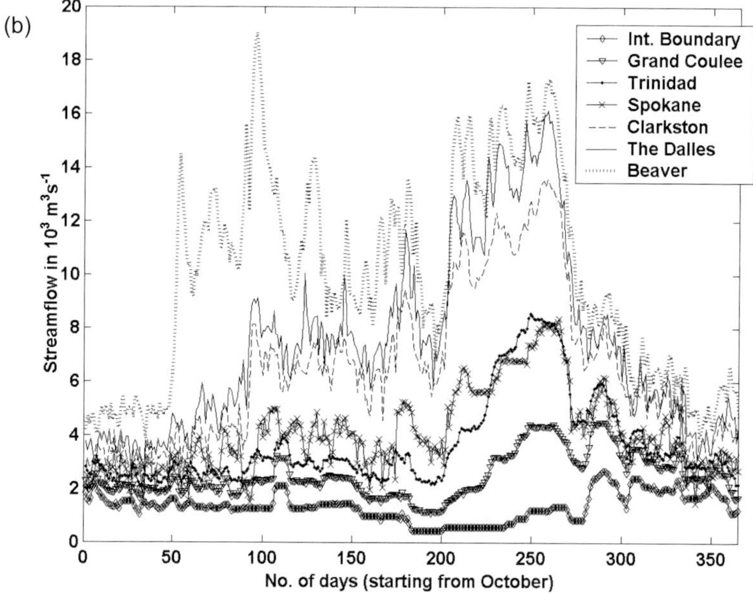

Fig. 4 (a) An example of a significant winter snowmelt event occurring before a major spring freshet (1894 water year); (b) an example of flow regulation decreasing spring freshet magnitude and increasing flows during the rest of the year (1997 water year); and (see opposite) (c) comparison of flow hydrographs for two low-flow years of the Columbia River at The Dalles (1926 and 1977). Regulation and power peaking effects are so prominent in 1977 that there is little annual cycle; unregulated flows in 1926 show a distinct, but attenuated, annual cycle.

snowmelt. All winter freshets, except the fourth type, are generated by rain-on-snow events. The largest known freshets (e.g. 1861, 1881 and 1892) involved both Interior and Western sub-basins. However, the Canadian part of the Interior sub-basin is not generally affected by these floods.

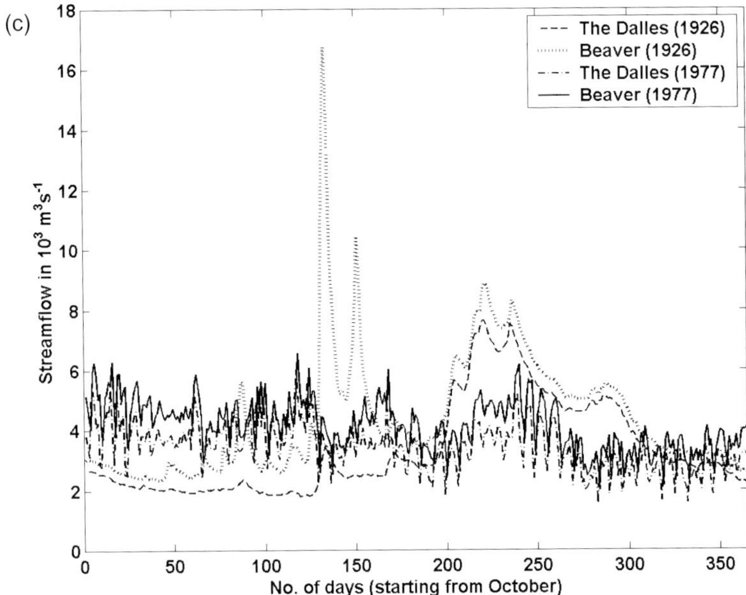

Fig. 4 Continued. (c) see caption opposite.

6 Downstream effects of hydrological alterations Changes in streamflow and sediment transport have exerted an important influence on the Columbia River estuary, but these changes are not yet well understood. Almost totally unknown are the effects of these hydrological changes on the Columbia River buoyant plume in the coastal ocean. Future studies should include the downstream effects of hydrological change by: salinity intrusion and salinity stratification, habitat availability, the fluvial tidal regime, estuarine sediment dynamics, Columbia River plume area, volume, turbidity, and seasonality.

7 Effects of future climate change Although major changes during the past 150 years in Columbia River hydrological processes have resulted primarily from human manipulation and secondarily from climate, it is still vital to consider how climate will constrain future management options in the river basin, including efforts to restore depleted salmon populations. Climate projections forecast gradual regional warming, possibly accompanied by higher winter precipitation. This may lead to increased incidences of winter freshets and lower natural spring freshet flows. These hydrological changes would aggravate conflicts over water supply during the critical spring freshet period, by decreasing natural flows and increasing water demand.

DISCUSSION

Spring freshets are especially important to downstream migration of juvenile salmons (Bottom *et al.*, 2005). Large freshets can also affect salmonid populations by modifying habitat structure and distribution. On the other hand, water quality and habitat availability may decline during very dry years, effects that are further exacerbated by human alterations like diking and irrigation diversion. Important lessons can be learnt from the flow histories of extreme years because salmon are most severely tested under stressful climatic conditions.

Very large freshets that occurred in the pre-regulation period before 1900 lasted 30–60 days, with the sharpness of the peak governed by the relative timing of snowmelt throughout the basin. Significant snowmelt events occur in many winters before a major spring freshet, e.g. 1862, 1881,

1894, 1948, 1956, 1974 and 1997, reducing the intensity of the following spring freshet (e.g. 1894, Fig. 4(a)) and subsequent summer flows. The occurrence of major freshets even after winter snowmelt events indicates the high magnitude of the snow pack in such years. Flow regulation decreases spring freshet magnitude and increases flows during the rest of year through winter drawdown of reservoirs, filling of the reservoirs during the freshet and de-synchronization of flow peaks throughout the basin. The result is a "spring" freshet in high-flow years like 1996 and 1997 that lasts from January to June (Fig. 4(b)). The effects of human manipulation (especially a weekly power-peaking cycle) are also very prominent during very low flow years like 1977 and 1992. The spring freshet in such years is now an almost totally artificial event. The closest pre-regulation analog to such years (1926) still showed a marked annual cycle, though it was of reduced intensity (Fig. 4(c)).

The Columbia River has historically been a major source of economic activity for the Pacific Northwest, and is one of the more heavily modified rivers in the USA today. Understanding human and climate-induced changes in its hydrological properties is, therefore, a topic of considerable interest. Long streamflow records are essential to determining how runoff has changed over time. Daily streamflow records of the Columbia River at The Dalles dates back to June 1878. However, the observed daily flow does not alone provide enough information to understand or separate anthropogenic and climate effects. It is also necessary to have an estimate of virgin (naturalized) flow of the river to provide a historical perspective of water resources development, to separate anthropogenic and climate effects, and to compare present water use scenarios with those of the past decades. The United State Geological Survey (USGS) has calculated a monthly averaged adjusted river flow at The Dalles since 1879 that accounts for the effects of flow regulation. The Bonneville Power Administration (BPA) has estimated the monthly averaged virgin (or naturalized) flow at The Dalles, i.e. the flow in the absence of both flow regulation and irrigation depletion for 1929–2000 (BPA, 2004). Naik & Jay (2005) have estimated the monthly virgin flow of the Columbia River at The Dalles from records of irrigated area for the missing years, i.e. for the period 1879–1928. Examination of the virgin flow record shows that climate change since the late 19th century has caused a decrease of 8–9% in its annual average flow volume. The decrease in flow due to irrigation diversion during the same period is also 7–8% (Naik & Jay, 2011). Broadly speaking, there are three periods of Columbia River flow management. Before 1900, mainstem dams were absent and flow diversions relatively small. Numerous dams were constructed between 1900 and 1970, and irrigation depletion increased 500%. Since about 1970, river flows have been managed on a system-wide basin, significantly affecting interannual transfers of flows for the first time.

REFERENCES

BPA (Bonneville Power Administration) (2004) *2000 Level Modified Streamflow*. Bonneville Power Administration, Portland, various paginations.

Bottom, D. L., Simenstad, C. A., Burke, J., Baptista, A. M., Jay, D. A., Jones, K. K., Casillas, E. & Schiewe, M. H. (2005) Salmon at river's end: the role of the estuary in the decline and recovery of Columbia River salmon. *NOAA Tech. Memo., NMFS-NWFSC-68*, 246 pp., US Dept. of Commerce.

Hamlet, A. F. & Lettenmaier, D. P. (1999) Effects of climate on hydrology and water resources in the Columbia River basin. *J. Am. Water Resour. Assoc.* **35**(6), 1597–1623.

Jay, D. A. & Naik, P. K. (2002) Separating human and climate impacts on Columbia River hydrology and sediment transport. In: *Southwest Washington Coastal Erosion Workshop Report 2000* (ed. by G. Gelfenbaum & G. M. Kaminsky), 38–48, *US Geol. Surv. Open-File Report 02-229.*

Naik, P. K. & Jay, D. A. (2005) Estimation of Columbia River virgin flow: 1879–1928. *Hydrol. Processes* **19**(9), 1807–1824.

Naik, P. & Jay, D. A. (2010) Human and climate impacts on Columbia River hydrology and salmonids. *J. River. Res. Applic.* doi: 10.1002/rra.1422.

Naik, P. K. & Jay, D. A. (2011) Distinguishing human and climate influences on the Columbia River: changes in mean flow and sediment transport. *J. Hydrol.* (submitted).

Sherwood, C. R., Jay, D. A., Harvey, B., Hamilton, P. & Simenstad, C. A. (1990) Historical changes in the Columbia River estuary. *Prog. Oceanogr.* **25**, 299–352.

Effect of streamflow regulation on mean annual discharge variability of the Yenisei River

SVETLANA STUEFER[1], DAQING YANG[1] & ALEXANDER SHIKLOMANOV[2]

1 *Water and Environmental Research Center, University of Alaska Fairbanks, Fairbanks, Alaska 99775, USA*
sveta.stuefer@alaska.edu

2 *Water System Analyst Group, University of New Hampshire, Durham, NH 03824, USA*

Abstract The magnitude of natural and anthropogenic changes in hydrological systems is one of the major scientific questions yet to be addressed. Relative to climatic effects, dam impacts are much more direct and often cause abrupt changes in the water regimes of rivers. We expect these changes to be evident and detectable in the mean annual discharge (MAD) records and discharge–precipitation relationship of the Yenisei River, Siberia, Russian Federation. We use statistical analysis to compare three periods: (a) natural streamflow (1936–1956), (b) filling of reservoirs (1957–1980), and (c) operation of reservoirs (1981–2006). Comparison of reconstructed and observed MAD suggests that streamflow regulation affects the homogeneity of the MAD between filling of reservoirs and operation periods. We conclude that dam regulation in the Yenisei River is strong enough to modify the MAD response to annual precipitation, particularly during the 1980–2004 period.

Key words discharge; precipitation; Yenisei River; dam; reservoir; streamflow regulation; Arctic

INTRODUCTION

Among the Arctic watersheds, the Yenisei River is most affected by anthropogenic impacts. The total storage of operating and constructed reservoirs in the Yenisei River watershed is 482 km^3 (Shiklomanov *et al.*, 2000). The Yenisei River is referred to as "strongly affected by flow regulation/fragmentation", based on a worldwide classification of fragmentation of the river channels by dams and by water regulation resulting from reservoir operation, interbasin diversion, and irrigation (Dynesius & Nilsson, 1994). Babkina (1988) reported that as of 1986, there were 64 reservoirs in the Angara-Yenisei basin. In addition to the large storage capacity, the cascading reservoirs upstream of the Yenisei and Angara rivers confluence were designed for water redistribution from year to year. The effect of streamflow regulation on seasonal hydrology has been clearly documented for the Siberian Rivers (Shiklomanov, 1978; McClelland *et al.*, 2004; Yang *et al.*, 2004; Berezovskaya *et al.*, 2005; Adam *et al.*, 2007). Seasonal streamflow regulation during spring and early summer reduces the peak discharge, and releases additional water from the reservoirs during the winter months. While it is evident that streamflow regulation has a marked effect on the seasonality of discharge, its effects on the mean annual discharge (MAD) is more complex and often neglected for the large rivers. The primary impacts of water regulation resulting from reservoir operation on MAD include: (a) filling of the reservoirs, and (b) long-term intra-annual streamflow regulation; that is, water is captured in the reservoir and then released in the following years. Such streamflow regulation directly affects the MAD variability.

The magnitude of natural and anthropogenic changes in hydrological systems is one of the major scientific questions yet to be addressed. Our research into this question has been prompted by the fact that, relative to climatic effects, dam impacts are much more direct and often cause abrupt changes in the water regimes of rivers. We expect these changes to be evident and detectable in the mean annual discharge (MAD) records and discharge–precipitation relationship of the Yenisei River. We analyse observed and reconstructed Yenisei River MAD and quantify the effect of streamflow regulation by reservoirs on the MAD and precipitation relationship. The reservoir effect was eliminated in reconstructed MAD using the approach presented by Shiklomanov & Lammers (2009). Our analysis mainly relies on historical measurements from the conventional network. New data on water use and water consumption is also presented.

DATA

Mean annual discharge

Mean annual discharge of the Yenisei River at the Igarka station (67.4°N, 86.5°E) with a drainage area of 2440 × 10^3 km^2 (Lammers *et al.*, 2001) was collected from 1936 to 2006 (Fig. 1). Mean annual discharge is the most accurate observational component of the water balance; its accuracy is 6.1% for the Igarka gauging station (Shiklomanov *et al.*, 2006).

Reconstructed discharge

The reconstruction of MAD for the Yenisei River basin from 1957 to 2006 was implemented using the hydrograph routing model (HRM). The HRM was developed at the University of New Hampshire (USA) in collaboration with the Arctic and Antarctic Research Institute (Russia) (Shiklomanov, 1996; Shiklomanov & Lammers, 2009). The model, based on the genetic formula or Duhamel integral, operates at the sub-basins between gauges. In the HRM, the sub-basin is considered a black box that transforms the inlet hydrograph, $Q'^{(t-\tau)}$, into the outlet hydrograph, $Q_{(t)}$:

$$Q_{(t)} = \int_0^t Q'^{(t-\tau)} P_{(\tau)} \mathrm{d}\tau \tag{1}$$

$P_{(\tau)}$ is the influence function (the travel curve), which is approximated with a two-parameter equation suggested by Nash (1958) and Kalinin *et al.* (1969):

$$P(\tau) = \frac{1}{\tau(n-1)} \left(\frac{t}{\tau}\right)^{n-1} e^{-\left(\frac{t}{\tau}\right)} \tag{2}$$

where n is the number of sub-basins with the same travelling time, τ. The HRM works on a daily time step and uses information from 35 major hydrometric sites across the Yenisei River basin. HRM provides more accurate hydrograph simulations (Shiklomanov & Lammers, 2009) than the approaches based on comparison of mean pre-dam and post-dam hydrographs (e.g. McClelland *et al.*, 2004; Yang *et al.*, 2004) and it reduces the uncertainty associated with water balance modelling approaches for hydrograph simulation (e.g. Adam *et al.*, 2007) as it uses only observed river discharge, which is the most accurately measured component of the hydrological cycle (Shiklomanov *et al.*, 2006).

Mean annual reconstructed discharge was calculated from daily data and used to represent MAD from 1957 to 2006 (Fig. 1). The cumulative difference between reconstructed and observed Yenisei MAD from 1957 to 2006 is calculated at 604 km^3. Adam *et al.* (2007) estimated 580 km^3 of runoff reduction due to streamflow regulation from 1956 to 1999. This estimate is comparable with HRM results over the same time period (567 km^3).

Annual precipitation

There are difficulties in using gauge-based precipitation data sets for water budget analyses and predicting streamflow in large Siberian river basins (e.g. Fekete *et al.*, 2005). These problems are associated with the construction of precipitation data sets in the northern regions, because of underestimation of solid precipitation by gauges, lack of observations in the mountain regions, and the varying numbers and locations of stations in the networks. On a temporal scale, station networks across Eurasia gave rise to an overestimation of annual precipitation during earlier years (Rawlings *et al.*, 2006). For the largest Siberian basins, Adam & Lettenmaier (2007) showed that there is close agreement among five available gauge-based gridded precipitation products. Those products are CRU TS 2.0 (Climate Research Unit of the University of East Anglia); UDel Arctic (University of Delaware; Willmott & Matsuura, 2005); PREC/L (precipitation over the land); Vasclimo (Variability Analyses of Surface Climate Observations at the Global Precipitation Climatology Centre); and UW (University of Washington). Among the five products, the best fit to the length of discharge records is represented by UDel from 1930 to 2004, whereas UW covers the period from 1930 to 1989. Pavelsky & Smith (2006) demonstrate that discrepancies in long-term precipitation variability between UDel and UW are small compared with the discrepancies that exist between river discharge and any of the precipitation products. Working with historical data, we realize that any long-term assessment of precipitation variability depends on conventional gauge observations. We chose to use the original UDel product at the resolution 0.5° × 0.5°, since it yields a similar variability to other products and yet best represents the period of discharge records. The UDel monthly precipitation was summed over the year, and the basin average was calculated (Fig. 1).

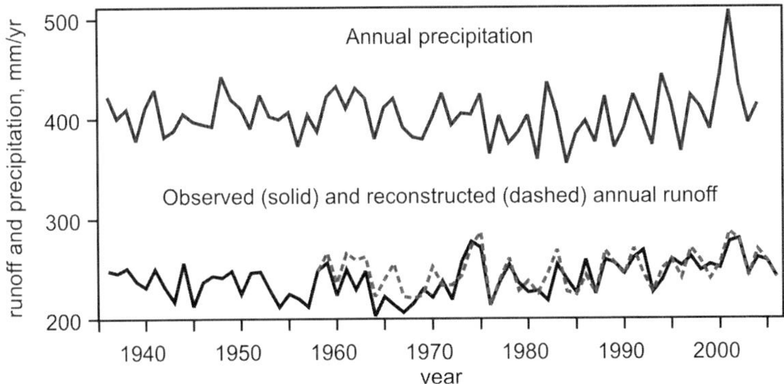

Fig. 1 Observed and reconstructed mean annual runoff and annual precipitation (mm/year).

METHODS

Assessment of reservoir impact on MAD

There are seven large (with capacity greater than 1 km^3) operational reservoirs in the Yenisei River watershed: Bratskoe (169.3 km^3), Irkutskoe (48.1 km^3), Krasnoyarskoe (73.3 km^3), Ust'-Ilimskoe (59.4 km^3), Sayano-Shushenskoe (31.3 km^3), Ust-Khantaiskoe (23.5 km^3) and Kureiskoe (9.9 km^3) (Malik *et al.*, 2000). After reviewing extensive Russian literature on streamflow regulation effects, we highlight a study by Shiklomanov & Veretennikova (1978), where decrease in the Yenisei River MAD was quantified based on reservoir parameters, the hydrological budget of each reservoir, and maps of mean evaporation. Evaporation increase was attributed to the additional evaporation from the reservoir's flooding zone and assumed no climate change. Shiklomanov & Veretennikova (1978) calculated actual MAD decrease from 1936 through 1970 and projected MAD decrease from 1971 through 2000 (Table 1). Their analysis was performed for all reservoirs with a volume greater than 0.05 km^3. Future projections were exaggerated; they assumed that total useful reservoir storage in the Yenisei and Lena River watersheds would increase by 900 km^3 by the end of the 20th century. In reality, only two large reservoirs in the Yenisei basin were filled after 1975: the Sayano-Shushenskoe (capacity 31 km^3) and Kureiskoje (capacity 10 km^3) (Malik *et al.*, 2000). The Bogucharskoe reservoir (58.6 km^3) is still under construction.

Table 1 Reduction in mean annual discharge (km^3/year) under the effect of economic activity (Shiklomanov & Veretennikova, 1978). Long-term (1936–2006) average MAD is 585 km^3/year.

Period of averaging	1936–1940	1941–1950	1951–1955	1956–1960	1961–1965	1966–1970	1971–1975	1976–1980	1981–1985	1986–1990	1991–2000
Evaporation	0	0	0	0.01	0.20	0.51	0.90	1.05	1.19	1.62	3.30
Storage	0	0	0	0.36	19.5	19.7	4.03	15.7	11.9	16.5	39.3
Total	0	0	0	0.37	19.7	20.2	4.93	16.7	13.1	18.1	42.6

RESULTS

Examination of mean annual discharge

Long-term MAD data from 1936 to 2006 allow us to separate the records into three similar periods in length, so as to represent different stages of streamflow regulation by reservoirs in the Yenisei River watershed. We examine the homogeneity of the mean, variance, and distribution function of MAD during these three periods: (1) natural regime (no streamflow regulation) from 1936 to 1956, (2) reservoir filling and initial operation (most of the dead volume was filled during this period) from 1957 to 1980, and (3) operation of reservoirs from 1981 to 2006 (Table 2). We assume that filling and operation of reservoirs with a capacity less than 1 km^3 cannot be detected in the annual discharge of the Yenisei River outlet because of its extremely large watershed area. The period of filling and initial operation in the Yenisei basin refers to 1957–1980, because six reservoirs with a total capacity of 404.9 km^3 were constructed and filled during this period. Afterward, only the Kureiskoje reservoir (9.9 km^3) was filled (from 1986 to 1990).

Table 2 Consistency analysis of mean, variance and distribution functions for natural streamflow regime, reservoir filling and operation periods.

Test	Natural	Filling and initial operation	Operation
Yenisei River	1936–1956	1957–1980	1981–2006
Length (years)	21	24	26
Mean (cubic km per year)	575	569	608
Coefficient of variation	0.06	0.09	0.06
Chi-square to normal distribution	h=0	h=0	h=0
F test		h=0 (p=0.066)	h=0 (p=0.251)
T test		h=0 (p=0.636)	h=1 (p=0.003)
Kolmogorov-Smirnov		h=0 (p=0.558)	h=1 (p=0.004)
Yenisei River Reconstructed	1936–1956	1957–1980	1981–2006
Length (years)	21	24	26
Mean (cubic km per year)	575	592	611
Coefficient of variation	0.06	0.08	0.07
Chi-square to normal distribution	h=0	h=0	h=0
F test		h=0 (p=0.088)	h=0 (p=0.601)
T test		h=0 (p=0.181)	h=0 (p=0.156)
Kolmogorov-Smirnov		h=0 (*p*=0.145)	h=0 (*p*=0.073)

The result h is 1 if the test rejects the null hypothesis at the 5% significance level (two-tailed tests); 0 otherwise. The p-value of a test is given in parentheses.

Hypothesis tests on equal mean (T test), variance (F test) and distribution function show that the biggest difference (23 km^3) between the reconstructed and observed mean MAD was detected during the second period "Filling and initial operation". Results show that streamflow regulation affected homogeneity of the mean in Yenisei MAD, causing lower observed MAD as if it would be without construction of reservoirs. Two tests (equal means and the same distribution functions) reject the null hypothesis at less than 5% significance level between "Filling and initial operation" (period 2) and "Operation" (period 3). Once the same hypothesis is tested on the reconstructed MAD, no tests are rejected (Table 2). This implies that the streamflow regulation does affect homogeneity of mean MAD and could influence trends detected in the observed MAD. Detailed trend analysis of the Yenisei River discharge (Shiklomanov & Lammers, 2009) showed that Sayano–Shushenskoe and Krasnoyarskoe reservoirs exert a considerable influence on MAD trend over the upper part of the river. Among others, Sayno-Shushenskoe and Krasnoyarskoe reservoirs are designed for long-term intra-annual streamflow regulation. These reservoirs are located upstream of confluence of the Yenisei and Angara rivers, in the upper part of the Yenisei River watershed.

Mean annual discharge and precipitation relationship

A linear fit between discharge and precipitation can explain 10 to 11% of the variability in the observed MAD (Fig. 2(a) and (b)) and 33 to 49% variability in the reconstructed MAD (Fig. 2(c) and (d)). The relationship between annual precipitation and MAD improves significantly for the reconstructed discharge, especially during the 1981–2004 period. Hydrological interpretations of discharge–precipitation scatter plots suggest that streamflow regulation in the Yenisei River watershed is strong enough to affect MAD response to precipitation. The reconstructed MAD reveals much higher R-squared in relation to precipitation, whereas observed records exhibit a weak linkage during reservoir operation. This shows that reconstruction is useful for describing river system response to the climatic forcing, particularly precipitation, in the Yenisei River watershed.

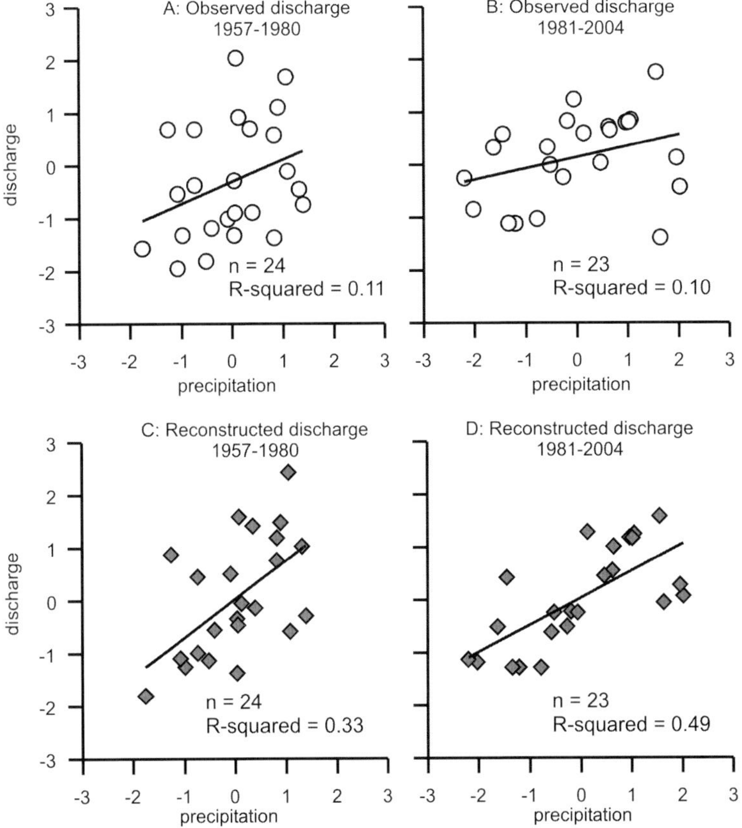

Fig. 2 Relationship between standardized annual precipitation and MAD for the "Filling & Operation" period (1957–1980) and "Operation" period (1981–2004). Length of "Operation" period is limited by precipitation data availability (1936–2004). We use the de-trended and standardized discharge and precipitation data (zero mean, unit standard deviation) so that they can be compared with each other.

CONCLUSIONS

Climate change and human activities affect watershed hydrology. We used reconstructed and observed discharge data to segregate climatic effects and the effect of streamflow regulation by reservoirs in the Yenisei River watershed. We demonstrate clear and visible changes in discharge records and in the discharge–precipitation relationship during three periods: (a) natural streamflow

(1936–1956), (b) filling of reservoirs (1957–1980), and (c) operation of reservoirs (1981–2006). Results show that the relationship between annual precipitation and mean annual discharge is distorted after 1957 due to construction and operation of large reservoirs. The interesting result is that long-term intra-annual streamflow regulation in the upper part of the Yenisei River watershed significantly affects mean annual discharge, even downstream at the outlet station Igarka, and distorts the relationship between discharge and basin-wide precipitation. We conclude that streamflow regulation resulting from construction and operation of reservoirs limits use of the observed Yenisei River discharge for climate change analysis not only at seasonal, but also at intra-annual time scale.

Acknowledgements The study was supported by the National Science Foundation grant ARC-0612334. Data support from the Arctic RIMS project is greatly appreciated.

REFERENCES

Adam, J. C. & Lettenmaier, D. P. (2007) Exploring the roles of precipitation and temperature changes on the variability of Eurasian arctic river discharge. *J. Geophys. Res.* **112**, D24114, doi:10.1029/2007JD008525.

Adam, J. C., I. Haddeland, F. Su & Lettenmaier, D. P. (2007) Simulation of reservoir influences on annual and seasonal streamflow changes for the Lena, Yenisei, and Ob' Rivers. *J. Geophys. Res.* **112**, D24114, doi:10.1029/2007JD008525.

Babkina, I. V. (1988) Economical use of reservoirs in the Angaro-Yenisei basin. In: *Siberian Water Resources: Study, Management and Control* (ed. by A. V. Petenkov), 94–99, Krasnoyarsk (in Russian).

Berezovskaya, S., Yang, D. & Kane, D. (2004) Compatibility of precipitation and runoff trends over the large Siberian watersheds. *Geophys. Res. Lett.* **31**, L21502, doi:10.1029/2004GL021277.

Berezovskaya, S., Yang, D. & Hinzman, L. (2005) Long-term annual water balance analysis of the Lena River. *Global Planet. Change* **48**, 84–95.

Dynesius, M. & Nilsson, C. (1994) Fragmentation and flow regulation of river system in the northern third of the world. *Science* **266**, 753–761.

Fekete, B. M., Vörösmarty, C. J., Roads, J. O. & Willmott, C. J. (2004) Uncertainties in precipitation and their impacts on runoff estimates. *J. Clim.* **17**, 294–304.

Kalinin, G. P., Kuchment, L. S. & Koren, V. I. (1969) Principles for elaboration of mathematical runoff models. In: *Floods and their Computation: Proceedings of the Leningrad Symposium*, vol.1, 60–65. UNESCO publication.

Malik, L. K., Koronkevich, N. I., Zaitseva, I. S. & Barabanova, E. A. (2000) Development of dams in the Russian Federation and NIS countries. A WCD briefing paper prepared as an input to the World Commission on Dams, Cape Town, www.dams.org.

McClelland, J. W., Holmes, R. M., Peterson, B. J. & Stieglitz, M. (2004) Increasing river discharge in the Eurasian Arctic: Consideration of dams, permafrost thaw, and fires as potential agents of change. *J. Geophys. Res*, **109**(D18), 10.1029/2004JD004583.

Nash, J. E. (1958) The form of the instantaneous unit hydrograph. In: *IAHS Darcy Symposium*, vol. 3, *Crues/Floods*, 114–118. Available at: www.iahs.info/redbooks/042.htm.

Pavelsky, T. M. & Smith, L. C. (2006) Intercomparison of four global precipitation data sets and their correlation with increased Eurasian river discharge to the Arctic Ocean. *J. Geophys. Res.* **111**, D21112, doi:10.1029/2006JD007230.

Rawlins, M. A., Willmott, C. J., Shiklomanov, A., Linder, E., Frolking, S., Lammers, R. B. & Vörösmarty, C. J. (2006) Evaluation of trends in derived snowfall and rainfall across Eurasia and linkages with discharge to the Arctic Ocean. *Geophys. Res. Lett.* **33**, L07403, doi:10.1029/2005GL025231.

Shiklomanov, A. I. (1996) Influence of human activity and climate change on runoff in the Yenisei River basin. PhD Thesis, Arctic and Antarctic Research Institute, St Petersburg, Russia.

Shiklomanov, A. I., Yakovleva, T. I., Lammers, R. B., Karasev, I. Ph., Vörösmarty, C. J. & Linder, E. (2006) Cold region river discharge uncertainty – estimates from large Russian rivers. *J. Hydrol.* **326**, 231–256.

Shiklomanov, A. I. & Lammers, R. B. (2009) Record Russian river discharge in 2007 and the limits of analysis. *Environ. Res. Lett.* **4**, 1–9.

Shiklomanov, I. A. (1978) Dynamics of anthropogenic changes in annual river runoff in the USSR. *Trudy GGI* **239**, 3–26.

Shiklomanov, I. A. & Veretennikova, G. M. (1978) Effect of reservoirs on the annual runoff of the rivers in the USSR. *Soviet Hydrology: Selected papers* **17**(1), 23–32.

Shiklomanov, I. A., Shiklomanov, A. I., Lammers, R. B., Peterson, B. J. & Vörösmarty, C. J. (2000) The dynamics of river water inflow to the Arctic Ocean. In: *The Freshwater Budget of the Arctic Ocean* (ed. by E. L. Lewis, E. P. Jones, P. Lemke, T. D. Prowse & P. Wadhamns), 281–296. Kluwer Academic Publishers, Dordrecht, The Netherlands.

Willmott, C. J. & Matsuura, K. (2005) Arctic land-surface precipitation: 1930–2004 gridded monthly time series (v. 1.03). Center for Climatic Research, University of Delaware, Newark.

Yang, D., Ye, B. & Kane, D. (2004) Streamflow changes over Siberian Yenisey River basin. *J. Hydrol.* **296**, 59–80.

Yang, D., Kane, D., Zhang, Z., Legates, D. & Goodison, B. (2005) Bias-corrections of long-term (1973–2004) daily precipitation data over the northern regions. *Geophys. Res. Lett.* **32**, L19501, doi:10.1029.

Hydrological process change with air temperature over the Lena Basin in Siberia

B. YE[1], D. YANG[2], T. ZHANG[3], Y. ZHANG[4] & Z. ZHOU[1]

1 *State Key Laboratory of Cryospheric Sciences, Cold & Arid Regions Environmental and Engineering Research Institute (CAS), Lanzhou, China*
yebs@lzb.ac.cn

2 *Water and Environment Research Center, University of Alaska Fairbanks, Fairbanks, USA*

3 *National Snow and Ice Data Center, Cooperative Institute for Research in Environmental Sciences, University of Colorado, Boulder, Colorado, USA*

4 *Institute of Tibetan Plateau Research, Chinese Academy of Sciences*

Abstract We use long-term monthly discharge and sub-basin air temperature data in the Lena River to examine the relationship between hydrological processes and permafrost change. The ratio of the maximum to minimum monthly discharge (Qmax/Qmin) decreased, while the recession coefficient in the cold season (Qapr/Qdec, discharge in April *vs* discharge in November) increased over the upper Lena and Aldan sub-basin during 1936 to 2000. The annual basin air temperature (AT) has increased from 1940 to 2000. There is a significant relationship between Qmax/Qmin, Qapr/Qdec and AT. The positive relationship between Qapr/Qdec and AT, and the negative relationship between Qmax/Qmin and AT became significant from a single year to 7-year running average. These results suggest that the Qmax/Qmin and Qapr/Qdec changes may be related to the basin warming and perhaps permafrost degradation.

Key words hydrology; permafrost; temperature; Siberia

INTRODUCTION

In cold regions, the hydrological regime is closely related to permafrost conditions, such as permafrost extent and thermal characteristics. Permafrost has a very low permeability and commonly acts as a barrier to infiltration or as a confining layer to aquifers. Because it is a barrier to infiltration, permafrost increases the surface runoff and reduces subsurface flow. Permafrost extent over a region plays a key role in the distribution of surface–subsurface interaction (Carey & Woo, 2001; Lemieux *et al.*, 2008; Woo *et al.*, 2008). Permafrost and non-permafrost rivers have very different hydrological regimes. Relative to non-permafrost basins, permafrost watersheds have higher peak flow and lower base flow (Woo, 1986; Kane, 1997). In the permafrost regions, watersheds with higher permafrost coverage have lower subsurface storage capacity and thus a lower winter runoff and a higher summer peak flow (Woo, 1986; Kane, 1997; Yang *et al.*, 2003). There exists a significant positive relationship between the ratio of maximum to minimum monthly discharge (Qmax/Qmin) and basin permafrost coverage over the Lena River. This relationship indicates that permafrost condition does not significantly affect streamflow regime over the low permafrost (less than 40%) regions, but strongly affects the discharge regime for regions with high permafrost (greater than 60%) (Ye *et al.*, 2009).

The ratio of Qmax/Qmin decreased during 1937–2000 for the Lena River. The recession coefficient (RC, ratio of April to December discharge) during the cold season increased from 1937 to 2000 in the main branches of the Lena River without reservoir regulation (Ye *et al.*, 2009). These changes may be related to permafrost degradation. This study analyses the relationship between the Qmax/Qmin, RC, and basin air temperature. It is difficult to accurately determine changes in permafrost distribution. We therefore use the basin air temperature to reflect permafrost condition changes. The objective of this study is to explore the effect of the permafrost degradation on hydrological processes and their changes.

BASIN DESCRIPTION, DATA SETS AND METHOD OF ANALYSIS

The Lena River originates from the Baikal Mountains in the south central Siberian Plateau and flows northeast and north into the Arctic Ocean (Fig. 1). Relative to other large rivers, this basin

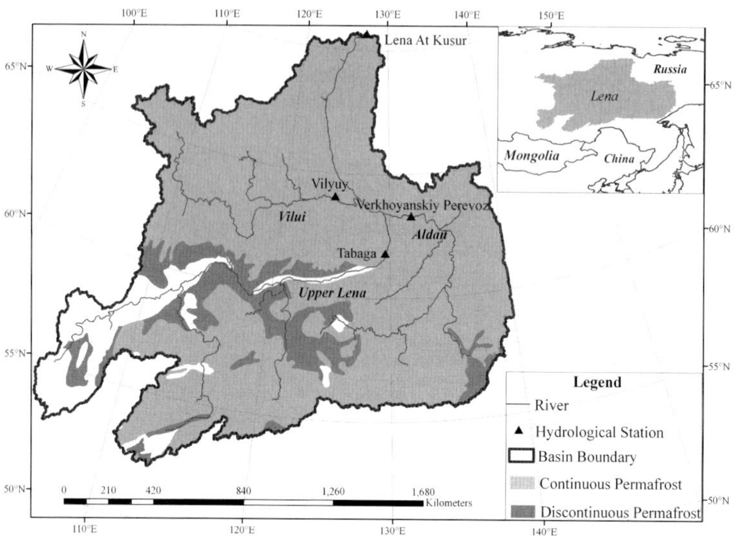

Fig. 1 Lena River basin map with the basin boundary and permafrost distribution.

has fewer human activities and much less economic development (Dynesius & Nilsson, 1994). There is only one large reservoir in the Vilui sub-basin; a large dam (storage capacity 35.9 km^3) and a power plant were completed in 1967 near the Chernyshevskyi (112°15′W, 62°45′N) (Ye *et al.*, 2003). The other two main branches, the upper Lena and Adam sub-basins, are affected little by human activities.

The drainage areas from the 1-km DEM match well to those reported by the Pan-Arctic River Discharge Database (Lammers *et al.*, 2001), with the relative errors being <15% for the sub-basins. The coverage of permafrost in a basin is defined as the weighed average of the four different types of permafrost. Considering the ranges of permafrost coverage, we use the mean coverage as representative coverage for each permafrost type, i.e. 95%, 70%, 30% and 5% permafrost coverage in continuous, discontinuous, isolated and sporadic areas, respectively. Variations from the mean CP are ±5%, ±20%, ±20% and ±5% for continuous areas, discontinuous areas, isolated areas, and sporadic permafrost, respectively. The upper Lena basin at Tabaga is covered by 72% permafrost and the Adam basin at Verkhoyanskiy Perevoz is covered by 92% permafrost (Table 1, Fig. 1).

Since the late 1930s hydrological observations in the Siberian regions, such as discharge, stream water temperature, river-ice thickness, dates of river freeze-up and break-up, have been carried out systematically by the Russian Hydrometeorological Services; the observational records were quality-controlled and archived by the same agency (Shiklomanov *et al.*, 2000). The discharge data are now available from the R-ArcticNet (v4.0) – a database of Pan-Arctic river discharge during 1936–2000 (Lammers *et al.*, 2001). In this analysis, we use the long-term monthly discharge records collected at Tabaga station in upper Lena and Verkhoyanskiy Perevoz station in Aldan sub-basin outlet (Fig. 1). Relevant information for these stations is given in Table 1.

Table 1 List of hydrological stations used in this study.

Station name/ Location	Latitude °N	Longitude °E	Data period		Drainage area (× 1000 km^2)	Annual runoff (km^3)	(mm)
Tabaga/Upper Lena	61.83	129.6	1936	1999	897	221.0	246.4
Verkhoyanskiy Perevoz/ Aldan sub-basin outlet	63.32	132.02	1942	1999	696	166.0	238.5

The methods of analyses include calculation of monthly mean discharge and hydrographs for the Tabaga station in upper Lena and the Verkhoyanskiy Perevoz station in Aldan sub-basin, determination of the ratio of monthly maximum to minimum flows (Qmax/Qmin), and the recession coefficient (RC, ratio of April to December discharge) during the cold season. We also use monthly air temperature (AT) data (Jones, 1994) for the Siberian regions, and calculate the basin mean values for the upper Lena, and Aldan sub-basins. Based on these data, we carry out analyses of changes in hydrological parameters (Qmax/Qmin and RC) and their relationships with basin mean temperature during 1937–2000.

RESULTS

It is difficult to directly investigate the linkage between discharge and permafrost change over a large basin, because of the difficulty of identifying the permafrost distribution and change. Here we use the basin mean air temperature, instead of the basin permafrost change to examine the linkage between hydrological processes and permafrost changes.

Relationship between Qmax/Qmin and basin air temperature

Figure 2 shows the Qmax/Qmin and annual mean basin air temperature (AT) during 1937–2000 at the upper Lena and Aldan basins. There is a significantly negative relationship in the annual scale and more significant one for the 5-year moving average. Usually, air temperatures directly affect evaporation and snow melt, and consequently the hydrological regime. Air temperature can also directly affect permafrost distribution, which influences the hydrological regime in the permafrost basin (Ye *et al.*, 2009). Increases in temperature lead to permafrost degradation, consequently more infiltration of surface water and a flat discharge regime. The temporal relationship between Qamx/Qmin and AT during 1937–2000 (Fig. 2) shows a similar result to the spatial comparison between Qmax/Qmin and permafrost coverage in the Lena basin (Ye *et al.*, 2009).

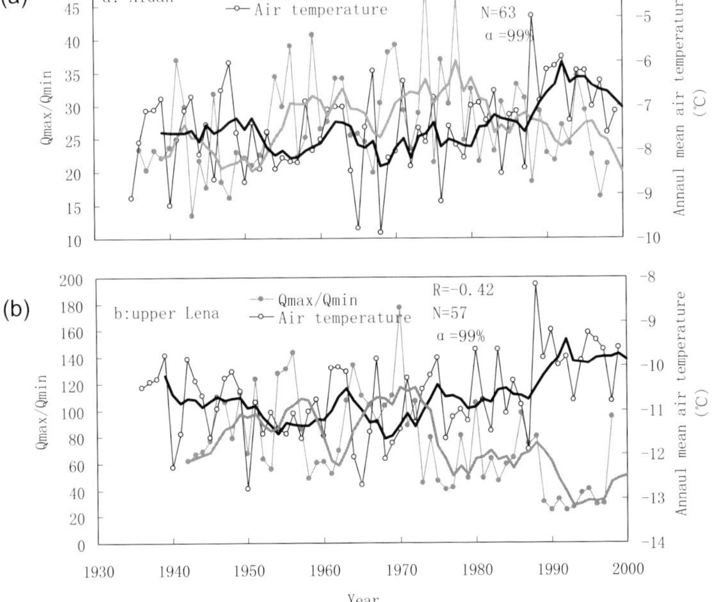

Fig. 2 The ratio of maximum *vs* minimum monthly discharge (Qmax/Qmin) and annual basin mean air temperature for: (a) Tabaga/Upper Lena and, and (b) Verkhoyanskiy Perevoz/Aldan sub-basin outlet, during 1936–2000 (The broad line is 5-year moving average).

Permafrost degradation is a slow response to climate change. Annual scale analyses may not reflect the relationship between climate change and permafrost.

The 1-year, 3-year, 5-year and 7-year moving average of basin temperature and Qmax/Qmin are used in the study. Table 2 shows the correlation coefficients of AT *vs* Qmax/Qmin, and AT *vs* Qapr/Qdec for the two sub-basins. The correlation coefficient increases as the moving average period increases. All relationships using the 7-year moving average are significant at 99% level. This result indicates slow responses of hydrological processes to climate warming and perhaps permafrost change.

Table 2 The correlation coefficient of annual basin mean air temperature with Qmax/Qmin and Qapr/Qdec, for 1-year, 3-year, 5-year and 7-year moving average.

Sub-basin	Parameter	1-year	3-year	5-year	7-year
Upper Lena	Qmax/Qmin	–0.33**	–0.37**	–0.33*	–0.37**
	Qapr/Qdec	0.26*	0.25*	0.27*	0.34**
	N	63	61	59	57
Aldan	Qmax/Qmin	–0.42**	–0.72**	–0.83**	–0.89**
	Qapr/Qdec	0.30*	0.65**	0.82**	0.88**
	N	57	55	53	51

Note: * and ** indicate 95% and 99% significant levels, respectively. N is sample number.

Relationship between recession coefficient and basin temperature.

The hydrological process is mainly a recession in the clod season from December to April without rain supply. The recession is controlled by the water released from the ground water reservoir. The ratio of April to December discharge (Qapr/Qdec) is defined as the recession coefficient. Figure 3 shows the annual basin mean temperature and Qapr/Qdec during 1936–2000. There is a good

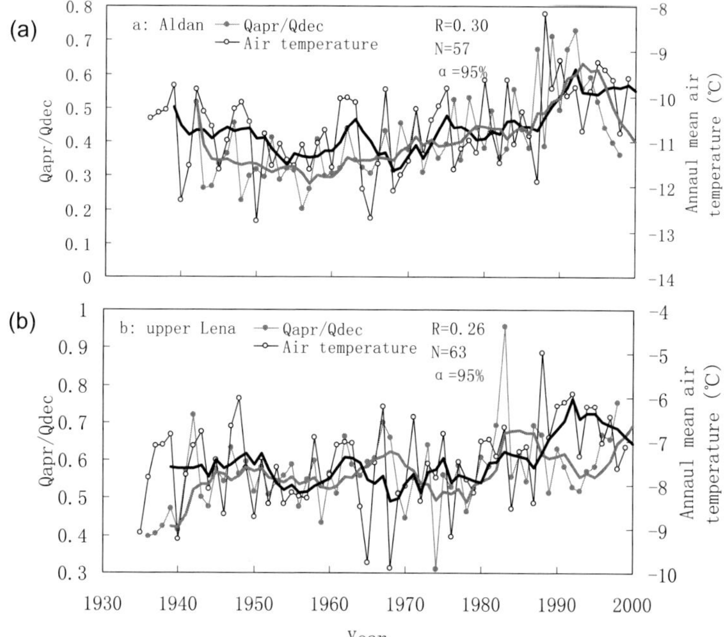

Fig. 3 The recession coefficient in cold season (Qapr/Qdec) and annual basin mean air temperature at: (a) Tabaga/Upper Lena, and (b) Verkhoyanskiy Perevoz/Aldan sub-basin outlet during 1936–2000 (the broad line is 5-year moving average).

relationship between them on annual scale, 95% significant level. The 5-year moving average is more consistent. This result indicates the discharge recession would become slow as climate is warming up. The large coefficient implies large capacity and regulation of the groundwater reservoir, which may be caused by permafrost degradation, especially permafrost disappearance in a warming climate. Permafrost degradation does not enhance the infiltration rates, but enlarges the infiltration area, and consequently the groundwater reservoir capability. Similar results have also been found in some rivers with permafrost distribution in northwest China (Niu *et al.*, 2011).

The 1-year, 3-year, 5-year and 7-year moving average of annual basin mean air temperature and Qapr/Qdec have been analysed (Table 2). The result is similar to that of Qmax/Qmin and AT. This again indicates the permafrost degradation is a slow response to climate change, and the hydrological parameters caused by permafrost change also slowly respond.

CONCLUSION

Monthly discharge and annual basin mean temperature data are used to examine the relationship between discharge process and climate change over the Lena River in Siberia. The ratio of maximum to minimum discharge (Qmax/Qmin) has decreased over time, while the recession coefficient (Qapr/Qdec) in the cold season has increased in the two branches – upper Lena and Aldan – during 1936–2000 (Ye *et al.*, 2009). These results suggest hydrological process change due to climate change and permafrost degradation. The annual basin mean air temperature increased from 1940 to 2000. There is a significant relationship between Qmax/Qmin *vs* AT, and Qapr/Qdec *vs* AT. The positive relationship between Qapr/Qdec and AT, and the negative relationship between Qmax/Qmin and AT gradually become significant from the annual scale to the 7-year mean. These results imply that both the Qmax/Qmin and Qapr/Qdec change may be related to climate warming and perhaps permafrost degradation. Permafrost degradation means that the impermeable stratum of the permafrost will disappear and lead to more surface water to infiltrate to groundwater. Permafrost degradation also extends the infiltration area and enlarges the groundwater reservoir, enhances reservoir regulation, and slows the recession process. A warmer climate will lead to a flatter discharge regime in cold regions.

Acknowledgements This work was supported by the National Key Basic Research program of China (2010CB951404), Hundred Talents Program of Chinese Academy of Sciences and US National Science Foundation grant (ARC-0612334).

REFERENCES

Bliss, N. B. & Olsen, L. M. (1996) Development of a 30-arc-second digital elevation model of South America. In: *Pecora Thirteen, Human Interactions with the Environment – Perspectives from Space* (Sioux Falls, South Dakota, August 1996).

Brown, J., Ferrians, O. J. Jr, Heginbottom, J. A. & Melnikov, E. S. (1997) Circum-Arctic map of permafrost and ground-ice conditions. US Geological Survey in Cooperation with the Circum-Pacific Council for Energy and Mineral Resources. Circum-Pacific Map Series CP-45, scale 1:10 000 000, 1 sheet. Washington, DC, USA.

Brown, J., Ferrians, O. J. Jr, Heginbottom, J. A. & Melnikov, E. S. (2001) Circum-Arctic map of permafrost and ground-ice conditions. National Snow and Ice Data Center/World Data Center for Glaciology. Digital Media. Boulder, Colorado, USA.

Carey, K., & Woo, M. (2001) Slope runoff processes and flow generation in a subarctic, subalpine catchment. *J. Hydrol.* **253**, 110–129.

Dynesius, M. & Nilsson, C. (1994) Fragmentation and flow regulation of river systems in the northern third of the world. *Science* **266**, 753–762.

Frauenfeld, O. W., Zhang, T., Barry, R. G. & Gilichinsky, D. (2004) Interdecadal changes in seasonal freeze and thaw depths in Russia. *J. Geophys. Res.* **109**, D05101, doi:10.1029/2003JD004245.

Jones, P. D. (1994) Hemispheric surface air temperature variations: a reanalysis and an update to 1993. *J. Climate* **7**, 1794–1802.

Kane, D. L. (1997) The impact of Arctic hydrologic perturbations on Arctic ecosystems induced by climate change. In: *Global Change and Arctic Terrestrial Ecosystems*, *Ecol. Studies* **124**, 63– 81. Springer-Verlag, New York, USA.

Lammers, R., Shiklomanov, A., Vorosmarty, C., Fekete, B. & Peterson, B. (2001) Assessment of contemporary arctic river runoff based on observational discharge records. *J. Geophys. Res.* **106**(D4), 3321–3334.

McClelland, J. W., Holmes, R. M., Peterson, B. J. & Stieglitz, M. (2004) Increasing river discharge in the Eurasian Arctic: Consideration of dams, permafrost thaw, and fires as potential agents of change. *J. Geophys. Res.* **109**, D18102, doi:10.1029/2004JD004583.

Niu, L., Ye, B., Li, J. & Yu, S. (2011) Effect of permafrost degradation on hydrological processes in typical basins with varying permafrost coverage in Western China. *Science China (Earth Science)* **54**(4), 615–624, doi: 10.1007/s11430-010-4073-1 .

Smith, L. C., Pavelsky, T. M., MacDonald, G. M., Shiklomanov, A. I. & Lammers, R. B. (2007), Rising minimum daily flows in northern Eurasian rivers: A growing influence of groundwater in the high-latitude hydrologic cycle. *J. Geophys. Res.* **112**, G04S47, doi:10.1029/2006JG000327.

Shiklomanov, I. A., Shiklomanov, A. I., Lammers, R. B., Peterson, B. J. & Vorosmarty, C. J. (2000) The dynamics of river water inflow to the Arctic Ocean. In: *The Freshwater Budget of the Arctic Ocean* (ed. by E. L. Lewis *et al.*), 281–296. Springer, New York, USA.

Woo, M.-K. (1986) Permafrost hydrology in North America. *Atmos. Ocean* **24**(3), 201–234.

Woo, K., Kane, D., Carey, S. & Yang, D. (2008) Progress in permafrost hydrology in the new millennium. *Permafrost and Periglacial Processes* **19**, 237–254.

Yang, D., Robinson, D., Zhao, Y., Estilow, T. & Ye, B. (2003) Streamflow response to seasonal snow cover extent changes in large Siberian watersheds. *J. Geophys. Res.* **108**(D18), 4578, doi:10.1029/2002JD003149.

Ye, B., Yang, D. & Kane, D. L. (2003) Changes in Lena River streamflow hydrology: human impacts *versus* natural variations. *Water Resour. Res.* **39**(7), 1200, doi:10.1029/2003WR001991.

Ye, B., Yang, D., Zhang, Z. & Kane, D. L. (2009) Variation of hydrological regime with permafrost coverage over Lena Basin in Siberia. *J. Geophys. Res.* **114**, D07102, doi:10.1029/2008JD010537.

Zhang, T., R. G. Barry, G., Knowles, K., Heginbottom, J. A. & Brown, J. (1999) Statistics and characteristics of permafrost and ground-ice distribution in the Northern Hemisphere. *Polar Geogr.* **23**(2), 132–154.

Streamflow analysis for the Yana basin in eastern Siberia

IPSHITA MAJHI & DAQING YANG

Water and Environmental Research Center, Institute of Northern Engineering, University of Alaska Fairbanks, Fairbanks, Alaska 99775-5860, USA

ipmajhi@alaska.edu

Abstract We analyse Yana River streamflow and climate data in order to understand climate change and its impact on basin hydrology. Basin temperature and precipitation records show little change during 1977–1999. Discharge data near the basin mouth suggest changes (increase and decrease) over the summer months. Basin precipitation has a positive correlation with discharge during June, July and August. The relationship between snow water equivalent and discharge follows an inverse relation; maximum snow water equivalent and discharge have a linear relation, with inconsistencies in some years. Further examination is needed to improve this relationship. The results of this study are useful for a better understanding of the hydrological regime and changes over the northern regions.

Key words Yana River, Siberia; discharge; snow cover

INTRODUCTION

Arctic climate and hydrology have changed significantly in the past decades. Recently we have studied hydrological regimes and changes over the Kolyma and Lena basins in order to quantify and understand human impact and climatic effects on regional hydrological changes (Yang *et al.*, 2003; Ye *et al.*, 2003; Majhi *et al.*, 2008). The Yana basin lies in eastern Siberia and drains into the Mesozoic continental collisional/accretionary zone of very complex geology. This region has a mountainous topography with the Verkhoyansk Range reaching 2000 m. The Yana River is an average sized basin along the coast of the Arctic Ocean. It has a drainage area of 238 000 km^2 (Fig. 1), a length of 1073 km (Huh *et al.*, 1998), and annual discharge of 34 km^3/year. There are no dams in the basin, which has a low population density. It thus provides ideal conditions to examine the effect of climatic variation on streamflow changes. The objective of this study is to better understand factors affecting discharge and its changes. We examine streamflow change and its relation with climatic variables, such as precipitation, temperature and snow cover. The results of this analysis will improve our understanding of hydrological response to climate change in the northern regions.

DATA AND METHODOLOGY

The Russian Federal Service for Hydrometeorology and Environment (Roshydromet) has monitored discharge of Russian rivers since the early part of this century. The discharge data for this study were obtained from the University of New Hampshire (www.r-arcticnet.sr.unh.edu). Stage height readings were made daily and cross-channel measurements of discharge were made 25–30 times for a rating curve. Estimates of daily discharge from the rating curves were accurate to ±5% (Shiklomanov, 2000). Discharge data are available over various parts of the basins; this analysis focuses on the basin scale. We use monthly and daily flow data collected near the mouth of the river during 1972–1999.

Passive microwave remote sensing in recent years has provided the ability to monitor various features of the Earth's atmosphere and surface, including snowpack properties. A previous study was done on snow covered area for the Lena basin (Yang *et al.*, 2003). A special sensor microwave imager (SSM/I) on the US Defense Meteorological Satellite Program (DMSP) has a daily temporal and good spatial coverage for most areas, which is an important feature for snow pack monitoring. The SSM/I data are from a seven channel microwave radiometer, which has dual polarized channels at 19, 37 and 85 GHz, and a vertically polarized channel at 22 GHz (Hilburn *et al.*, 2010). The daily SWE data were compiled by the University of New Hampshire from 1987 to

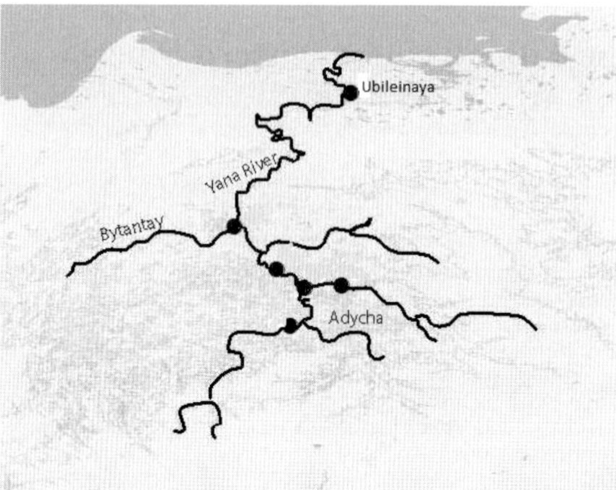

Fig. 1 Yana basin with all the stations listed from downstream to upstream.

2003, and are available through their website (www.r-arcticnet.sr.unh.edu). We use these data for snow analysis in this study.

The results of the analysis can be briefly summarized in three parts. First we define the climatic regime and change, using monthly precipitation and data. In the second section, we quantify the discharge regime and change at monthly and annual timescales over the basin. Finally, we compare the snow water equivalent and discharge data at various time scales, in order to examine their compatibility and the effect of SWE change on discharge.

RESULT AND DISCUSSION

Basin climatology

We analysed temperature and precipitation data for the Yana basin during 1972 to 1999 – a common data period for this study. The mean annual temperature ranged from –14°C to –18°C. The cold temperature is characteristic of regions with continuous permafrost. The basin has a long cold season of eight months, with temperatures ranging from 0°C in September to around –1°C in May. The coldest month is January, with a mean monthly temperature of –45°C. The brief warm season has a temperature range of 9°C in June to 8°C in August; July is the warmest month with a mean temperature of 12°C. Both the seasonal variation and the linear trend have low values.

Mean annual precipitation over the basin ranged from 171 mm to 300 mm, with an average of 217 mm. Trend analysis of the precipitation data showed no significant change for any month; the highest change was for July, about 4 mm during 1972–1999. The rest of the months had very low trend. The month with the most significant trend was April, with an α value of 0.03. It is interesting to note that the precipitation changes are mostly during the winter months, except for August. We compared the relationship between precipitation and temperature during 1972–1999, and found that they were strongly correlated, and significant at an α value of ±0.05 for a few months. The winter months (October to April) showed positive correlations, implying warmer temperatures associated with higher precipitation.

Basin hydrology

The Ubileinaya station, situated on the main river valley, (70.77°N, 136.08°E) (Fig. 1), is the closest station to the basin outlet. It has a drainage area of 224 000 km^2, with a mean annual flow of 1020 m^3/s from 1972 to 1999. The highest discharge takes place in June (4300 m^3/s), followed

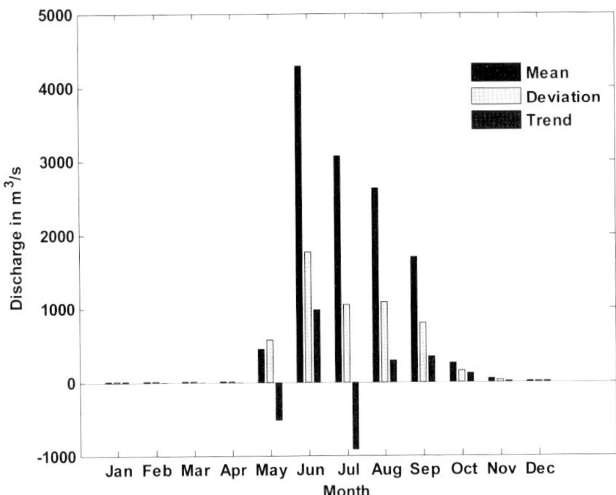

Fig. 2 Ubileinaya mean discharge, trend and standard deviation from 1972 to 1999.

by July (3055 m^3/s), August (2615 m^3/s), and September (808 m^3/s). Low flows were from December (12 m^3/s) to April (0.1 m^3/s). Seasonal flow fluctuations showed a high standard deviation for the months with high flows, i.e. June (1771 m^3/s), July (1051 m^3/s), August (1075 m^3/s), and September (805 m^3/s). Flow records showed a decrease of 500 m^3/s in May, and an increase in June by approximately 1000 m^3/s. The increase in June was compensated by July, which had a decrease of 1000 m^3/s. Flows in August, September and October also increased by 100 m^3/s (Fig. 2).

Flow changes in March (–1.5 m^3/s) and April (2.2 m^3/s) were significant. There is a positive change in the baseflow for the rest of the winter months. Annual discharge increased by 492 m^3/s (57%) during 1972 to 1999 (Fig. 2), significant at 83%; this significant change is most likely due to natural causes.

SWE *vs* runoff

The snow water equivalent (SWE) data used for this work relate to the period 1988–2000. We examine the relationship between SSM/I snow water equivalent and discharge data collected near the basin outlet.

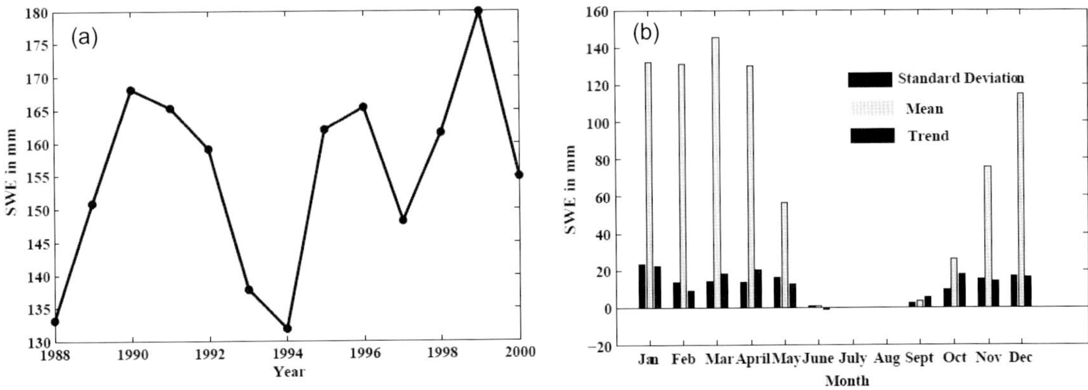

Fig. 3 (a) Mean March SWE during 1988–2000; (b) monthly mean, STD and trend for SWE for 1988–2000.

Snow starts to accumulate from September (5 mm) and continues to build up through October (26 mm), November (76 mm), December (116 mm), January (130 mm) and February (Fig. 3(b)). The maximum SWE occurs in March, ranging from 133 mm to 180 mm. Max SWE has an increasing trend during 1988 to 1999, with a sudden drop in 2000 (Fig. 3(a)).

Snow ablation starts in April, when SWE reduced from 146 mm in March to 130 mm in April. With the increase in air temperature in May, SWE reduced to 66 mm. There are variations in snowmelt processes among the years. Earlier melt was observed in the springs of 1997, 1998 and 1999, when the snow had gone by Julian day 142. For the other years, snow disappeared around Julian day 150. In spring 2000, snowmelt was complete by day 150.

SWE *vs* discharge

Discharge and SWE follow an inverse relationship, with the advent of snowmelt at about day 86 (last week of March) and its final disappearance at day 150, around the last week of May. The discharge subsequently peaks on Julian day 160 (Fig. 4). This is typical for the Arctic regions with continuous permafrost. The time series of discharge and snow water equivalent emphasizes the inverse relationship (Fig. 5). A high value of SWE does not always lead to a high peak discharge; this could be due to different ablation rates, which are very variable from year to year. We calculated the ablation rates around mid May to the first week of June, and it varies from as low as 2 mm/d to as high as 16 mm/d at the start of snowmelt. The melt rates change with increase in temperature.

CONCLUSION

Yana basin is a pristine and permafrost basin in eastern Siberia. It drains into the Laptev Sea. The basin has a long cold season of eight months, with temperatures ranging from 0°C in September to around −1°C in May. The basin has warmed up over the past three decades. Mean precipitation for the Yana basin ranged from 171 to 300 mm. Precipitation and temperature during 1972–1999 were strongly correlated (significant at α value of ± 0.05) for a few months. The winter months (October to April) showed positive correlations, implying warmer temperatures related with high precipitation. Monthly flow increased during the low flow months and decreased in June. These changes are significant and consistent with the slight increase in temperature over the last three decades.

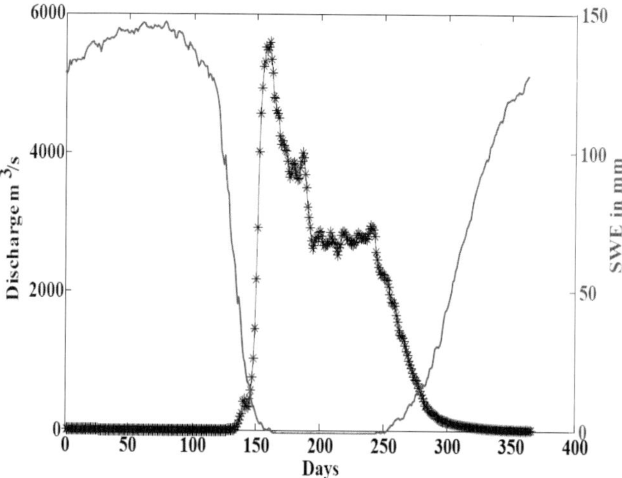

Fig. 4 Mean daily discharge and SWE for the basin, 1988–2000.

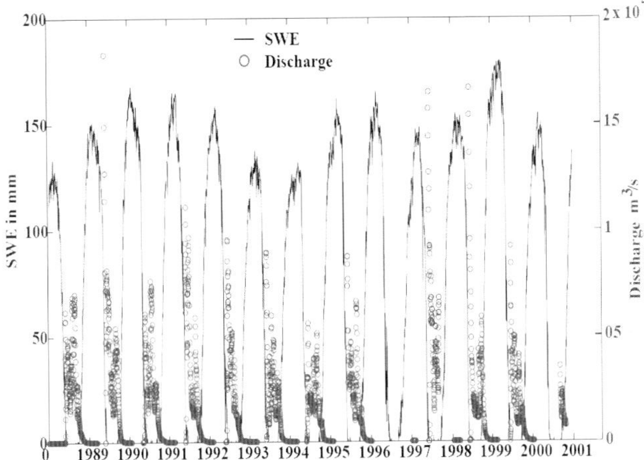

Fig. 5 Time series of SWE and discharge for the basin, 1988–2000.

The snow water equivalent shows the maximum accumulation in March; the interannual variability in the maximum SWE shows an increasing trend during 1988–2000. The SWE and discharge relationship does not always show consistency, that is, the years with high SWE do not always have peak discharge. Different melt rates could result in this discrepancy. There is a logarithmic relation between daily maximum SWE and discharge. Similar results exist for monthly flow and SWE. It is necessary to continue this research so as to better understand the discharge dynamics and hydrological cycle for the Arctic regions.

REFERENCES

Hilburn, K., Wentz, F., Mears, C., Meissner, T. & Smith, D. (2010) Description of Remote Sensing Systems, version 7. Geophysical Retrievals, Remote Sensing Systems. http://www.ssmi.com/papers/hilburn/Hilburn_V7_Poster_AMS_SatMet_2010_Annapolis.pdf.

Huh, Y., Ponteleyev, G., Babich, D., Zaitsev, A. & Edmond, J. M. (1998) The fluvial geochemistry of the rivers of Eastern Siberia: II Tributaries of the Lena, Omoloy, Yana, Indigirka, Kolyma and Anadyr draining the collisional/accretionary zone of the Verkhonyonsk and Cherskiy ranges. *Geochimica et Cosmochimica Acta* **62**, 2053–2075.

Majhi, I. & Yang, D. (2008) Streamflow characteristics and changes in Kolyma Basin in Siberia. *J. Hydrometro.* **9**, 267–279.

Shiklomanov, I. A. & Shiklomanov, A. I. (2003) Climatic change and the dynamics of river runoff into the Arctic Ocean. *Water Resour.* **30**, 593–601.

Yang, D., Robinson, D., Zhao, Y., Estilow, T. & Ye, B. (2003) Streamflow response to seasonal snowcover extent changes in Siberian watersheds. *J. Geophys. Res.-Atmos.* **108**(D18):4578, doi:10.1029/2002 JD003149.

Ye, B., Yang, D. & Kane, D. L. (2003) Changes in Lena River streamflow hydrology: human impacts *versus* natural variations. *Water Resour. Res.* **39**(7), 1200, doi 10.1029/2003 WR 001991.

Temperature effects on seasonal streamflow and variation at different spatial scales in cold regions

GENXU WANG, GUANGSHENG LIU & LIN YUN

Institute of Mountain Hazards and Environment, Chinese Academy of Sciences, Chengdu 610041, China
wanggx@imde.ac.cn

Abstract A typical permafrost watershed and alpine cold forest watershed in the Qinghai-Tibet Plateau were selected to analyse the effects of soil and air temperature on runoff processes. The primary factors influencing surface runoff processes during different seasons were analysed by Principal Component Analysis (PCA), statistical regression, and the power spectrum fractal methods. The results indicated that regarding hydrological processes, different factors are dominant in different seasons, but temperature is probably the main controlling factor to be considered for runoff processes analysis in permafrost watersheds and cold alpine forest watersheds. Some statistic relationships illustrating the effect of temperature on runoff processes in different season and its variation at different spatial scales were developed in this study. These relationships provide a practical way for estimating the effects of temperature on runoff processes and the variation patterns at different spatial scales.

Key words runoff processes; spatio-temporal variability; temperature effects; cold region; Tibet

INTRODUCTION

Over the last decade, a number of studies have focused on diagnostic hydrological processes and their seasonal variation in arctic permafrost regions. Because of the presence of frozen soil, runoff in the arctic permafrost regions is commonly characterized by greater water yield and larger direct runoff ratios for both rain and snowmelt than those in the non-frozen or temperate regions (Hayashi *et al.*, 2003). McNamara *et al.* (1998) reported that the thawing active layer was an important factor influencing seasonal direct runoff in the Subarctic Wolf Creek watershed in Canada. However, in spite of significant progress in some separate hydrological processes (snow cover formation and snowmelt, freeze–thaw of the ground), the attempts to understand a detailed mechanism of hydrological cycle and runoff generation for these regions have had little success (Kuchment *et al.*, 2000; Hayashi *et al.*, 2003; Yamazaki *et al.*, 2006).

Recent catchment hydrological studies on scaling issues indicate that the scale effects vary significantly in different contexts and experimental methodologies. When monitoring surveys involve catchments larger than 1 km^2, the scale effect is even more obvious (Cerdan *et al.*, 2004). Based on studies of the scale issue in hydrology in the last decade, we conclude that there is no single unanimous scale effect that dominates all hydrological processes, and that the effect is either site- or context-specific, or dependent on the size of the monitored plots or catchments (Cerdan *et al.*, 2004). One of the factors that make scaling difficult is the heterogeneity of catchments and the variability of hydrological processes. Therefore, it is necessary to study the scale issue in different catchments with different geographical and climate or vegetation conditions. Rainfall and runoff relationships have been widely used as a diagnostic tool for studies of runoff processes, as well as an important input parameter in hydrological design (Merz *et al.*, 2006). The scale effect on the rainfall–runoff relationship is crucial in hydrological studies. The objectives of the present study are: (1) to understand the effects of soil freeze–thaw variation on runoff processes in a permafrost watershed of QTP, and (2) to determine and quantify the nature of the scale effect for seasonal hydrological processes in the typical permafrost watershed.

STUDY SITE AND METHODOLOGY

The Zuomaokong watershed (127.63 km^2) and Hailuogou watershed (80.5 km^2), were selected as the study area (Fig. 1). The Zuomaokong watershed belongs to a permafrost region and the vegetation is dominated by *Kobresia pygmaea* C. B. Clarke and *Kobresia humilis* Serg (Wang *et*

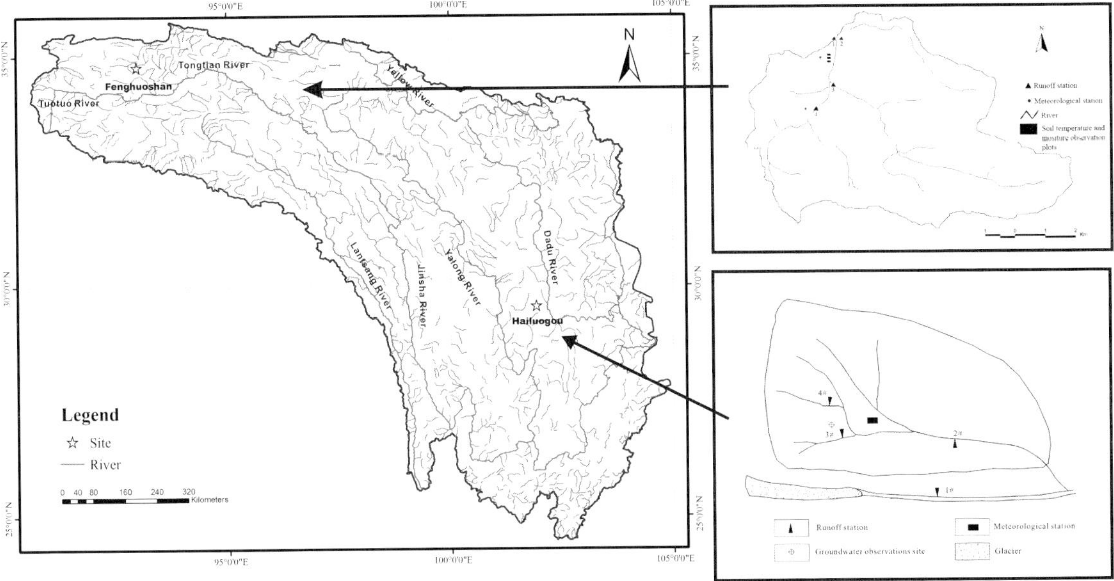

Fig. 1 The location of the research area and the river distribution.

al., 2001b; Zhou, 2001). The Hailuogou watershed is situated in an alpine forest region. *Abies* fabric forest is the main vegetation type in the subalpine area; it is a type of subalpine dark coniferous forest in southwestern China. There are four sub-catchments ranging from 0.41 km^2 to 80.5 km^2 in the research area.

In the Zuomaokong watershed, there were five runoff observation points (one at the outlet of the entire catchment, and the other four at the outlets of sub-basins). The discharge was measured at each site twice a day. Two soil temperature and moisture observation locations were located near the outlet section and the middle section of the watershed (Fig. 1). Soil moisture was determined by a frequency domain reflectometer (FDR) using a calibrated soil moisture sensor equipped with a Theta-probe (Holland, Eijkelamp Co.). Soil temperature was monitored using a thermal resistance sensor. In the Hailuogou watershed, the observation system at the Gongga Alpine Ecosystem Observation Station was established in 1988. The alpine hydrological observation system is one of the most important sections of the observation. This observation system contains two groundwater observation sites, one meteorological station, and four runoff observation sections. There are also additional instruments to observe air temperature and precipitation near the outlet of each sub-basin. At each runoff observation point, the discharge was measured by an automatic water level gauge placed at the outlet of each basin.

Fractal models have parameters that correlate features at one scale to those at all others, and as such they present an appealing methodology for linking processes across scales. To determine the scaling exponent for the transformation of rainfall–runoff dynamics across different spatial scales, annual runoff processes were separated into two periods based on the variation of runoff coefficients. One is from late June to early September, when the soil thawed out completely, and the runoff coefficients were lower. This is the summer season in the permafrost watershed and wet season in the alpine cold forest watershed. The other is the combined autumn, winter and spring seasons, when the soil freezes and thaws alternately. This is the cold season for the permafrost watershed and dry season in the alpine cold forest watershed. Principal component analysis (PCA) and statistical regression analysis were used to identify the main factors influencing the runoff formation and its seasonal variation.

RESULTS AND DISCUSSION

Impacts of temperature on seasonal runoff in permafrost region

In the permafrost watershed, our results revealed a significant linear relation ($R^2 \geq 0.45$, $P \leq 0.001$) between the discharge of the spring flood period and topsoil (0–20 cm) temperature (Fig. 2(a)). Surface runoff increased as the temperature of the active layer elevated. In particular, when the soil temperature was above 0.5°C, the runoff coefficient rose significantly. However, a weak exponential relation existed between air temperature and runoff (Fig. 2(b), $R^2 \leq 0.28$, $P = 0.01$). There was no clear relationship between precipitation and runoff, suggesting that precipitation had a weaker influence on runoff during the spring period. An exponential relation existed between the upper ground (0–60 cm) temperature, air temperature and autumn runoff (Fig. 3, $R^2 \geq 0.58$, $P \leq 0.001$). The autumn runoff declined exponentially with the decrease in soil and air temperature. Precipitation only played a minor role in spring flood runoff and autumn runoff, and exhibited a limited effect on direct runoff. Active soil thawing and freezing changed the soil water storage capacity, soil water infiltration capacity, and soil hydraulic conductivity, redistributing water in the soil profile. Consequently, seasonal variations in freeze–thaw of the active layer were the main cause for the seasonal changes in interflow and groundwater discharge, thus influencing the surface runoff process.

Air temperature and soil temperature at different depths of the active layer constituted the first component, with a level of influence on runoff of 41.3%, while soil moisture under the shallow active layer and spring precipitation were secondary factors, with a level of influence of 27.1%. Therefore, thawing processes of the active layer (including soil temperature and moisture) and air temperature were the primary factors influencing spring runoff. In autumn, the temperature and moisture of active layer and, air temperature explained over 82% of the runoff variation.

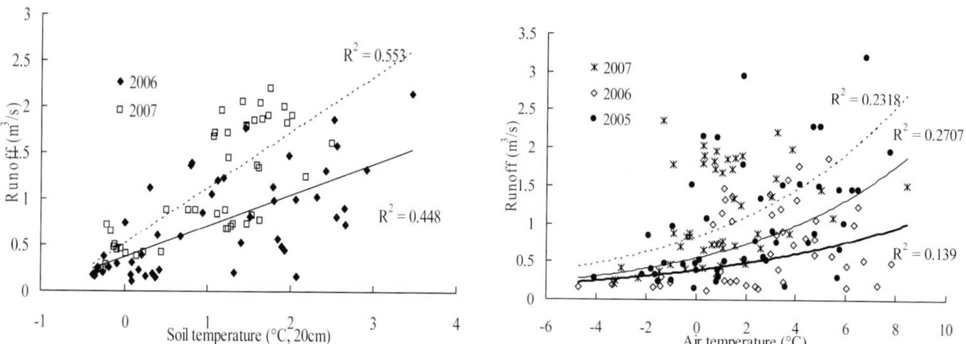

Fig. 2 Relationship between year-to-year spring flood-season river runoff (monthly) and soil temperature at 20 cm depth and air temperature, for a permafrost watershed on the Qinghai-Tibet plateau.

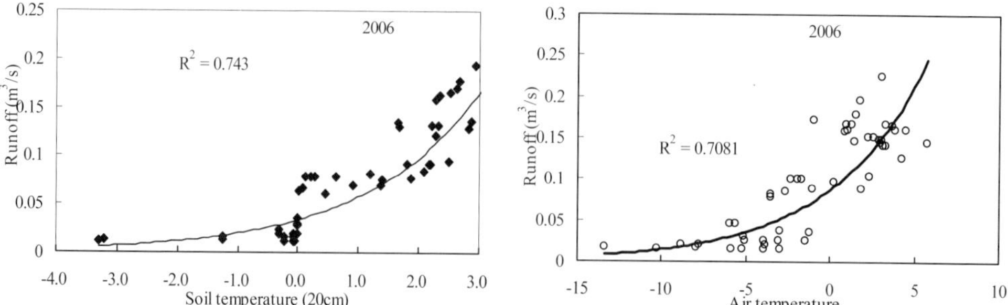

Fig. 3 Relationship between daily runoff and soil and air temperatures temperature of a permafrost watershed on the Qinghai-Tibet plateau during the autumn period.

Relationship between temperature and seasonal runoff in alpine forest region

Mean monthly temperature, rainfall, and runoff data for the dry season from 1990 to 2006 in the four sub-basins were collected and analysed. The primary factor affecting runoff in the dry season was the air and soil temperature. Air and soil temperature and runoff were significantly correlated, i.e. a significant exponential relationship for each gauging site during the dry season (Fig. 4, $R^2 \geq 0.57$, $P \leq 0.001$). In the alpine cold forest watershed, the dry season runoff increased exponentially with the soil and air temperature. Similar to in the permafrost watershed, when air temperature was above 0.5°C, the runoff coefficient rose significantly. Precipitation had a weak influence on runoff during the dry period ($R^2 \leq 0.21$, $P = 0.09$). Based on the PCA results, air temperature and surface soil temperature (0–20 cm) constituted the first component, with a level of influence on runoff of 56.7%. Soil moisture and precipitation were secondary factors, with a level of influence of 25.4%.

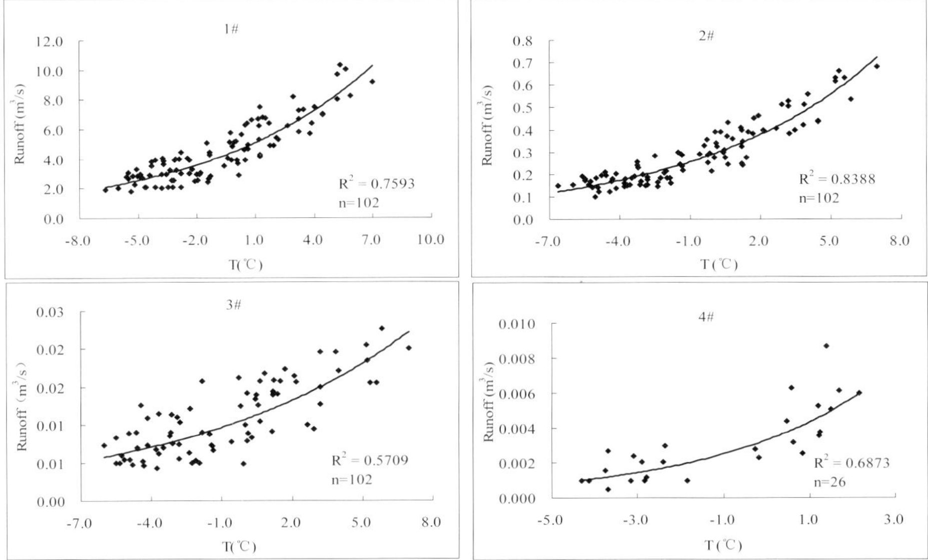

Fig. 4 The statistical relationship between the monthly scale runoff and temperature in the dry season in four sub-basin of the alpine forest watershed. Runoff had a significant exponential relationship with temperature in the dry season.

Variation of temperature effects on runoff at different spatial scales

1 Permafrost grassland watershed In the permafrost watershed, the daily mean runoff processes in the autumn and spring seasons had a significant power-function relation with the daily mean air temperature between September and October for different spatial scales ($R^2 > 0.71$, $P < 0.01$). The air temperature and precipitation are the controlling factors on runoff processes. The following equation was obtained for daily runoff processes in the autumn season, taking the two factors of air temperature and precipitation into account:

$$Q = 0.0296\mathrm{e}^{0.028k}(\mathrm{e}^{(0.0248\ln k + 0.0912)T_a}) + (0.0007k)P \quad (1)$$

where k is the spatial scale (sub-basin area), P is precipitation (mm), and T is temperature (°C). Based on equation (1), the effects of temperature on the discharge of the basin and its variation at different spatial scales were estimated. In the autumn and spring seasons, air temperature had a significant influence on the soil freeze–thaw cycle, as mentioned above, which further affected the rainfall–runoff relationship. The correlation coefficient between the observed and simulated runoff using equation (1) was greater than 0.82, and the Nash value was greater than 0.67. In the permafrost watershed, there are many different dominant factors of hydrological processes in different seasons, but temperature is probably the main controlling factor to effect the variation of runoff at different spatial scales.

2 Alpine cold forest watershed Temperature and runoff were significantly correlated in an exponential relationship at the monthly scale for each gauging site during the dry season (Fig. 4). Based on this relationship, an exponential regression for each basin was built for the dry season, and the data series of spatial scales (e.g. catchment area) and temperature can be built. Then, the scale relationship was derived after the catchments area data were normalized. For the dry season, the scale relationship formula between temperature and runoff was as follows:

$$Q = (0.0532\,k - 0.0183\,)\mathrm{e}^{0.1891\,k^{(-0.1765\,)}T} \tag{2}$$

This relation was applied to all the catchments to test its general applicability for dry season runoff processes from 1 November 2007 to 30 April 2008, and the results are summarized in Fig. 4. The results are in agreement with the observations. In general, the results showed a significant scale effect between different catchments. For the wet season, a nonlinear regression was conducted for the runoff, rainfall, and temperature data was established as follows:

$$Q = 0.0007\,\mathrm{e}^{(0.01k)T} + (0.0004k - 0.0002)P + (0.001k^2 + 0.04k - 0.01) \tag{3}$$

Using the above relation (3), the wet season runoff processes in 2007 and 2008 were simulated for the different spatial-scale catchments. The results were also in agreement with the observations. It was inferred from equations (2) and (3) that temperature was the main controlling factor for runoff processes in the dry season. In the wet season, however, precipitation and temperature were coupling factors affecting the runoff. In different seasons, the effects of temperature on surface runoff varied with not only spatial scales, but the spatial pattern of precipitation. Based on equations (2) and (3), the effect of temperature and its variation on surface runoff were estimated. The R^2 between the simulated and observed runoff in different seasons and at different time scales were more than 0.75, and the Nash coefficients were more than 0.65. These results demonstrated the acceptability and vitality of the relationships, which suggest variations in temperature effects on runoff in different seasons across various spatial-scales (catchments).

CONCLUSION

In the two watersheds, air temperature is the primary factor controlling runoff processes in the dry season over the study area. In the wet season, precipitation, soil moisture and temperature were the primary factors to affect the seasonal discharge and its variation at different spatial scales. In the permafrost watershed, temperature and runoff showed an exponential relationship during the spring and autumn seasons, but not for the summer season. In the alpine forest watershed, similar to the permafrost watershed, there was a significant exponential relationship between air temperature and runoff during the dry season. However, in the wet season, there was a complex relationship among temperature, precipitation and discharge.

Overall, in the two watersheds, temperature is probably the main controlling factor for runoff process variation. A statistical relation illustrating the effect of temperature on runoff processes and its variation at different spatial scales was developed in this study. The scale issue is of central concern in hydrological processes for understanding the potential of upscaling or downscaling methodologies. In the permafrost watershed, many different factors are dominant of hydrological processes in different seasons, but temperature is probably the main controlling factor to consider in scaling models.

Acknowledgements This study were funded by the Natural Science Foundation of China (no. 40730634 and no. 40925002)) and the knowledge innovation project of the Chinese Academy of Sciences (KZCX2-YW-331).

REFERENCES

Cerdan, O., Le Bissonnais, Y., Govers, G., Lecomte, V., van Oost, K., Couturier, A., King, C. & Dubreuil, N. (2004) Scale effect on runoff from experimental plots to catchments in agricultural areas in Normandy. *J. Hydrol.* **299**, 4–14.

Hayashi, M, van der Kamp, G. & Schmidt, R. (2003) Focused infiltration of snowmelt water in partially frozen soil under small depressions. *J. Hydrol.* **270**, 214–229.

Ishikawa, M., Zhang, Y., Kadota, T. & Ohata, T. (2006) Hydrothermal regimes of the dry active layer. *Water Resour. Res.* **42**, W04401, doi:10.1029/2005WR004200.

Kuchment, L. S., Gelfan, A. N. & Demidov, V. N. (2000) A distributed model of runoff generation in the permafrost regions. *J. Hydrol.* **240**, 1–22.

McNamara, J. P., Kane, D. L. & Hinzman, L. D. (1998) An analysis of streamflow hydrology in the Kuparuk River Basin, arctic Alaska: a nested watershed approach. *J. Hydrol.* **206**, 39–57.

Merz, R., Blöschl, G. & Parajka, J. (2006) Spatio-temporal variability of event runoff coefficients. *J. Hydrol.* **331**, 591–604.

Walker, D. A., Jia, G. J. & Epstein, H. E. (2003) Vegetation-soil-thaw-depth relationships along a low-arctic bioclimate gradient, Alaska: synthesis of information from the ATLAS studies. *Permafrost and Periglacial Processes* **14**, 103–123.

Wang, G., Li, Q. & Cheng, G. (2001) Climate change and its impact on the eco-environment in the source regions of Yangtze and Yellow Rivers in recent 40 years. *J. Glaciology and Geocryology* **23**(4), 346–352.

Yamazaki, Y., Kubota, J., Ohata, T., Vuglinsky, V. & Mizuyama, T. (2006) Seasonal changes in runoff characteristics on a permafrost watershed in the southern mountainous region of eastern Siberia. *Hydrol. Processes* **20**, 453–467.

Zhou, X. M. (2001) *Chinese Kobresia pygmaea Meadow*. Science Press, Beijing (in Chinese).

Zhou, Y., Guo, D., Qiu, G. & Cheng, G. (2000) *Geocryology in China*. Science Press, Beijing (in Chinese).

Impact of human activity on streamflow in the Huaihe River Basin, China: analysis and simulation

CHUANGUO YANG[1], ZHENCHUN HAO[1], ZHONGBO YU[1], ZHAOHUI LIN[2] & SHAOFENG LIU[2]

1 *State Key Laboratory of Hydrology-Water Resources and Hydraulic Engineering, Hohai University, Nanjing 210098, China*
cgyang@hhu.edu.cn

2 *International Center for Climate and Environment Sciences, Institute of Atmospheric Physics, Chinese Academy of Sciences, Beijing100029, China*

Abstract A distributed hydrological model coupled with a coarse grid land surface model is set up to simulate hydrological processes in the Huaihe River Basin, China. Parameters of the land surface model are interpolated from global soil and vegetation data sets. The characteristics of the basin, including topography, river networks and aquifer geology, are derived from a digital elevation model (DEM) and a national geological survey atlas. The NCEP/NCAR re-analysis data set and observed precipitation data are used as meteorological inputs. The coupled model is firstly calibrated and validated by using observed streamflow over the period 1980–1987. A long-term continuous simulation is then carried out for 1980–2003 forced with observed rainfall data. Results indicate that streamflow is over-estimated for dry years since the 1990s when water withdrawal increased substantially due to the growing industrial activities and the development of water projects. Two methods are proposed to study the human dimension in the hydrological cycle. One is to reconstruct the natural streamflow series using local volumes of withdrawals. The simulated results are consistent with the reconstructed hydrographs. The other method is to integrate a designed modular into the coupled model to represent the impact of human activities. This method can significantly improve the model's performance in streamflow simulation. This study shows that the coupling of hydrological and atmospheric models is a powerful tool for studying the human impact on the hydrological cycle.

Key words streamflow; human activity; hydrology model; withdrawal; Huaihe River, China

INTRODUCTION

Hydrology models and land surface models have been used to study the terrestrial water cycles. The basic theories of hydrological systems were developed on the basis of numerous laboratory and *in situ* experiments before the 1950s and 1960s. Thereafter, lumped conceptual hydrological models have been widely used in various climate regions (e.g. Zhao, 1992). In the last few decades, distributed hydrological models, which account for spatial heterogeneities with discretization by grids or representative elemental areas (REA), have been used for better understanding of hydrological processes and water resource policy development (Beven & Kirkby, 1979; Abbott *et al.*, 1986; Wood *et al.*, 1988; Yu *et al.*, 2006). Few of the lumped models and distributed models integrate the human activity dimension. Research on large-scale distributed hydrological models and its coupling with meso-scale meteorological models and global circulation models (GCMs), are attracting more attention than ever (Benoit *et al.*, 2000; Nijssen *et al.*, 2001; Yu *et al.*, 2006). Some parameterization schemes of hydrological processes at fine scales are being introduced into land surface models to improve water flux description (Yang & Niu, 2003; Koster *et al.*, 2004).

The human dimension in the hydrological cycle has two major aspects: land-use/land-cover change (LUCC) and direct withdrawals activities (e.g. dams, pumping wells). Globally, large reservoirs have a total storage capacity of 7000 km^3 (ICOLD, 1998), which accounts for three times the annual average water storage in river channels, or one-sixth of the global annual river discharge (Hanasaki *et al.*, 2006). This raises the question about how water withdrawals due to human activities alter streamflow in river channels. Human activities such as construction of dams and sluice gates, water withdrawal for agricultural, industrial and urban needs, and land-use/land-cover changes (Isik *et al.*, 2008; Yang *et al.*, 2008) bring new challenges in the spatio-temporal variation analysis of the water cycle and distributed modelling at the basin scale. Some observed hydrological data, e.g. streamflow and groundwater level, become less representative of natural processes due to such human activities. Not only must hydrological data series be well analysed and reconstructed, but also the structure, parameterization and calibration of hydrological models needs more support from new theories and experiments.

In this study, a fine-grid hydrological model coupled with a coarse-grid land surface model is used to study the impacts of human activity on streamflows in recent years. First, the model structure, parameters, forcing data, and calibration are described. Then the impact of human activity on streamflow is discussed using a long-term continuous simulation. Finally methods of studying human activity impacts are proposed.

MODEL AND CALIBRATION

Model descriptions

A coupled coarse land surface model and finer-grid hydrological model (LSX-HMS) (Yu *et al.*, 2006) was used in this study. The single column land surface model (LSX) consists of a six-layer soil module, a two-layer vegetation module (trees and grass) and a three-layer snow module. River/lake and groundwater are explicitly predicted with the components of HMS. The volume of groundwater in an assumed single-layer aquifer is described in the two-dimensional (2-D) Boussinesq equation with Darcy's law representing groundwater flow between grid cells. In each hydrological grid cell, there exists one major river channel conceptualized with a rectangular cross-section. Surface water flow, including river and lake flow, is resolved as 2-D diffusion waves with eight probable orientations, and the flow velocity is parameterized with the Manning equation. With the predicted elevation of surface water and the groundwater table, water fluxes between the river, lake and groundwater or the vadose zone are calculated using Darcy's law.

Parameters and forcing data

The LSX-HMS has been used in the Asia continent (Yang *et al.*, 2010). The parameters of the land surface model include soil texture and vegetation type. The spatial distribution and physical attribution of vegetation types are as prescribed in Dorman & Sellers (1989). Soil textures for the upper six layers (~4.25 m) are interpolated into global T62 (~1.9°) grid resolution using a bilinear method from an initial global data set. The coupled model system employs 10 × 10 km hydrological grids. The USGS HYDRO1K DEM data set (1 × 1 km) is used to derive the parameters of the hydrological model required for the description of basin characteristics with a newly developed DEM algorithm. River banks and width are determined with the DEM algorithm and some empirical relationships. The China national 1:4 000 000 geological survey data set is gridded to the hydrological grids using ArcGIS software. Hydrogeological parameters of the single-layer aquifer are then obtained for each lithological type with a look-up table method.

The 6-hour NCEP/NCAR re-analysis data are the basic meteorological forcing data for the coupled model system. Daily precipitation gauge data from 833 meteorological stations across China for the period 1951–2006 are gridded with a revised method described by Xia (2008) which accounts for the topographic effects on rainfall. The gridded observed daily precipitation is disaggregated into hourly values according to a random statistical function.

Model calibration

The model was calibrated over a three-year period (1980–1982) and validated for a five-year period (1983–1987) by comparing the simulated and observed streamflows at Wangjiaba (drainage area of 29 844 km^2), Lutaizi (88 630 km^2) and Bengbu (132 220 km^2) in the Huaihe River basin. Three hydrogeological parameters, i.e. aquifer thickness, hydraulic conductivity and porosity of one assumed aquifer were calibrated by a set of sensitivity analysis experiments (Yang *et al.*, 2010). Monthly simulated streamflow series are compared with the observed hydrographs of the three stations in both the calibration period and validation period in Fig. 1. Both high and low values of the simulated river flow are consistent with the observed values. For the monthly

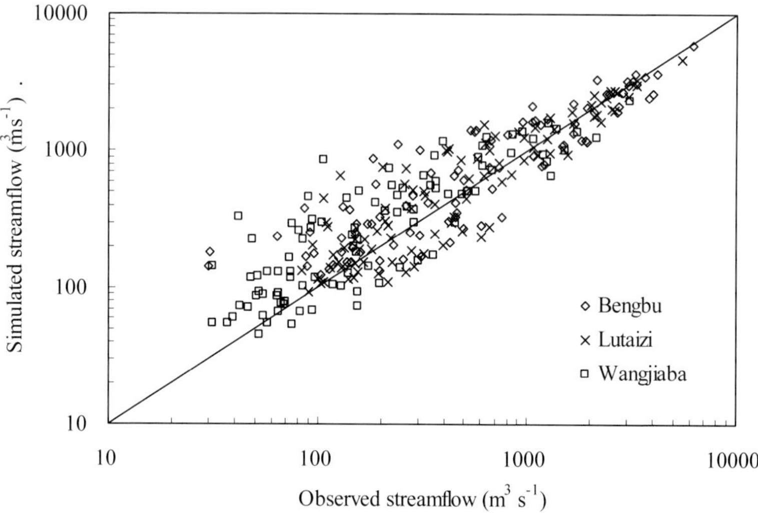

Fig. 1 Monthly streamflow at the Bengbu, Lutaizi and Wangjiaba stations from 1980 to 1987: the observed *versus* the simulated.

streamflow series, water balance index (WBI), Pearson's product moment correlation coefficient (PMC) and the Nash-Sutcliffe coefficient of efficiency (NSI) are 1.046, 0.952 and 0.902 at the Bengbu Station, respectively, 1.029, 0.968 and 0.926 at Lutaizi Station, and 1.048, 0.910 and 0.817 at Wangjiaba Station in the calibration period. For the validation period, WBI, PMC and NSI are 1.026, 0.921 and 0.848 at Bengbu Station, 1.020, 0.922 and 0.849 at Lutaizi Station, and 1.082, 0.845 and 0.640 at Wangjiaba Station, respectively.

Snow simulation

The basin has an average temperature of 14.9°C, with a trend of temperature increase at a rate of 0.23°C per ten years since 1951. Flat plain accounts for 2/3 of the basin area. There exist small volumes of snow mainly in the winter season, i.e. December, January and February, and there was no snow coverage from May to October (Fig. 2). The maximum of daily average snow depth is 0.8 mm from 1980 to 1987. Due to the Asian monsoon climate characteristics, floods occur in the wet season of July, August and September. The contribution of snow melt is limited to river flows of the basin.

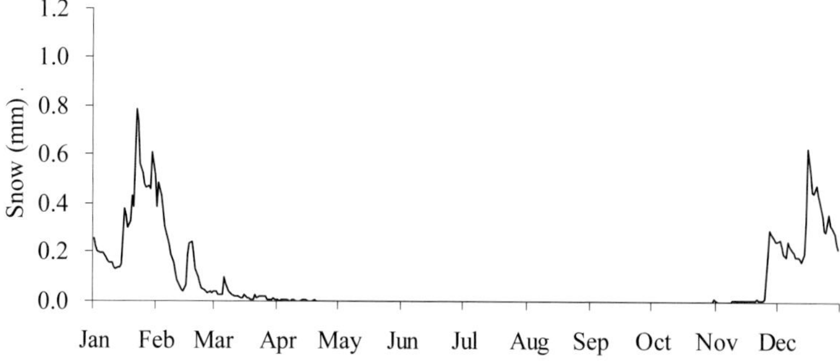

Fig. 2 Daily snow depth in the basin, average from 1980 to 1987.

ANALYSIS OF HUMAN ACTIVITY IMPACT

Comparison with the observed streamflow

In the Huaihe River Basin, agriculture and industry have developed rapidly since the 1980s. Frequent droughts and floods spurred on the new construction of water supply and management projects, and the full use of the existing projects, including more than 5700 reservoirs and 5000 sluice gates. A long-term continuous simulation from 1980 to 2003 was conducted with the coupled model system forced by the gauged rainfall data. Long-term monthly simulated and observed streamflows at the Bengbu Station are compared in Fig. 3. The annual and seasonal variations of the simulated streamflows are generally consistent with the observations. However, there is a remarkably high estimation for the low flows in dry seasons and dry years. The intercept of the fitting linear line is 362.77 $m^3 s^{-1}$.

Results for different periods, including the periods of 1980–1989, 1990–2003, 13 wet years and 11 dry years, indicate that the model has a good performance in the 1980s and in the wet years, and high estimates of streamflows in the years (1990–2003) and in the dry years (Yang *et al.*, 2010). Modelling of the dry years becomes more complicated due to various uncertainties. Small fluctuations of the forcing data may result in a relatively large bias of surface runoff in dry years; meanwhile more surface water is used at the developed basin in the dry seasons.

Trend analysis

Surface water supplies and changes due to human activities are regarded as important error sources for the coupled hydrological model simulations since the 1980s, which is confirmed by both the trend analysis of streamflow and rainfall, and the monthly observed streamflow with abnormal values in the dry years, e.g. no streamflow in wet seasons (June and July) of 2001. According to the analysis of precipitation and streamflows (Yang *et al.*, 2010), rainfall has a slightly increasing trend in the period, while observed streamflow shows a remarkable decreasing trend. The model simulated streamflow provides a hydrograph with a similar trend compared to the rainfall series. The difference between the observed and simulated streamflows can not be negligible for the dry years after 1990.

Surface water withdrawals

The uses of surface water decrease and redistribute local surface runoff directly, and thus affect the basin's streamflow hydrographs. Table 1 gives the annual volumes of surface water withdrawal

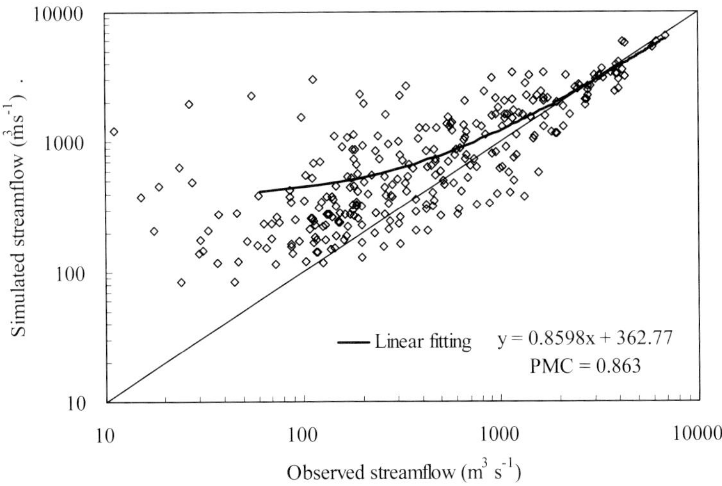

Fig. 3 Monthly observed and simulated streamflows at the Bengbu Station from 1980 to 2003.

Table 1 Annual surface water withdraws in the upstream of Bengbu Station and its percentage of simulated streamflows.

Year	1997	1998	1999	2000	2001	2002	2003
Withdraws ($\times 10^9$ m^3)	14.81	11.87	13.75	11.26	14.16	12.54	9.96
Percentage of simulated streamflow (%)	55.09	24.35	54.75	23.88	64.61	43.25	16.96

upstream of Bengbu station from 1997 to 2003, provided by the Water Resources Bulletin of the Huaihe River Basin (Huaihe River Committee, China). A huge volume of surface water, annual average 12.46×10^9 m^3, was supplied for local agriculture, industry and urban uses, i.e. about 395.2 m^3 s^{-1} of streamflow, or 109.0% of the observed streamflow of the 11 dry years at Bengbu station. Comparing with the simulated streamflow indicates that more than half of the streamflow was withdrawn due to local human activities in the dry years, which could have detrimental effects on the river ecology and the local environment. Human activities play an important role in the deviation between the simulated and observed streamflows in recent years, especially in the dry years.

STUDY METHODS OF HUMAN ACTIVITY IMPACTS

Reconstruction of natural streamflow

According to the above analysis, surface water withdrawals account for high percentages of the streamflows in the Huaihe River Basin. Therefore, it is necessary to reconstruct the natural streamflow with the information on water withdrawals. In this study, the difference between the observed streamflow and the simulated value was used as weights of the annual water withdrawal to adjust the monthly observed streamflow for each year. Some negative weights in winters were changed to be a small weight of 10 m^3 s^{-1} in the adjusting process. The simulated streamflows are in good agreement with the observed streamflows adjusted for the withdrawal. Performance indices PMC and NSI reach high values of 0.934 and 0.940, increasing by 0.075 and 0.298 compared to the original observed values, respectively. A better water balance is obtained with a WBI of 0.948 when the withdrawal is included. WBI fell below 1.0 for the first time in this study, which indicates the probable existence of return flow from the withdrawals.

Simulation of human activity impact

The other method to study the impacts of human activity is to integrate a new module into the coupled model. In the new module (Fig. 4), a fraction of surface runoff (R) which is directly affected by human activities is dammed by a withdrawal factor zw (0~1), and stored in a fictitious "reservoir". Due to irrigation and industrial water use, water in the reservoir returns to the natural water cycle with a rate of WR at the next time step. In Fig. 4, E stands for evapotranspiration, and I

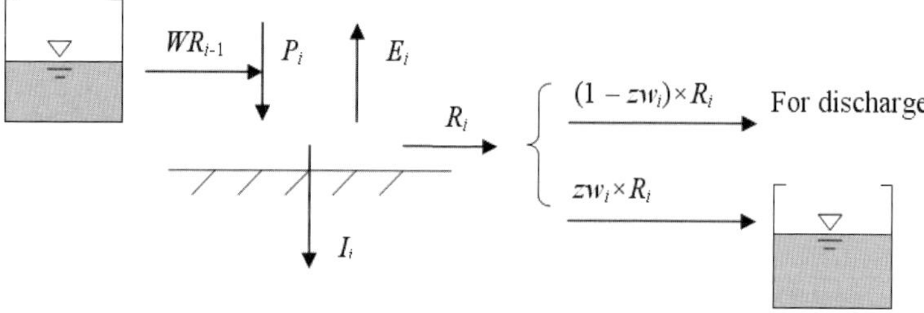

Fig. 4 Sketch of the human activity module newly introduced into the coupled model.

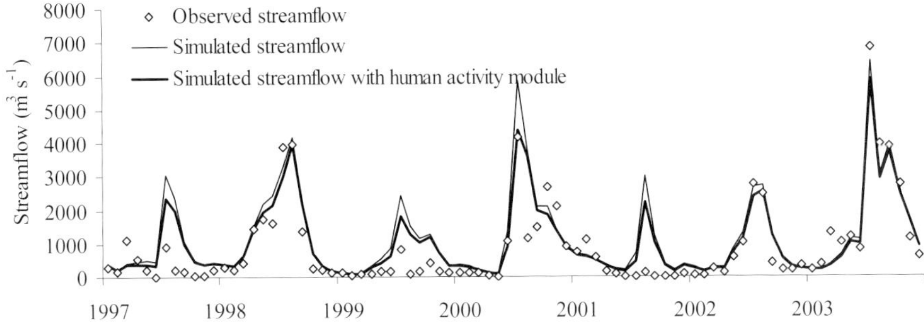

Fig. 5 Monthly observed and simulated streamflows at Bengbu Station from 1980 to 2003.

for infiltration at the land surface. In this study, *WR* has a constant value of 100 kg m^{-2} d^{-1} and *zw* is set as 0.2 in wet years and 0.4 in dry years according to the percentage of surface water withdrawal accounted for using the simulated streamflow, as shown in Table 1.

Compared with the original simulated result, monthly streamflow simulated using the model with the human activity module is obviously smaller in dry years. The annual average of simulated streamflow decreases by 141.1 m^3 s^{-1} in 1997 and 106.1 m^3 s^{-1} in 2001. The model performance was improved for the last decade (Fig. 5). The results indicate that the designed module represents well the effect of human activities on the local land water cycle.

CONCLUSIONS

Hydrological simulation and prediction are facing new challenges due to the expansion of human activities and recent climate change, and are particularly important in long-term simulations. In this study, how to detect the impacts of human activities in Huaihe River Basin was discussed. Results show that more than half of natural streamflow was withdrawn for human use in recent dry years. Reconstruction of the natural streamflow series is necessary for a basin with over-exploitation of water resources, especially for the dry years. The reconstruction method used for streamflow could be improved if monthly withdrawal data are used. Integrating a human activity module can help to improve the performance of the hydrological model in the basin. A more physical and functional module to represent human activities is planned for future study, for example, the water withdrawal factor *zw* and the return rate *WR* vary with soil moisture and annual crop growth stage. It is also planned to assimilate operation records of local water conservancy projects into hydrological models.

Acknowledgements This work is jointly supported by the National Basic Research Program of China (973 Program) (2010CB951101), the National Natural Science Foundation of China (NSFC) Projects (40830639), China Postdoctoral Science Foundation funded project (20100471374) and the Fundamental Research Funds for the Central Universities (2010B24914).

REFERENCES

Abbott, M. B., Bathurst, J. C., Cunge, J. A., O'Connell, P. E. & Rasmussen, J. (1986) An introduction to European hydrological system – systeme hydrologique Europeen (SHE) Part 1. History and philosophy of physically based distributed modeling system. *J. Hydrol.* **87**, 45–59.

Benoit, R., Pellerin, P., Kouwen, N., Ritchie, H., Donaldson, N., Joe, P. & Soulis, E. D. (2000) Toward the use of coupled atmospheric and hydrologic models at regional scale. *Mon. Weather Rev.* **128**, 1681–1706.

Beven, K. & Kirkby, M. J. (1979) A physically based, variable contributing area model of basin hydrology. *Hydrol. Sci. Bull.* **24**, 43–69.

Hanasaki, N., Kanae, S. & Oki, T. (2006) A reservoir operation scheme for global river routing models. *J. Hydrol.* **327**, 22–41.

Huaihe River Committee, Ministry of Water Resources, China. *Water Resources Bulletin of the Huaihe River Basin* (1997–2003). http://www.hrc.gov.cn/.

ICOLD (1998) *World Register of Dams*. International Commission on Large Dams, Paris, France.

Isik, S., Dogan, E., Kalin, L., Sasal, M. & Agiralioglu, N. (2008) Effects of anthropogenic activities on the lower Sakarya River. *Catena* **75**, 172–181.

Koster, R., *et al.* (2004) Regions of strong coupling between soil moisture and precipitation. *Science* **305**, 1138–1140.

Nijssen, B., O'Donnell, G. M., Hamlet, A. F. & Lettenmaier, D. P. (2001) Hydrological sensitivity of global rivers to climate change. *Climatic Change* **50**, 143–175.

Wood, E. F., Sivapalan, M., Beven, K. J. & Band, L. E. (1988) Effects of spatial variability and scale with implications to hydrologic modeling. *J. Hydrol.* **102**, 29–47.

Xia, Y. (2008) Adjustment of global precipitation data for orographic effects using observed annual streamflow and the LaD model. *J. Geophys. Res.* **113**, D04106, doi: 10.1029/2007JD008545.

Yang, C., Lin, Z., Yu, Z., Hao, Z. & Liu, S. (2010) Analysis and simulation of human activity impact on streamflow in the Huaihe River Basin with a large-scale hydrology model. *J. Hydromet.* **11**(3), 810–821.

Yang, T., Zhang, Q., Chen, Y.D., Tao, X., Xu, C. & Chen, X. (2008) A spatial assessment of hydrologic alteration caused by dam construction in the middle and lower Yellow River, China. *Hydrol. Processes* **22**, 3829–3843.

Yang, Z. & Niu, G. (2003) The versatile integrator of surface and atmosphere process. Part 1. Model description. *Global Planet. Change* **38**, 175–189.

Yu, Z., Pollard, D. & Cheng, L. (2006) On continental-scale hydrologic simulations with a coupled hydrologic model. *J. Hydrol.* **331**, 110–124.

Zhao, R. (1992) The Xinanjiang model applied in China. *J. Hydrol.* **135**, 371–381.

Distributed modelling of snow- and ice-melt in the Lhasa River basin from 1971 to 2080

MONIKA PRASCH[1], MARKUS WEBER[2] & WOLFRAM MAUSER[1]

1 *Department of Geography, Ludwig-Maximilians-Universität Munich, Luisenstraße 37, 80333 Munich, Germany*
m.prasch@lmu.de

2 *Commission for Glaciology, Bavarian Academy of Sciences, Alfons-Goppel-Straße 11, 80539 Munich, Germany*

Abstract The contribution of melt water release from snow and ice to water availability in mountain regions and adjacent forelands can often only be roughly assessed with simple models, because only sparse data are accessible. The impact of global climate change on water availability thus is afflicted with large uncertainties. We present a distributed modelling approach to determine the contribution of snow- and ice-melt to runoff at a regional scale in the Himalayan basin of the Lhasa River in Tibet under past and future climatic conditions. To fulfil the complex input data requirements, publicly available data are used. The successful validation of the model results for the past proves the application of the approach even in remote regions. Under IPCC SRES A2 climatic conditions with constant precipitation snowmelt will clearly decrease, whereas changes in ice-melt are small, although glacier retreat continues. However, runoff is reduced because of increasing evapotranspiration.

Key words snowmelt; glacier ice-melt; mountain hydrology; climate change; sparse data; Lhasa River, Tibet

INTRODUCTION

Future freshwater availability is a key factor for regional development, as not only drinking water but also food and energy production, health, and industrial development are all, to a certain extent, based on sufficient freshwater supply. In order to mitigate the water-related effects of Global Climate Change (GCC) for societies, appropriate adaptation strategies and Integrated Water Resources Management (IWRM) options for potential changes must be developed. This in turn requires comprehensive knowledge of the future impacts of GCC on the water balance.

Mountainous alpine headwaters play a very significant role in the water balance of many large river basins. The temporal storage of water as snow and ice in the high mountains, and its release during subsequent melting periods to adjacent forelands, makes the water release of snow and glaciers extremely important (Viviroli & Weingartner, 2004). However, the amount of melt water release often can only be roughly assessed with simple models, because sufficient data are seldom accessible in remote mountain regions. The impact of GCC on snow- and ice-melt and accordingly on water availability is thus afflicted with large uncertainties. Particularly in the Himalayas, the timing of the melt-out of glaciers and the adjacent melt water release under future climate conditions is controversial (IPCC, 2007, 2010; Kaltenborn *et al.*, 2010).

To determine the contribution of snow- and ice-melt to runoff in complex, meso- to large-scale watersheds, the spatial heterogeneity of complex basins with mountainous head-watersheds and forelands with large river valleys need to be captured synchronously by models. This includes the simulation of all relevant processes under varying climatic conditions and hydrological regimes. The modelling approach should also be applicable under future climatic conditions. As a consequence, the use of universal parameters, the conservation of mass- and energy, a foundation on physical principles and the abandonment of calibration to measured runoff are required for the modelling approach (Mauser & Bach, 2009). Finally, it should be possible to apply regional climate model (RCM) outputs as future meteorological drivers. Therefore, a complex modelling approach was developed within the framework of the GLOWA-Danube project (www.glowa-danube.de) for the Upper Danube River basin, with an excellent data basis describing the natural and socio-economic characteristics of the watershed. In this paper we present the transferability of the approach to the High Asian basin of the Lhasa River in Tibet with sparse data availability, and solely using publicly available data; the work was carried out within the framework of the Brahmatwinn project (www.brahmatwinn.uni-jena.de).

THE LHASA RIVER BASIN

The Lhasa River basin (LRB) is located in the southern part of the Tibet Autonomous Region of the People's Republic of China in Central Asia (Fig. 1). It is the largest tributary in the middle reach of the Brahmaputra in Tibet, with a basin area of 32 500 km^2. The location on the Tibetan Plateau characterizes the physio-geographic conditions of the watershed. The relief ranges from 3535 m a.s.l. at its junction with the Brahmaputra, up to 7162 m a.s.l. at the peak of the Nyainqêntanglha Mountains, forming the northwestern watershed. Climatic conditions are determined by a strong seasonal course of the precipitation (630 mm), which falls during the monsoon months in summer. Mean annual air temperatures vary between –9°C in the Nyainqêntanglha Mountains to +10°C in the Lhasa River valley near the river's mouth.

Due to the pronounced wet season during the summer monsoon and the dry season lasting the rest of the year, the runoff shows a clear seasonal cycle, with the flood peak reached in August. About 90% of the mean annual runoff is observed between May and November, whereas in the winter season, runoff is low. Most glaciers in the LRB are found along the Nyainqêntanglha Mountains due to their altitude and orientation to the southeast from where the monsoon comes, triggering orographic precipitation. Altogether 670 km^2, equating to 2% of the total basin area, are glacierized (WDC 2009).

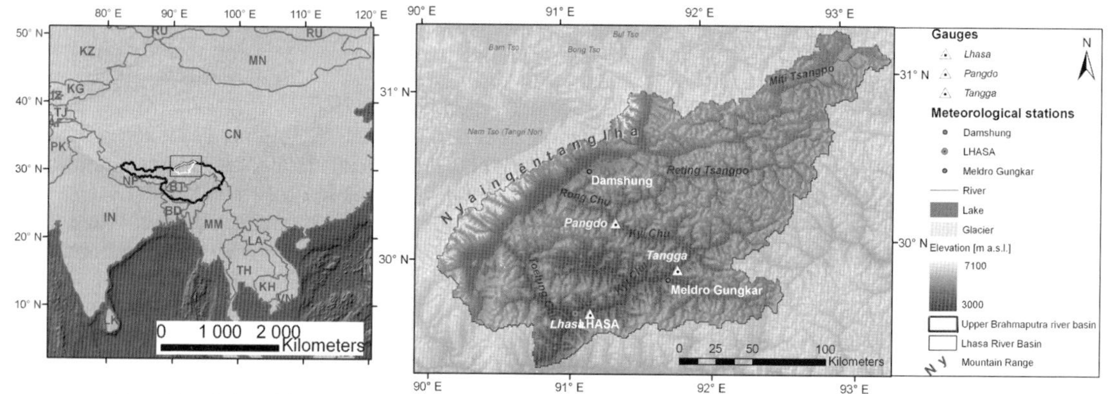

Fig. 1 The Lhasa River basin.

THE MODELLING APPROACH

The models

For the distributed modelling of the hydrological processes in a heterogeneous, large-scale river basin, the PROMET (Processes of Radiation, Mass and Energy Transfer) hydrological model (Mauser & Bach, 2009) was applied. PROMET, which has already been successfully used with changing climatic conditions in the Upper Danube basin (Mauser & Bach, 2009), is a fully spatially distributed raster model with a resolution of 1 × 1 km, and runs on an hourly time step. For simulation of hydrological processes, it uses distinct components for meteorology, land surface energy and mass balances, vegetation, snow and ice, soil hydraulics and soil temperature. Groundwater and channel-flow components with hydraulic structures are also included. As part of the snow and ice component, the SURGES glacier model (Subscale Regional Glacier Extension Simulator) (Weber *et al.*, 2010) was implemented to capture the small-scale processes for simulating ice-melt on alpine glaciers. SURGES uses a subscale parameterization in approximating the heterogeneous glacier surface by subscale units with similar surface elevations. Since the ice-melt water release depends on the snow-free area of the glacier, the duration of the snow cover on all parts of a glacier surface must be calculated. Therefore, the mass and energy balance on the glacier surface, including snow accumulation and ablation, snow metamorphism and glacier geometry changes are modelled for all elevation levels. The meteorological drivers are provided by the meteorology component of PROMET and extrapolated to all subscale elevation levels.

The components of PROMET fully interact and consider all processes related to water and energy fluxes. Within, as well as throughout all its components and feedbacks, PROMET strictly conserves mass and energy. The model characteristics allow the simulation of all relevant processes under varying climatic conditions and hydrological regimes. Furthermore, PROMET is not calibrated in using measured discharge data to fully exploit the predictive capabilities of the physically-based approach. It can therefore be used under past as well as future climatic conditions (Mauser & Bach, 2009). Additionally, these characteristics form the basis for coupling with RCMs, which are also usually raster-based. Since the spatial resolution of RCM outputs is in the range of 100 to 1000 km^2, the gap between this scale and the resolution of 1 km^2 of PROMET must be closed. Therefore the SCALMET (Scaling Meteorological variables) (Marke, 2008) coupling tool refines the coarse resolution of the RCM outputs in using different scaling techniques. They range from direct interpolation methods, such as bilinear interpolation, to statistical approaches using empirical relations. A combination of physically-based and statistical methods is also applied. Not only air temperature and precipitation are downscaled with these methods, but also wind speed, longwave and shortwave radiation, humidity and surface pressure, which are especially important for the simulation of snow and ice processes.

Derivation of input data

For the detailed simulation of the water flow considering snow and ice-melt water, vegetation growth, soil texture, groundwater flow and finally the interaction in routing the water in the river channel by the PROMET components, air temperature, precipitation, air humidity, wind speed and incoming short- and longwave radiation are required as meteorological drivers for past and future climate conditions. RCM outputs of the COSMO-CLM ERA 40 and ECHAM 5 model runs are used in this study as input data for SCALMET. Within the framework of the Brahmatwinn project, bias correction methods were applied to the CLM air temperature as well as to precipitation during the monsoon months in the whole Upper Brahmaputra basin, as described in Dobler & Ahrens (2008).

Additionally, data describing topography, land use and land cover, soil texture, river bed characteristics and glacier geometry are required to run the models. Gridded GIS layers describe the spatial distribution of the required parameters in the basin, whereas plant and soil parameters are provided in data tables for each land cover and soil texture class, because they are assumed to be homogenous throughout the whole catchment. Since sparse data is available for the LRB, publicly accessible data is used.

For topographic properties, the digital elevation model of the Shuttle Radar Topography Mission SRTM (Jarvis *et al.*, 2006) was applied in the LRB. Slope and aspect were deduced from the digital elevation model using the terrain analysis tool TOPAZ (Gabrecht & Martz, 1999), which was also applied to derive the watershed of the catchment, the main channel network with the channel slope and the flow direction. They, in turn, are required to run the river routing component of PROMET. Channel width is also needed. It was derived under the assumption that it correlates with the accumulated upstream area, also provided by TOPAZ, consistent with Mauser & Bach (2009). To determine the flow velocity, Manning's roughness parameter, which was deduced by field data, was applied. For the simulation of groundwater flows, the distance to the main river channel is required, which is also provided by TOPAZ. Depending on the distance, the storage time constant is set between one hour and one year.

The NASA TERRA/MODIS land cover product (Boston University, 2004) provides information about land cover and land use in the LRB. The parameterization of the land-use classes was adopted from Europe and adapted to the conditions of Tibet according to literature and field data. In order to model the water processes in the soil, soil texture classes were taken from the Harmonized World Soil Database (FAO *et al.*, 2009). The required parameters for the classes were adapted from the parameterization according to Mauser & Bach (2009).

In order to run SURGES, the area–elevation distribution of all glaciers and the ice thickness for all elevation levels is required as input data for all glacierized grid cells in the test basin. In general, the availability of glacier data is poor in the LRB, but the Chinese Glacier Inventory (CGI) provides polygons of the glacier areas of 1970 and additional information of their mean ice thickness (WDC, 2009). By intersecting the glacier boundaries with the ASTER Global Digital elevation model GDEM (ERSDAC, 2009) and aggregating the elevation values to levels at intervals of 100 m, the elevation levels for all glaciers in the LRB are deduced. Within the aggregation steps, the slope is averaged by weighting of the area. In this process, each glacier is considered separately, and so the elevation levels are not aggregated among different glaciers. Next, the area per elevation level is calculated. To consider the different ice thickness distribution of a glacier in more detail than assuming a homogenous ice block, the mean ice thickness provided by the CGI is modified for all elevation levels of the glacier. Thereby a correlation between surface slope and ice thickness (Haeberli & Hoelzle, 1995), and the thinning out of the glacier to its edges and the glacier front are considered.

Validation

While the uncertainty of the glacier area is in the 5% range (Wang *et al.*, 2009), the accuracy of the GDEM varies between ±7–14 m (ERSDAC, 2009). Because the elevation model was recorded in the year 2000 and is combined with the glacier boundaries of 1970, on average about 6–10 m w.e. (water equivalent) (Frauenfelder & Kääb, 2009) has melted away. Consequently, the elevation data are slightly too low. However, this average value is approximately within the range of accuracy of the elevation model. To check the quality of the deduced ice thickness for the elevation levels as described above, the derivation method was applied in the Eastern Alps, where area and geometry of the glaciers are similar to glaciers in the LRB, because no data are available in the LRB. The deduced ice volumes were compared to the data of the Austrian Glacier Inventory (Lambrecht & Kuhn, 2007). The volume deduced is 2% smaller, which corresponds to a deviation of 1 m ice thickness. Finally, the volumes of the elevation levels of a mountain and a valley glacier were compared. The overall deviation varies between 3 and 6%, which corresponds to a difference of 2 m or 3 m, respectively, of mean ice thickness. In comparison to uncertainties of ice thickness values for the Austrian Alps of between 5 and 10% (Fischer, 2009), the deviation is small. Additionally, the deviations are also within the accuracy range of the digital elevation model.

Detailed validation studies of the PROMET model components, including SURGES and SCALMET, have proven the quality of the model (Marke, 2008; Mauser & Bach, 2009). Its performance in the LRB is thus shown here. First, the downscaled ECHAM 5-driven CLM data are compared to average recordings of the three meteorological stations Lhasa, Damshung and Meldro Gungkar (see Fig. 1) from 1980 to 2000 in the LRB. While air temperature is reproduced, the deviations of the precipitation sum are between 10 and 22%. Next, glacier changes are validated. Glacier mass balances of –0.2 to –0.3 m w.e. per year between 1970 and 2000 and an area retreat of 21% in the LRB (Frauenfelder & Kääb, 2009) are in accordance to the simulated –0.3 m w.e. per year and an area loss of 20%. Finally, the mean simulated runoff is compared to three station observations of Lhasa, Pangdo and Tangga between 1996 and 2000. The deviation is below 5%. Since the ECHAM 5-driven CLM data can only reproduce the climate signal, ERA 40-driven CLM data are applied for the validation of monthly runoff data at the three gauges. Figure 2 shows the correlation between modelled and observed runoff with coefficients of determination of 0.87 to 0.89 and Nash-Sutcliffe efficiency coefficients (Nash & Sutcliffe, 1970) between 0.84 and 0.88.

The validation results prove the reproduction of glacier and hydrological dynamics in the LRB during past climatic conditions by the modelling approach. In order to determine the impact of future climate change on the melt water release of snow and ice and accordingly on runoff, results of the CLM ECHAM-5 driven model runs under IPCC SRES scenario A2 conditions are presented until 2080.

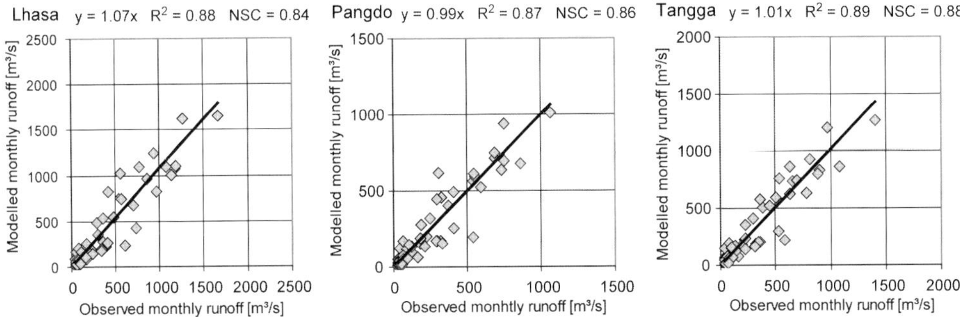

Fig. 2 Comparison of monthly modelled and observed runoff at the Lhasa, Pangdo and Tangga gauges from 1996 to 2000.

RESULTS

The mean annual air temperature in the LRB of –0.1°C between 1971 and 2000 will increase by 4.3 K until 2080 according to the A2 scenario. Although no significant trend is simulated for the amount of precipitation of 630 mm in the LRB, the percentage of snowfall is clearly reduced from 28% (1971–2000) to 15% . The spatial comparison of the percentage of snowfall between the past and the future periods from 2011–2040 and 2051–2080 is shown in Fig. 3(a). The highest proportion of snowfall is reached along the Nyainqêntanglha Mountains and in the northeastern parts of the basin, with 40–100%. From 2011 to 2040, the reduction rises to 10% for almost the entire basin area, and reaches 20% in the northeastern Nyainqêntanglha Mountains. The spatial pattern of the trend continues in the 2051–2080 period in accordance with the increasing air temperature. A maximum reduction of 30% is reached in the high mountain regions. This means a reduction of up to 3 months in the snow cover duration throughout the LRB.

The considerable changes in the meteorological conditions result in a proceeding glacier retreat under A2 scenario conditions in the LRB. The spatial distribution of the ice water reservoir is shown for 1970, 2000, 2040 and 2080 in Fig. 3(b). The changes in the past period are small, but glaciers did melt, particularly at their edges. In the scenario simulations, the retreat until 2040 is again small, but some glaciers in the northeastern parts disappear. Afterwards, an accelerating melt is simulated, and large storages with reservoirs greater than 50 m w.e. per km^2 decrease. More grid cells lose glacierization. Only around the Nyainqêntanglha Mountain peak is an amount larger than 50 m w.e. per km^2 stored as glacier ice in 2080, whereas a level of 25 m w.e. is rarely exceeded in all the other regions. The glaciers in the northeast melt completely.

The simulated changes in snow and ice reservoirs impact the amount of snow- and ice-melt released. The amount of snowmelt has, in accordance with snow precipitation, been decreasing since 1970. This negative trend continues until the end of the simulation period in 2080. As a result, the 215 mm mean amount of the past is reduced by 101% in the A2 model until 2080 (Fig. 4). In contrast to the uniformly negative trend of the amount of snowmelt, the future development of ice-melt water release varies in the scenario. The development of ice-melt shows a slight decrease from 12 to 8 mm until 2040. Afterwards, the amount increases and exceeds 15 mm. The total amount of ice-melt water is small, however, and accounts only for one fifth of the snowmelt (Fig. 4). The clear decrease of glacial coverage and stored ice water equivalent is therefore not pronounced in the course of the ice-melt. Shorter snow periods, because of higher air temperatures, enable the melting of ice at higher elevations. As a consequence, the reduction of the ice water reservoir is balanced by larger or longer snow-free glacier areas.

In assessing future water availability, important water balance values include not only precipitation, snow- and ice-melt amounts, but also the development of evapotranspiration. In accordance with the rising air temperature, evapotranspiration increased slightly during the period 1971–2000. This positive trend continues in the future A2 simulations and reaches an increase of 22% by 2080.

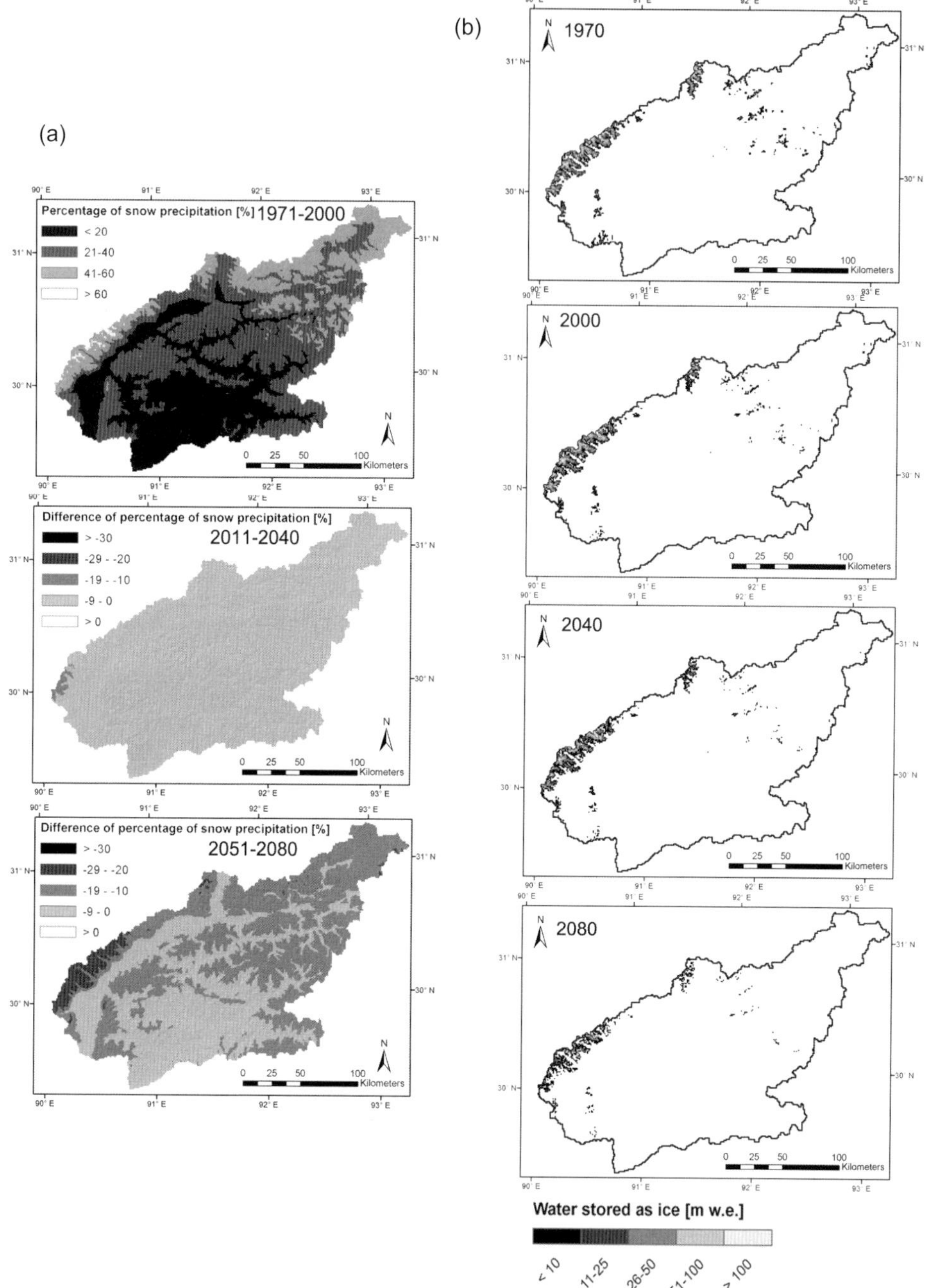

Fig. 3 (a) Percentage of snowfall. (b) The simulated ice water equivalent distribution in 1970, 2000, 2040 and 2080.

All the factors described above, together with soil water content and groundwater changes, determine the river runoff. Figure 4 shows the course of the mean annual discharge at the catchment outlet of the Lhasa River, together with precipitation, evapotranspiration and snow- and ice-melt. Since no trend is simulated for soil water content and groundwater changes, they are not

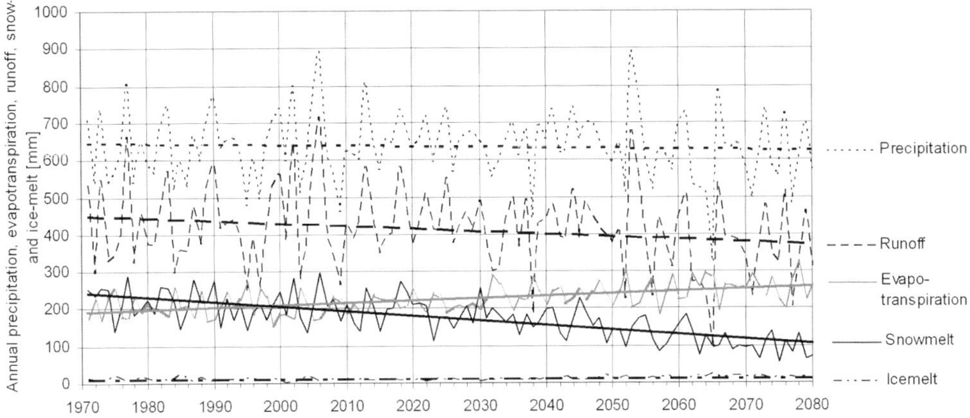

Fig. 4 Development of annual water balance parameters in the LRB from 1971 to 2080.

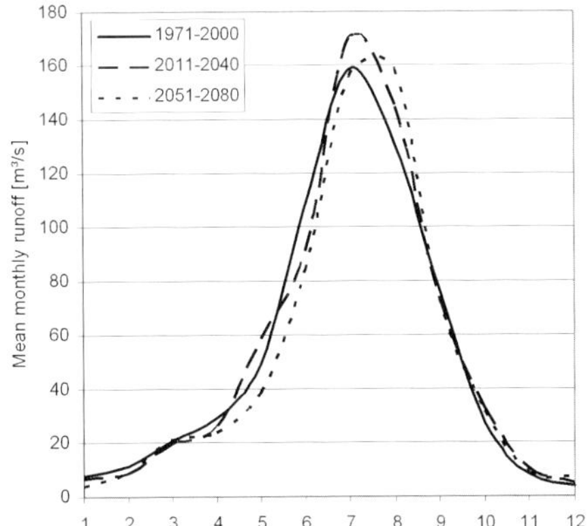

Fig. 5 Development of mean monthly runoff at the LRB outlet gauge for three climate periods.

illustrated. Runoff significantly decreases under A2 scenario conditions. While the amount of precipitation almost stays constant, the changes in runoff can be traced back to the increasing evapotranspiration.

Although the amount of ice-melt increases after 2050, as analysed above, the total amount is too small in comparison to other water balance components to influence runoff at the outlet gauge of the LRB. Thus, the mean annual runoff is determined by precipitation and follows its amount.

Snowmelt is of minor influence for the annual values, but it is relevant for the seasonal course of runoff. Accordingly, the comparison of mean monthly runoff for three climate periods (Fig. 5) shows a decrease from April to June, which is significant in May and June due to the trend analysis based on Mann (1945) and Kendall (1970). Since less precipitation is stored as snow than in the past period, less water is released as runoff in spring. The decrease of 18 to 22% (2051–2080 to 1971–2000) in these months is within the model accuracy with coefficients of determination of 0.87 to 0.89 and biases between 1 and 7% (see Fig. 2).

CONCLUSION

The presented approach enables the analysis of the future impact of GCC on water availability in the LRB and thus provides a basis for developing IWRM adaptation strategies. Although the models require a broad range of input data due to the complexity of the subject to be modelled, this study demonstrates their applicability in remote regions by using publicly available data. The analysis of the model results shows the dominant dependency of the Lhasa River runoff on precipitation and evapotranspiration. The amount of snowfall and the subsequent duration of snow cover determine the seasonal course, which follows the monsoon precipitation, peaking in summer. Under average climatic conditions, the amount of melt-water contribution from glacier ice, considered independently of snowmelt release, is too small to influence runoff at the catchment outlet, although glacier retreat continues according to the IPCC SRES A2 scenario run. Accordingly, runoff decrease is mainly caused by increasing evapotranspiration in the LRB in this scenario.

Acknowledgements Financial support by the European Commission Sixth Framework Programme (FP6) within the project "Brahmatwinn" and the Federal Ministry of Education and Research (BMBF) within the "GLOWA-Danube" project are gratefully acknowledged. We thank the Institute of Tibetan Plateau Research ITP and the International Centre for Integrated Mountain Development ICIMOD for data provision.

REFERENCES

Boston University (2004) NASA TERRA/MODIS HDF-EOS MOD12Q1 V004 Land Cover Product Binary Data, Eurasia subset, IGBP Class Scheme, available from http://duckwater.bu.edu/lc/mod12q1.html.

Dobler, A. & Ahrens, B. (2008) Precipitation by a regional climate model and bias correction in Europe and South Asia. *Met. Z.* **17**(4), 499–509.

ERSDAC (Earth Remote Sensing Data Analysis Center) (2009) The Ministry of Economy, Trade, and Industry (METI) of Japan and the United States National Aeronautics and Space Administration (NASA): Aster Global Digital Elevation Model (GDEM), available from http://www.ersdac.or.jp/GDEM/E/2.html.

FAO, IIASA, ISRIC, ISSCAS & JRC (2009) *Harmonized World Soil Database* (version 1.1). FAO, Rome, Italy and IIASA, Laxenburg, Austria.

Fischer, A. (2009) Calculation of glacier volume from sparse ice-thickness data, applied to Schaufelferner, Austria. *J. Glaciol.* **55**(191), 453–460.

Frauenfelder, R. & Kääb, A. (2009) Glacier mapping from multi-temporal optical remote sensing data within the Brahmatwinn river basin. *Proc. of the 33rd International Symposium on Remote Sensing of Environment* (May 2009, Stresa, Italy).

Garbrecht, J. & Martz, L. W. (1999) *TOPAZ Version 3.1*. USDA, Agricultura Research Service Grazinglands Research Laboratory, Oklahoma, USA.

Haeberli, W. & Hoelzle, M. (1995) Application of inventory data for estimating characteristics of and regional climate-change effects on mountain glaciers: a pilot study with the European Alps. *Annals of Glaciology* **21**, 206–212.

IPCC (2007) Asia. In: *Climate Change 2007: Impacts, Adaptation and Vulnerability*. Contribution of Working Group II to the Fourth Assessment Report of the Intergovernmental Panel on Climate Change (ed. by M. L. Parry, O. F. Canziani, J. P. Palutikof, P. J. Van Der Linden & C. E. Hanson), Cambridge University Press, Cambridge, UK.

IPCC (2010) IPCC statement on the melting of Himalayan glaciers. Geneva, 20 January 2010, Switzerland. Available from: http://ipcc-wg2.gov/publications/AR4/himalayastatement-20january2010.pdf

Jarvis, A., Reuter, H. I., Nelson, A. & Guevara, E. (2006) Hole-filled seamless SRTM data V3, International Centre for Tropical Agriculture (CIAT), available from http://srtm.csi.cgiar.org.

Kaltenborn, B. P., Nellemann, C. & Vistnes, I. I. (eds) (2010) *High Mountain Glaciers and Climate Change. Challenges to Human Livelihoods and Adaptation*, UN Environment Programme, GRID-Arendal, Birkeland Trykkeri AS, Norway.

Kendall, M. G. (1970) *Rank Correlation Methods* (4th edn). Griffin, London, UK.

Lambrecht, A. & Kuhn, M. (2007) Glacier changes in the Austrian Alps during the last three decades, derived from the new Austrian glacier inventory. *Annals of Glaciology* **46**, 177–184.

Mann, H. B. (1945) Nonparametric test against trends. *Econometrica* **13**, 245–259.

Marke, T. (2008) Development and application of a model interface to couple regional climate models with land surface models for climate change risk assessment in the Upper Danube Watershed. PhD Thesis, LMU München, Germany, available from http://edoc.ub.uni-muenchen.de/9162/ .

Mauser, W. & Bach, H. (2009) PROMET – Large scale distributed hydrological modelling to study the impact of climate change on the water flows of mountain wastersheds. *J. Hydrol.* **376**, 362–377.

Nash, J. E. & Sutcliffe, J. V. (1970) River flow forecasting through conceptual models. Part I – A discussion of principles. *J. Hydrol.* **10**(3), 282–290.

Viviroli, D. & Weingartner, R. (2004) The hydrological significance of mountains: from regional to global scale. *Hydrol. Earth System Sci.* **8** (6), 1016–1029.

Wang., Y., Hou, S.& Liu, Y. (2009) Glacier changes in the Karlik Shan, eastern Tien Shan, during 1971/72 – 2001/02. *Annals of Glaciology* **50**(53), 39–45.

WDC (2009) Chinese Glacier Inventory of the World. Data Center For Glaciology and Geocryology, Lanzhou, China. Available from http://wdcdgg.westgis.ac.cn/DATABASE/Glacier/glacier_inventory.asp.

Weber, M., Braun, L., Mauser, W. & Prasch, M. (2010) Contribution of rain, snow- and icemelt in the Upper Danube discharge today and in the future. *Geogr. Fis. Dinam. Quat.* **33**(2), 221–230.

Simulating discharge time series in regions with contrasting seasons using duration curves

VLADIMIR SMAKHTIN & NISHADI ERIYAGAMA

International Water Management Institute, PO Box 2075, Colombo, Sri Lanka

n.eriyagama@cgiar.org

Abstract Continuous discharge time series in ungauged basins where winter and summer flow generation mechanisms are distinctly different are simulated from limited observed meteorological data (rainfall, snow, temperature). Duration curves are used to convert the precipitation data from source gauges into a continuous hydrograph at an ungauged destination site. Temperature data is used as a control variable which determines whether precipitation is in a liquid (rainfall) or solid (snow) state, and whether the catchment is currently "active" to generate flow. The method is tested in several small catchments in Ontario, Canada, and is designed primarily for application at ungauged sites in data poor regions where the use of more complex and information consuming techniques of data generation may be difficult to justify.

Key words flow time series; flow duration curve; spatial interpolation; observed records; active storage; passive storage; precipitation index; ungauged basins

INTRODUCTION

Daily streamflow time series are required for a variety of hydrological analyses and engineering applications. In data poor regions and/or cases where flow time series *only* are sought, either the use of very simple deterministic models, or application of observed limited data transfer techniques may be justified. One such technique – a non-linear spatial interpolation of observed flow time series – was suggested by Hughes & Smakhtin (1996). Its key characteristic is a flow duration curve (FDC) which gives a summary of flow variability at a site and is interpreted as a relationship between any discharge value and the percentage of time that this discharge is equalled or exceeded. The underlying principle in this technique is that flows occurring simultaneously at sites in reasonably close proximity to each other correspond to similar percentage points on their respective FDCs. The site at which a streamflow time series is generated is referred to as a *destination site*. The site (or sites) with available time series, which is used for generation, is called a *source site*. In essence, the procedure is to transfer the streamflow time series from the location where data are available to the destination site.

The method was originally developed only for patching or extension of observed flow data. Subsequently, Smakhtin *et al.* (1997) illustrated how a FDC may be established at ungauged sites and translated into a complete flow time series. Smakhtin & Masse (2000) developed a modification of the method that allowed rainfall data to be utilized in the frequent cases when no source flow records are available. The modification entailed defining a current precipitation index (CPI), a continuous function of daily rainfall which would abruptly increase on rainy days and exponentially decay during dry periods, thus mimicking the general pattern of streamflow variability. In the modified algorithm, both source flow time series and source FDC were replaced by CPI time series and its duration curve, respectively. The method, in its various forms, was used by Metcalfe *et al.* (2005) to simulate non-regulated flow regimes at numerous hydropower facilities' sites in Ontario, Canada; by Lee *et al.* (2007) to simulate daily flow at ungauged locations in South Korea through the use of a GIS interface; by Archfield *et al.* (2010) to assess water availability at ungauged stream locations in Massachusetts, USA; and in some other applications.

One problem that remains unresolved to date was the use of the CPI based technique in regions, where precipitation falls in the form of both rain and snow, and where snowmelt dominates the spring flow. This paper examines one possible pragmatic solution to fill this gap.

METHOD

The dynamics of the accumulated catchment wetness at any precipitation observation point (*source site*) can be described by the CPI (Smakhtin & Masse, 2000). The CPI for any day is calculated as:

$$\mathrm{CPI}_t = \mathrm{CPI}_{t-1} \times \mathrm{REC} + R_t \quad (1)$$

where CPI_t is the current precipitation index (mm) on day t; R_t is the catchment precipitation for day t and REC is the daily recession coefficient. CPI is also referred to in this paper as "active catchment storage/wetness" – accumulated wetness that determines the current discharge from the catchment. On any day with no rain ($R_t = 0$) the CPI is equal to the CPI of the previous day multiplied by REC. However, if it rains, the daily rainfall depth too is added to the present day's CPI. CPI is therefore similar to the antecedent precipitation index, reflecting the rate of soil moisture depletion during a period of no rainfall, but it also represents the effects of the current precipitation as well. The range of the REC value is the same as that of the baseflow recession constant, and can be estimated by means of regional regression models with catchment characteristics. Transferring CPI time series at source sites into flow time series at destination sites includes several steps:

(a) *Source site selection for data transfer* In data poor regions, source site selection is limited/ obvious, e.g. meteorological stations within the basin or immediately adjacent to it shall be used. If more than one source site is identified, weights may be assigned to each of them.
(b) *Generation of tables of CPI values* These discrete tables are produced for each source site and each month of the year for fixed percentage points on the CPI duration curve, i.e. if one source site is used, the total number of required tables is 12. The tables use 17 arbitrarily-selected fixed percentage points, which cover the entire range of probabilities. Another alternative is to generate an annual 1-day CPI duration table using the entire CPI record instead of one for each month, especially in cases where only an annual regionalised FDC is available/can be established at the destination site. The differences between the two alternatives do not lead to significant differences in results (Metcalf *et al.*, 2005).
(c) *Calculation of FDC tables at ungauged destination site* The set of FDCs for each month of the year (or entire year only) should be established prior to the simulation of the actual time series. A FDC at an ungauged site obviously cannot be calculated from an observed record, as the latter does not exist. It has to be established using various procedures of hydrological regionalisation (e.g. Smakhtin *et al.*, 1997; Castellarin *et al.*, 2004; Natural Resources Canada: http://www.retscreen.net/ang/home.php). The destination FDCs are approximated by tables of values, similarly to CPI above.
(d) *Data transfer from individual source sites and final estimate* This is the main computational step during which the percentage point of each day's CPI at each source site is identified and the flow value for the equivalent percentage point from the destination site's FDC is read off. The discharge tables are used to "locate" the CPIs and flows on corresponding curves. L-interpolation is used between fixed percentage points. The procedure is repeated for each source site. Finally, a weighted averaging of all flow estimates for the destination site (obtained using individual source CPIs) is performed.

The last step is repeated for each day during the calculation period. The beginning of the period corresponds to the earliest start date of the selected meteorological records, and the end date to the latest of all end dates.

In regions with pronounced differences between seasonal precipitation types, adjustments need to be made to the way that CPI is calculated. To account for these differences, the additional variable – temperature – has to be included. Changes in precipitation type can then be related to temperature; this approach would help to avoid seeking observed snow data, which often simply do not exist.

During the warm part of the year, when daily mean temperatures are continuously positive ($T_t > 0$), the CPI calculation is the same as in equation (1). In the late autumn and winter periods, when the daily temperature becomes and continuously stays negative ($T_t < 0$), precipitation is assumed to fall as snow. This snow does not immediately contribute to "active" catchment storage, which is equivalent to CPI, but accumulates until temperatures become positive again. Therefore, precipitation which falls during negative temperature days is assumed to increment what is referred to here as "passive catchment storage" (PS, mm). PS is incremented by the amount of daily precipitation and no water is released from it at negative temperatures. Active catchment storage, CPI, however, recedes in line with REC:

$$\text{If } T_t < 0: \ \text{CPI}_t = \text{CPI}_{t-1} \times \text{REC} \ \text{ and } \ \text{PS}_t = \text{PS}_{t-1} + R_t \qquad (2)$$

When the temperatures become positive again in spring and PS is non-zero, melting occurs:

$$\text{If } T_t > 0 \ \text{ and } \ \text{PS}_{t-1} > 0: \ \text{MELT}_t = \text{DD} \times T_t \qquad (3)$$

where MELT_t is snowmelt on day t (mm/d) and DD is the degree-day factor (mm/°C per day). The "daily" check is made to ensure that there is still "water" in passive storage. If precipitation occurs during the melting stage, it is considered to be in the form of rain. The required CPI values during the melting period are then calculated as:

$$\text{CPI}_t = \text{CPI}_{t-1} \times \text{REC} + R_t + \text{MELT}_t \qquad (4)$$

The DD parameter is the amount of melting which occurs per one degree of positive air temperature within 24 hours – per one degree-day. The DD varies depending on the month, terrain and weather conditions and snow density. In the current study, the DD parameter value has been fixed at 3.5 mm/°C per day as per Kuusisto (1984). Region-specific DD values may be obtained from the literature. Once the continuous CPI time series and their corresponding duration curves are calculated for each selected source meteorological station, they may be used as described above in steps (c) and (d) of the algorithm to calculate the destination flow time series.

APPLICATION TO ONTARIO CATCHMENTS

The approach described above is generic. Therefore, the choice of a "testing ground" is determined by (naturally) the climate of the area and ease of access to input data on which to test the method. The Canadian province of Ontario satisfies both criteria. The catchments selected have been drawn mainly from the southern part of Ontario (Fig. 1). The exact geographical location of these catchments is, however, largely irrelevant, but the amount and quality of available observed source data are of primary importance. To test the algorithm under the "harshest" possible conditions, only one source meteorological station, nearest to each selected test catchment outlet, was used in the present study.

All input data are referred to in the coding system followed by the Canadian National Climate Data Archive (CDCD: http://climate.weatheroffice.gc.ca/prods_servs/index_e.html) and the Canadian National Water Data Archive (HYDAT: http://www.ec.gc.ca/rhc-wsc/default.asp?lang=En&n=9018B5EC-1, Table 1). The data selection procedure took into account such factors as the length of record, amount of missing data, existence of concurrent precipitation and temperature data, closeness of meteorological station to the catchment, catchment size, and flow regulation (unregulated flow records must be selected).

All catchments are gauged; yet two different scenarios of observed data availability were considered. The first scenario represents the case where limited historical flow time series may be available to construct a FDC for a destination site. The second scenario is that the catchment is completely ungauged. For the first scenario, 12 FDC tables were generated (one for each calendar month) and it was assumed that they did not change with time and remain a representative flow "signature" of each selected catchment. In the second scenario use was made of Regional Normalized FDCs and specific runoff maps (runoff/km^2) developed by Natural Resources Canada

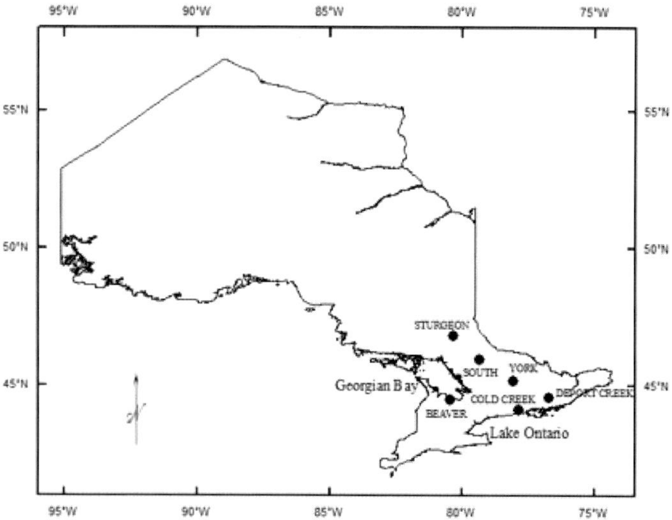

Fig. 1 Locations of the study catchments (black circles) on the map of Ontario Province.

for the entire country as part of the RETScreen software (a software for evaluating renewable energy projects: http://www.retscreen.net/ang/hydrology_data.php). Within each FDC region and specific runoff region, normalized FDCs have been provided for several HYDAT gauge locations (but not for the ones simulated in this study). Three types of normalized FDCs were derived by averaging the nearest available FDCs (from the same FDC region and specific runoff region) to each simulated gauge location. Annual FDCs for each site were then estimated by multiplying each averaged non-dimensional FDC by catchment area and specific runoff. The corresponding annual CPI duration curve for source sites were used in the simulation in this case.

The purpose of the first scenario is to examine how the method performs in conditions when an accurate FDC can be established for the ungauged location. In this case, most, if not all, estimation uncertainty will be related to the amount, quality and temporal pattern of the source meteorological data. The second scenario then represents the case where uncertainty is associated with both the estimation of regional duration curves and source meteorological data.

DISCUSSION AND CONCLUSIONS

Table 1 presents the results of simulations in all catchments considered so far, and Fig. 2 gives a snapshot of some individual, arbitrarily selected simulations. No attempt has been made to "calibrate" the proposed "model". On the other hand, calibration options of this approach are limited to changing the recession parameter value, the degree-day factor and the meteorological stations' number and weights. Therefore the results illustrate the performance of the approach under very stringent conditions. The recession REC value for each catchment was assumed to be equal to the median recession ratio value of the flow time-series (FREND, 1989), but generally, recession characteristics of streams may be estimated from regional relationships with catchment parameters (Tallaksen, 1995).

Fit statistics in Table 1 suggest that flow hydrographs obtained under Scenario 1 for all catchments are of better quality than those under scenario 2. This indicates that the results of the simulation depend on the accuracy of the established FDC for the destination site. In conditions where there is a reliable FDC, the algorithm yields good results.

The use of just one meteorological station (as done in this study intentionally to represent the most limited data condition) is not normally recommended (although it might be unavoidable in

Table 1 Statistics of fit between observed daily flows, and flows simulated by the proposed model for the simulation period.

Catchment and simulation period	Gauged area (km^2)	Data	Mean (m^3/s)	SD (m^3/s)	Max (m^3/s)	Min (m^3/s)	R^2	CE
South: 02DD005	787	Obs	14.8	13.3	121	0.52		
1981–2006		Scenario1	11.8	12.4	89.7	0.55	0.72	0.67
		Scenario2	14.2	21.4	179	0.03	0.53	–0.88
Sturgeon: 02DC004	2980	Obs	37.4	34.6	271	6.12		
1990–2002		Scenario1	38.1	35.2	345	6.07	0.70	0.67
		Scenario2	40.6	59.5	410	3.01	0.46	–0.63
Beaver: 02FB009	572	Obs	8.85	7.70	62.0	2.13		
1967–1969		Scenario1	7.92	6.98	57.9	1.19	0.60	0.56
		Scenario2	8.62	20.1	166	2.13	0.16	–4.79
York: 02KD002	839	Obs	11.9	13.9	102	0.58		
1958–1983		Scenario1	11.8	12.7	104	0.05	0.60	0.58
		Scenario2	11.4	16.7	116	0.83	0.37	0.01
Cold Creek: 02HK007	159	Obs	1.95	1.91	24.7	0.45		
		Scenario1	2.02	1.97	28.6	0.47	0.44	0.30
1991–2006		Scenario2	2.14	3.10	21.8	0.16	0.25	–1.02
Deport Creek: 02HM002	189	Obs	2.11	1.63	11.2	0.02		
		Scenario1	1.99	1.92	20.6	0.00	0.48	0.25
1986–2006		Scenario2	2.54	3.64	26.0	0.19	0.18	–3.15

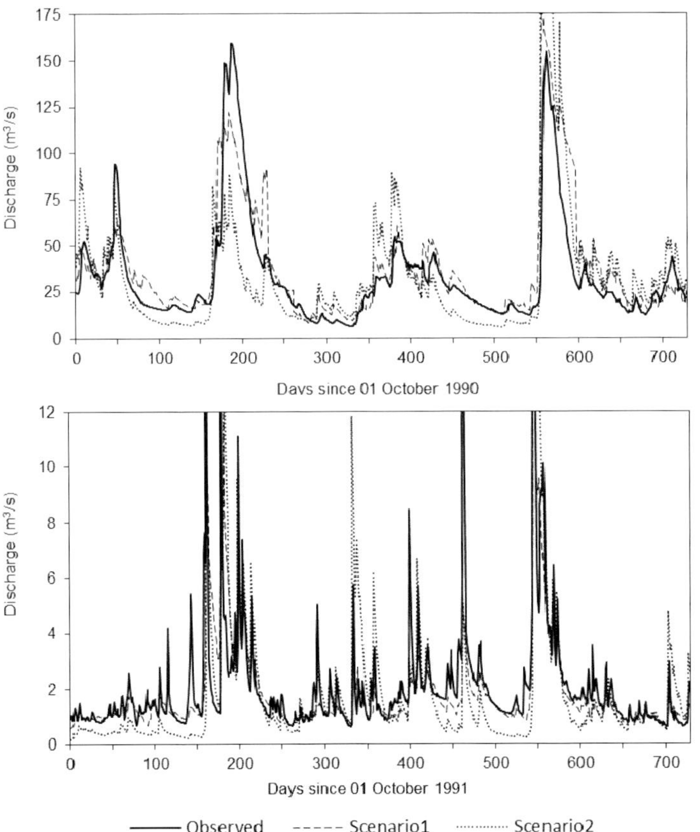

Fig. 2 Observed and simulated hydrographs for Sturgeon River (top) and Cold Creek (bottom).

some cases), because all the deficiencies of its data are effectively transferred to the simulated destination flow time-series. At the same time, one gauge may not always be representative of the spatial variability of catchment precipitation and temperature, even in small catchments, especially when it is located outside the catchment boundary. This is one of the reasons for the poor R^2 and CE values (Table 1) obtained for Cold Creek and Deport Creek, where the source meteorological station is a considerable distance outside the catchment (12–25 km) when compared with catchment dimensions (159 and 189 km^2, respectively).

Overall, comparison of observed and generated hydrographs, as well as fit statistics, shows that the approach is capable of reproducing the general pattern of daily streamflow variability, although low flows tend to be underestimated in Scenario 2 (Fig. 2). However, this can be explained by the low accuracy of the destination sites' FDCs estimation – from a rather coarse nation-wide regional study.

The paper has illustrated the application of the method only for the generation of *daily* streamflow hydrographs. At the same time a similar approach may be applied to generate monthly flow time-series from monthly precipitation data. In this case, the actual monthly step precipitation data may be used instead of the CPI time-series, and the recession coefficient will need a new interpretation.

The performance of the method has been illustrated using examples of predominantly small catchments. However, it may be anticipated that the method, in principle, is applicable to large river catchments as well. In this case, the number of initially available meteorological stations with suitable data will be large and it may be appropriate to generate the weighted average catchment wetness first. Depending on the type of data used, it could be daily or monthly weighted average wetness. Also it should be possible to split the large catchment into a set of "homogeneous" subareas and calculate weighted average time-series for each of them. Subareas then will replace the source meteorological sites. On the other hand, the large catchments are more likely to have suitable streamflow source gauges that can be used for data transfer between the sites and the original version of the spatial interpolation algorithm (which uses streamflow data *only*) may apply. The significant amount of remotely sensed data presently available, for a variety of hydrological analyses, may also be used for such simulations in the future.

REFERENCES

Archfield, S. A., Vogel, R. M., Steeves, P. A., Brandt, S. L., Weiskel, P. K. & Garabedian, S. P. (2010) The Massachusetts Sustainable-yield Estimator: a decision support tool to assess water availability at ungaged stream locations in Massachusetts. *Scientific Investigations Report 2009-5227*. US Department of the Interior, US Geological Survey.

FREND (1989) *Flow Regimes from Experimental and Network Data. I: Hydrological Studies; II: Hydrological Data.* Institute of Hydrology, Wallingford, UK. 345 pp

Hughes, D. A. & Smakhtin, V. Y. (1996) Daily flow time series patching or extension: a spatial interpolation approach based on flow duration curves. *Hydrol. Sci. J.* **41**, 851–871.

Kuusisto, E. (1984) Snow Accumulation and snowmelt in Finland. *Water Research Institute Publ. no. 55*. National Board of Waters, Finland, Helsinki, 120 pp.

Lee, J. U., Yang, J. & Choi, J. (2007) Development of Integrated GIS Interface for characteristics of regional daily flow. *World Academy of Science, Engineering and Technology* **28**, 180–185.

Metcalfe, R. A., Chang C. & Smakhtin V. (2005) Tools to support the implementation of environmentally sustainable flow regimes at Ontario's waterpower facilities. *Canadian Water Resources J.* **30**(2), 97–110.

Natural Resources Canada. Retscreen Clean Energy Project Analysis Software Website (http://www.retscreen.net/ang/home.php). Accessed on 17 December 2010.

Smakhtin, V. U. & Masse, B. (2000) Continuous daily hydrograph simulation using duration curves of a precipitation index. *Hydrol. Processes*. **14**, 1083–1100.

Tallaksen, L. M. (1995) A review of baseflow recession analysis. *J. Hydrol.* **165**, 349–370.

Siberian Lena River heat flow regime and change

BAOZHONG LIU[1] & DAQING YANG[2]

1 *Water and Environmental Research Center, University of Alaska Fairbanks, Fairbanks, Alaska 99775, USA*

2 *National Hydrology Research Center (NHRC), 11 Innovation Boulevard, Saskatoon S7N 3H5, Canada*

daqing.yang@ec.gc.ca

Abstract Heat flow, as a synthetic measure of discharge and water temperature, is useful to define the characteristics of a watershed's response to climate change. In this research, based on monthly discharge and water temperature data collected during 1950–1990, we defined the heat flow regime and quantified its change over the Lena watershed. Results show that near the Lena basin outlet, stream temperature is the dominant factor for the seasonal maximum heat flow in July. Trend analysis shows that the Lena River heat flow in June increased by 888 HU (41%) during 1950–1990 due to the stream temperature increase. This result may indicate a greater thermal impact of the Lena River on the local ecology over the Lena delta and on the land-fast sea-ice of the Laptev Sea.

Key words Lena River, Siberia; heat flow regime and change

1 INTRODUCTION

River discharge and water temperature are two major measures for characterizing a watershed undergoing climate change in terms of mass and energy, respectively. The water cycle that involves the ground, surface and atmosphere is essentially driven by energy. Thus, there is a need to combine mass and energy to investigate watershed response to climate change. Heat flow is not a new hydrological term; Mackay & Mackay (1975) mathematically defined it as a function of discharge and water temperature, and furthermore, they defined the mean open season river heat flow regimes at the Fort Providence and Fort Norman in the Mackenzie River basin. Elshin (1981) calculated the heat runoff (heat flow) for rivers in the European part of the former USSR.

Studies showed that Lena River runoff has experienced significant changes, i.e. an increase in winter, spring, and summer seasons, and a decrease in the autumn season, in the past few decades due to climate change and human activities (Yang *et al.*, 2002; Ye *et al.*, 2003). In addition, Yang *et al.* (2005) reported that the Siberian Lena watershed has experienced a basin-wide stream temperature rise in the early warm season during 1950–1992. From a perspective of the Earth system in which the hydrological cycle involves processes in and between the atmosphere, ground surface, and underground, the streamflow and water temperature changes demonstrate the responses of the permafrost-underlain Siberian Lena watershed to climatic fluctuations in terms of mass and energy. Thus, the heat flow regime and change are of great significance for understanding hydrological responses to climate change over the northern regions.

In this study, based on long-term (1950–1990) Lena River discharge and stream temperature data, we define the heat flow seasonality, interannual variability, and examine the long-term changes of the heat flow over the Lena basin as a whole. The purpose of this research is to present a new perspective regarding how large Arctic watersheds respond to climate change.

2 METHODS, DATA AND BASIN DESCRIPTION

The total heat flow in a given month is calculated for the open water season (May to October) using the following equation (Elshin, 1981):

$$H = kQTn \qquad (k = c_1 c_2 c_3 = 0.3615) \qquad (1)$$

where, H is the monthly total heat transported by a river in a given month (10^6 MJ); k is a constant, 0.3615 (10^6 MJ·s/(m^3 day °C)); Q is discharge (m^3/s); T is stream temperature (°C); and n is number of open water days in a given month; c_1 is specific heat of water, 4.184 J/(g °C); c_2 is a conversion coefficient for discharge, 10^6 g/m^3; c_3 is conversion coefficient for time, 88 400 s/day.

As the magnitude of the heat flow is very large for some months, we define 10^9 MJ/month as a heat unit (HU). The heat flow in the cold season (November to April) is negligible because the stream temperature is usually very close to 0°C when the river is ice-covered. In the calculation of heat flow, the number of open water days (n) for June to September is 30 or 31, and for May and October is 10 and 20, respectively. This is a conservative means to avoid potential overestimation of the heat flow calculation in May and October, because the river usually opens in mid May and freezes in late October.

Discharge data are available from the R-ArcticNet (v. 2.0) (a database of pan-Arctic river discharge, www.r-arcticnet.sr.unh.edu/main.html). Yang *et al.* (2005) have reported the source, method of observation, and quality of stream temperature data. The raw stream temperature data were measured three times a month (10th, 20th, and 30th). To derive the monthly mean stream temperature, we averaged different combinations of the water temperature measurements, for instance, averaging the 10th, 20th, and 30th day records in a month, averaging the 30th day of the previous month and the other three records in the month, or just averaging the 10th and 20th days. Finally, we selected the last method to estimate the monthly mean stream temperature. The estimate made by this approach is closer to the value in mid month, which is representative and conservative. This approach can be consistently applied to each month, including May and October. In data processing, when one of the 10th and 20th measurements is missing, we use a long-term mean value to fill the gap. If both records are missing, the monthly mean stream temperature in that year will be taken as a missing record.

The Lena River is one of the largest rivers in the Arctic. It originates from the Baikal Mountains in the south central Siberian Plateau and flows northeast and north, emptying into the Arctic Ocean via the Laptev Sea (Fig. 1). The drainage area of the Lena basin is about 2 430 000 km^2, approximately 78–93% of which is underlain by permafrost (Zhang *et al.*, 1999). The Lena River contributes 524 km^3 of freshwater per year, or about 15% of the total freshwater into the Arctic Ocean (Prowse & Flegg, 2000; Shiklomanov *et al.*, 2000). The drainage is covered mainly by forest (84%), shrub (9%), grassland (3%), cropland (2%), and wetland (1%) (Revenga *et al.*, 1998). The basin's total population is about 2.3 million people, with one city (Yakutsk) having a population of more than 270 000. Compared with other large Siberian rivers, such as the

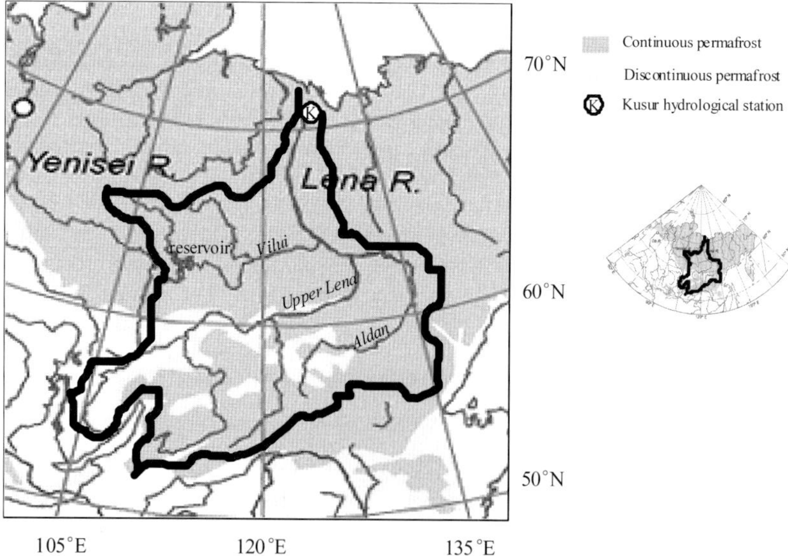

Fig. 1 The Lena watershed in Siberia. Also shown are permafrost distribution, major tributaries, basin boundaries, and the Kusur hydrologic station near the basin outlet.

Ob and Yenisei, the Lena basin has less human activity and much less economic development (Dynesius & Nilsson, 1994). There is only one large reservoir (capacity greater than 25 km^3), in west Lena basin, which was built during the late 1960s.

This study examines the response of the Lena basin as a whole to the climate change, thus data near the basin outlet are used (Fig. 1). Technically, we use the long-term mean and standard deviation to define the regime and interannual variation, and carry out trend analysis to identify a total change ("trend" in this paper) during a period by linear regression (Ye *et al.*, 2003). The standard *t*-test is used to indicate the statistical significance (or confidence level) of trend. In terms of confidence level, in this study we regard over 90% confidence as "significant", 60–89% as "considerable", and lower than 59% as "weak". Percentage of change (total trend to long-term mean) is also used to indicate the degree of change.

3 DATA ANALYSES AND RESULTS

3.1 Discharge

The seasonal cycle of monthly discharge near the Lena basin outlet (station K in Fig. 1) is presented in Fig. 2(a). It generally shows a low-flow (1378–3522 m^3/s) period during November to April and a high-flow (6392–74 657 m^3/s) season from June to October, with maximum discharge usually occurring in June due to snowmelt floods. Generally, watersheds with a high percentage of permafrost coverage have low subsurface storage capacity and thus a low winter baseflow, and a high summer peak flow (Kane, 1997). In the Lena River basin, which is mostly underlain by continuous permafrost (78–93%), the peak flow in June is about 54 times the lowest discharge in April and about 12 times the May runoff. The streamflow in July sharply drops to 55% of the June peak flow. Streamflow drops slowly to about 13 586 m^3/s in October due to less rainfall. The runoff in November and December (3522 and 3008 m^3/s, respectively) is relatively larger than that in January to April (1378–2814 m^3/s). The interannual variation of monthly runoff near the Lena River outlet is generally small in the cold season (standard deviation around 428–829 m^3/s), and large (standard deviation between 3314 and 10 390 m^3/s) in the warm season.

Trend analysis of monthly discharge records near the Lena basin outlet reveals a discharge increase during most months (Fig. 2(a)). Discharge increases over the period 1950–1990 are found of between 747 and 1009 m^3/s (21–73%) during November to April. These positive changes are statistically significant at a 92–99% confidence level. Strong streamflow increase near the Lena River outlet is found to be about 9280 m^3/s (145%) in May, which is statistically significant at the 99% confidence level. A weak discharge decrease is found of about 4600 m^3/s in June, but is statistically insignificant (59% confidence level). The negative trend in June is a reasonable consequence of the strong runoff increase in May due to earlier snow cover melt. Weak discharge increases are found, about 2078 m^3/s (5%) in July, about 1825 m^3/s (7%) in August, and about 484 m^3/s (2%) in September. These positive changes are statistically insignificant (below 39% confidence level). The streamflow decrease is found to be very weak in October (about –137 m^3/s, 1%), and statistically insignificant (6% confidence level). As the result of the monthly streamflow changes, yearly mean discharge shows a considerable (75% confidence level) upward trend, 1266 m^3/s (or 8%) over the period 1950–1990.

3.2 Stream temperature

The seasonal cycle of monthly stream temperature near the Lena basin outlet (Fig. 1) shows an extended ice-covered period from November to May, and an open water period from June to October (2.7–13.5°C), with higher stream temperature in both July and August (Fig. 2(b)). According to the original stream temperature measurements as taken every ten days, stream temperature records on 30 May near the Lena basin outlet are not available during 1950–1966, while mostly available during 1967–1990. The available stream temperature measurements for 30 May are usually very low (around 0.1°C), thus we count May as part of the ice-covered period.

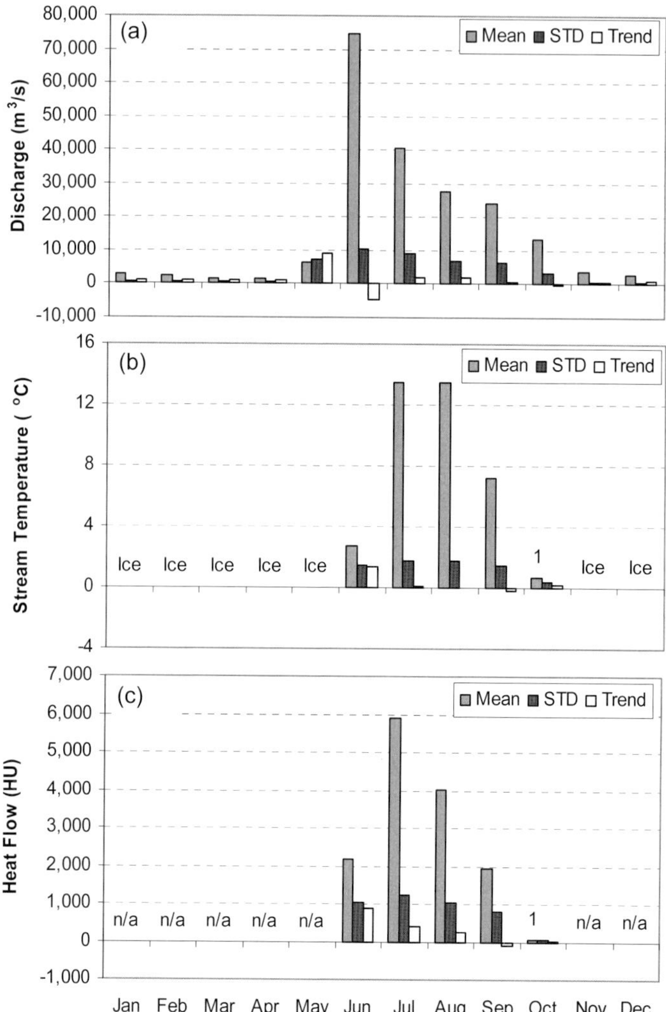

Fig. 2 Long-term (1950–1990) mean, standard deviation, and total trend of monthly discharge (a), stream temperature (b), and heat flow (c) near Lena basin outlet. Number of missing records indicated above bars, 1 HU = 10^9 MJ/month.

Water temperature measurements indicate that the Lena basin outlet becomes frozen in late October. The long-term (1950–1990) mean monthly stream temperature in June is 2.7°C. It increases sharply to 13.5°C in June and remains higher (around 13.5°C) in August. Monthly stream temperature drops sharply to 7.3°C in September, and continuously decreases to 0.7°C in October when the Lena River outlet gradually becomes frozen. The interannual variation of monthly stream temperature near the Lena basin outlet is generally large from June to September (standard deviation 1.4–1.8°C) and small in October (0.4°C).

Trend analysis of the monthly stream temperature over the period 1950–1990 reveals the strongest water temperature increase in June, 2.3°C, which is statistically significant at 93% confidence level. The stream temperature increases are found to be very weak in July and August (less than 0.1°C), and statistically insignificant (below 5% confidence level). A weak stream temperature decrease/increase is detected in September (–0.2°C)/October (0.2°C), both statistically insignificant (23% and 49% confidence levels, respectively).

3.3 Heat flow

According to the stream temperature measurements as taken every ten days over the period 1950–1990, the zero-heat transportation period near the Lena basin outlet is defined from November to May when the final part of Lena River is generally ice-covered. The heat transportation near the Lena basin outlet is active in the open water season, roughly from June to October (Fig. 2(c)). The heat transport in June near the Lena River outlet is large (2181 HU) due mainly to the large monthly discharge, although the monthly stream temperature is quite low (2.7°C) in this month. Heat flow increases sharply to the seasonal maximum in July (5888 HU). This indicates that the maximum July monthly stream temperature overwhelms the maximum June monthly discharge. The heat transport near the Lena River outlet in August drops to 4016 HU, due mainly to the decrease of monthly discharge in August. The heat flow decreases continuously to 1154 HU in September mainly due to the decrease of monthly stream temperature in this month. The heat transport in October before the upper Lena becomes frozen is about 71 HU. The annual total heat flow near the Lena River's outlet is 14 028 HU. The interannual variation of heat flow near the Lena basin outlet is generally large from June to September (standard deviation 787–1236 HU) and small in October (56 HU).

Trend analysis over the period 1950–1990 reveals heat flow increases during June to August and October, and decrease only in September (Fig. 2(c)). The largest heat flow increase, 888 HU (41%), is found in June, and is statistically significant at the 90% confidence level. This strong heat flow increase is due mainly to the strong upward trend of stream temperature in June, as the discharge trend is downward in this month. The positive changes are found to be 415 HU (7%) in July and 269 HU (7%) in August, both statistically insignificant (47 and 37% confidence levels, respectively). These positive changes are mainly due to runoff increases in these two months as the temperature changes are very weak. A weak heat flow decrease of about –86 HU (4%) is detected in September, but is statistically insignificant (16% confidence level). This weak decrease is mainly due to the weak downward trend of stream temperature as the discharge trend is weakly upward in this month. The heat flow in October has slightly increased (7 HU, 10%), statistically insignificant (19% confidence level), which is due mainly to the slight stream temperature increase in this month. As the result of monthly heat flow changes, the annual total heat transport near the Lena River outlet shows a considerable increase (1521 HU) over the period 1950–1990, although statistically it is less significant (68% confidence level).

4 SUMMARY

The heat flow regime over the Lena watershed is estimated in this research. Results show that the heat flow near the Lena River's outlet has considerably increased during 1950–1990. The strongest heat flow increase (41%) is detected in June and attributed to the stream temperature increase in this month. This heat flow increase may have a great impact on the land-fast sea-ice dynamics of the Laptev Sea (Bareiss *et al.*, 1999; Alexandrov *et al.*, 2000). There is a need to evaluate the impact of the June heat flow increase on the lives of both birds and aquatic species in the Lena delta. Weak heat flow increases are also detected in July and August, and attributed to the increased runoff. This may indicate increased sediment in the Lena River due to the strong thermal erosion on the riverbed and banks.

Acknowledgement This study was supported by the US NSF (Award no. ARC- 0612334).

REFERENCES

Alexandrov, V. Y., Martin, T., Kolatschek, J., Eicken, H. & Kreyscher, M. (2000) Sea ice circulation in the Laptev Sea and ice export to the Arctic Ocean: results from satellite remote sensing and numerical modeling. *J. Geophys. Res.* **105** (C), 17143–17159.

Bareiss, J., Eicken, H., Helbig, A. & Martin, T. (1999) Impact of river discharge and regional climatology on the decay of sea ice in the Laptev Sea during spring and early summer. *Arctic, Antarctic & Alpine Res.* **31**, 214–229.

Dynesius, M. & Nilsson, C. (1994) Fragmentation and flow regulation of river systems in the northern third of the world. *Science* **266**, 753–762.

Elshin, Y. (1981) River heat runoff in the European part of Russia. *Meteorology and Hydrology* **9**, 85–93.

Kane, D. L. (1974) A review of dam construction techniques in permafrost regions. *Northern Engineer* **6**, 25–29.

Mackay, J. R. & Mackay, D. K. (1975) Heat energy of the Mackenzie River. In: *Further Hydrologic Studies in the Mackenzie Valley, Canada* (Environmental-Social Committee Northern Pipelines. Task Force on Northern Oil Development report, Report 74-35), 1–23. Ottawa, Information Canada.

Prowse, T. D. & Flegg, P. O. (2000) Arctic river flow: A review of contributing areas. In: *The Freshwater Budget of the Arctic Ocean* (ed. by E. L. Lewis, E. P. Jones, P. Lemke, T. E. Prowse & P. Wadhams) (Proc. NATO Advanced Research Workshop, Tallin, Estonia, 27 April–1 May 1998), 269–280. Kluwer Acad., Norwell, Massachusetts, USA.

Revenga, C., Murray, S., Abramovitz, J. & Hammond, A. (1998) *Watersheds of the World: Ecological Value and Vulnerability.* World Resour. Inst. and World Watch Inst., Washington, DC, USA.

Shiklomanov, I. A., Shiklomanov, A. I., Lammers, R. B., Peterson, B. J. & Vörösmarty, C. J. (2000) The dynamics of river water inflow to the Arctic Ocean. In: *The Freshwater Budget of the Arctic Ocean* (ed. by E. L. Lewis, E. P. Jones, P. Lemke, T. E. Prowse & P. Wadhams) (Proc. NATO Advanced Research Workshop, Tallin, Estonia, 27 April–1 May 1998), 281–296. Kluwer Acad., Norwell, Massachusetts, USA.

Yang, D., Kane, D., Hinzman, L., Zhang, X., Zhang, T. & Ye, H. (2002) Siberian Lena River hydrologic Regime and recent change. *J. Geophys. Res.* D **107**, 4694, doi:10.1029/2002JD002542.

Yang, D., Liu, B. & Ye, B. (2005) Stream temperature changes over Lena River Basin in Siberia. *Geophys. Res. Lett.* **32**, doi:10.1029/2004GL021568.

Ye, B., Yang, D. & Kane, D. (2003) Changes in Lena River streamflow hydrology: human impacts vs. natural variations. *Water Resour. Res.* **39**, 1200, doi:10:1029/2003WR001991.

Zhang, T., Barry, R. G., Knowles, K., Heginbottom, J. A. & Brown, J. (1999) Statistics and characteristics of permafrost and ground-ice distribution in the Northern Hemisphere. *Polar Geogr.* **23**, 132–154.

2 Snow Cover, Permafrost and Glaciers

Changes in North American snow packs for 1979–2004 detected from the snow water equivalent data of SMMR and SSM/I passive microwave and related climatic factors

THIAN YEW GAN[1,2], ROGER BARRY[1] & ADAM GOBENA[2]

1 *National Snow and Ice Data Center (NSIDC), University of Colorado at Boulder, Colorado, USA*
tgan@ualberta.ca; adam.gobena@bchydro.com

2 *Department of Civil & Environmental Engineering, University of Alberta, Canada*

Abstract Changes to the North American (NA) snow packs for 1979–2004 were detected from snow water equivalent (SWE) retrieved from SMMR and SSM/I passive microwave data using the non-parametric Kendall's test. In NA, about 30% decreasing trends in SWE for 1979–2004 are statistically significant, or about three times more than significant increasing trends of SWE. Significant decreasing trends in SWE are more extensive in Canada than in the USA. The overall mean trend magnitudes are about –0.4 to –0.5 mm/year, which translates to an overall reduction of snow depth of about 5–6 cm in 26 years. From detected increasing (decreasing) trends of gridded temperature (precipitation) based on the North American Regional Reanalysis (NARR) and the University of Delaware data set for NA, and their respective correlations with SWE data, it seems that the extensive decreasing trends in SWE detected mainly in Canada are caused more by increasing temperatures than by decreasing precipitation.

Key words snow water equivalent; SMMR and SSM/I passive microwave data; North America; Kendall's non-parametric trend test; surface temperature; precipitation; climate anomalies

1 INTRODUCTION

This study on detecting changes in snow packs of North America (NA) is based on snow water equivalent (SWE) data retrieved from the brightness temperature (T_B) in K (Kelvin) of passive microwave remote sensing platforms, such as the Scanning Multichannel Microwave Radiometer (SMMR) from 25 October 1978 to 20 August 1987, and the Special Sensor Microwave Imager (SSM/I) since 7 September 1987. The SMMR sensor flew on NASA's Nimbus 7 while SSM/I are mounted on the Defense Meteorological Satellite Program (DMSP) satellites of the USA. Since May 2002, T_B retrieved from the Advanced Microwave Scanning Radiometer-EOS (AMSR-E) sensor aboard the Aqua satellite has also been used to estimate SWE. The frequencies (resolution) of SSM/I are 6.6 GHz (150 km), 19 GHz, 22 GHz, 37 GHz (25 km) to 85 GHz (12.5 km), while that of AMSR-E are 6.9 GHz (50 km) to 89 GHz (5 km). These T_B values are either horizontally (H) or vertically (V) polarized. Because of limitations of passive microwave data, retrieval of SWE from such data based on equations (1) and (2) applied for NA is expected to have varying accuracy, depending on the vegetation types, snowpack characteristics, frozen and unfrozen water, topography, mountains *versus* flat plains, possible effect of wet snow, lake ice, ice lenses, depth hoar and deep snowpacks. The traditional T_B difference between 19 and 37 GHz has been shown to be inappropriate for lake-rich environments in the Arctic, and retrieving tundra SWE can be challenging. In view of these limitations, this trend analysis of SWE data derived from such data should be more reliable in regions dominated by grassland (e.g. Canadian Prairies) than in the Canadian Arctic with many frozen lakes, in forested or mountainous regions, and snowpacks that consist of depth hoar and wind slabs. Even though such SWE data are subject to uncertainties, it is the only data available to analyse SWE trends at the continental scale. Also, by using a non-parametric approach, we have overcome some of the limitations of the aforementioned uncertainties. The observed Northern Hemisphere snow cover since the 1920s had remained quite steady until around the 1980s, after which a significant decrease in snow cover has been observed (Lemke *et al.*, 2007). With about 26 years of continuous SWE data from SMMR (late 1978 to 1987) and SSM/I (summer of 1987 until 2004), it is now feasible to perform a trend analysis of SWE over NA for the winter and spring seasons to detect possible changes.

2 RESEARCH OBJECTIVES

With the above background information, this study has the following objectives: (1) to analyse monthly monotonic trends for the October–April SWE over NA for the 1979–2004 period; (2) to compute trend magnitudes, trend homogeneity, and the spatial distributions of trends and variability in SWE across NA from 1979 to 2004; (3) to perform principal component analysis (PCA) on SWE, and to correlate PCAs of SWE with climate anomalies; and (4) to relate SWE to major climate variables, and from the results, identify possible cause(s) for the detected changes of the snowpacks of NA from 1979 to 2004.

3 SWE RETRIEVED FROM SMMR AND SSM/I PASSIVE MICROWAVE DATA

On the basis of volumetric scattering, which is the dominant loss mechanism for microwave radiation greater than 15 GHz incident on a snowpack, it is possible to empirically relate the brightness temperature (T_B) of a certain frequency and polarization (H or V) to the SWE of snow packs. The NSIDC at the University of Colorado, Boulder, USA, provides such passive microwave T_B data at Equal-Area Scalable Earth, or EASE Grid format. The 1979–1987 SMMR SWE data of NSIDC were retrieved from equation (1) (Chang *et al.*, 1987):

$$\text{SWE (mm)} = 4.77(T_{B18H} - T_{B37H}) \tag{1}$$

where T_{B18H} and TB_{37H} are the horizontally polarized T_B at 18 GHz and 37 GHz, respectively. The 1987–2004 SSM/I SWE data of NSIDC were retrieved from equation (2) (Armstrong *et al.*, 2001):

$$\text{SWE (mm)} = 4.77(T_{B18H} - T_{B37H} - 5) \tag{2}$$

which is slightly different from equation (1), probably because of sensitivity differences between SMMR and SSM/I sensors in detecting shallow snow. Daily SWE is adjusted for the surface forest cover using the BU-MODIS (NSIDC, 2005) land cover data so that SWE (mm) = SWE/(1 – Forest%). Further, to ensure snow packs are detectable by passive microwave data, SWE less than 7.5 mm is considered unreliable and set to zero. Given that the SMMR and SSM/I data are at 25-km resolution, we can only analyse meso- to regional-scale variability of snowpack from these data. Furthermore, some studies found errors in the SWE record from passive microwave observations suggesting systematic biases related to grain size, not SWE, e.g. in northern Siberia, the approach of Chang *et al.* (1987), because thin snowpack and a strong thermal gradient promoted large crystals that enhanced scattering, grossly overestimated SWE. For NA, a 7-month (October to April) running mean (solid curve) and the monthly (dotted curve) SWE anomalies based on SMMR and SSM/I data of 1979–2004 generally show negative anomalies since the late 1980s (see Figure 2.20 of Barry & Gan, 2011).

4 NON-PARAMETRIC, MANN-KENDALL'S TEST AND TREND MAGNITUDE (β)

The non-parametric, Mann-Kendall test (Mann, 1945; Kendall, 1975) for testing trends in seasonal time series should be more robust than parametric methods because it can handle non-normality, censoring, and seasonality. To safeguard against non-representative results, only data sets with a few missing values were tested. Since Kendall's statistic, S_k, is not based on the magnitudes of individual SWE, it is expected to be more robust than say, the simple linear regression. However, we expect some degree of uncertainties in estimating SWE from two different satellite sensors. From the perspective of detecting trend of SWE data, S_k for season k will be first based on SWE estimated from equation (1) from 1979 to 1987. In changing to SWE of SSM/I based on equation (2) from 1987 to 2004, S_k could be affected in one step change, because after the transition, it will be based on SWE values estimated from equation (2) only. Albeit SWE estimated from equations (1) and (2) based on SMMR and SSM/I T_B data, respectively, may not be exactly the same, this non-parametric approach merely accounts for whether SWE has been increasing (or decreasing) from one time step to the next, so the effect of using SWE data estimated from two satellites on S_k should be minimal.

5 DISCUSSION OF RESULTS

In terms of SWE, about 30% of detected decreasing trends in SWE for 1979–2004 are statistically significant at $\alpha/2 = 0.05$, which is about three times the detected increasing trends in SWE (see Table 1). Significant decreasing trends in SWE are more extensive in Canada than in the USA, where such decreasing trends are mainly found along the American Rockies (Fig. 1(a)). Scattered increasing trends are found mainly near Hudson Bay in Canada, but not near the Great Lakes where large increases in lake-effect snowfall since 1951 had been reported. Apparently in Canada statistically significant decreasing trends are mainly detected east of the Canadian Rocky Mountains. We did not detect much change to SWE along the coastal areas of NA likely because SWE data retrieved from passive microwave data suffer from mixing signals emitted from relatively dense vegetation along the coastal areas. Albeit few significant trends have been detected, we believe SWE along the western coastal areas of NA had also undergone decreasing trends. The SWE data along the American and Canadian Rockies could be affected by terrain features of mountainous areas, but the net effects may be more or less averaged out by the relatively coarse resolution of such data. This could be partly why significant trends are detected in these areas, but not along the western coastal areas generally covered with dense forest covers. A vigorous approach to compute the trend magnitude β_k (mm/year) used is:

$$\beta_k = median \frac{(X_{jk} - X_{ik})}{(j-i)} \tag{3}$$

where $k = 1, 2, q$ where $1 < I < j < n$. β_k estimated using equation (3) are about –0.4 to –0.5 mm/year, which are of smaller range than that of simple regressions, while the mean (median) of negative trends alone is about –1.5 (–1.2) mm/year (Table 2). This means that from 1979 to 2004, the overall SWE of NA has decreased by about 10 to 13 mm or possibly more, and in terms of snow depth could range from about 4 to 5 cm (depending on the average snow density which typically ranges between 200 and 250 g/cm^3). Based on the negative trends, apparently the decrease in SWE mainly occurred in the area extending from the western part of the Canadian high Arctic to north of the five Great Lakes; the decrease in SWE amounts to 25–30 mm.

Table 1 Results of Kendall's test on statistically significant increasing (+ to +****) and decreasing (– to –****) trends of monthly SWE for 1979–2004 at different significance levels.

Mon	+ $\alpha/2$ <.95	+* $\alpha/2$ >.95	+** $\alpha/2$ <.975	+*** $\alpha/2$ <.99	+**** $\alpha/2$ <.995	– $\alpha/2$ ≥.05	–* $\alpha/2$ <.05	–** $\alpha/2$ <.025	–*** $\alpha/2$ <.01	–**** $\alpha/2$ <.005	Total pixels analysed	SWE >7.5mm
Jan	4374	393	327	151	165	5148	1015	1007	566	695	13841	12091
Feb	4387	473	481	296	562	4090	985	1226	743	1598	14841	12147
Mar	5886	470	469	287	401	4303	1076	1197	650	967	15706	11982
Apr	7613	439	395	186	234	3368	806	928	494	1069	15532	9645

Table 2 Means and medians of trend magnitudes (β in mm/year) for NA SWE between 1979 and 2004.

Month	Overall mean of negative and positive trends	Mean of negative trends only (mm/year)	Median of negative trends only
January	–0.441	–1.216	–1.0
February	–0.541	–1.779	–1.5
March	–0.482	–1.506	–1.28
April	–0.376	–1.371	–1.18

6 INFLUENCE OF CLIMATIC ANOMALIES ON SWE

The possible effects of the Pacific Decadal Oscillation (PDO), Pacific/North American (PNA) teleconnection pattern, and the El Nino-Southern Oscillation (ENSO) on SWE are examined using the PDO, PNA, Southern Oscillation Index (SOI) and Nino3 indices (Tables 3 to 4) since these

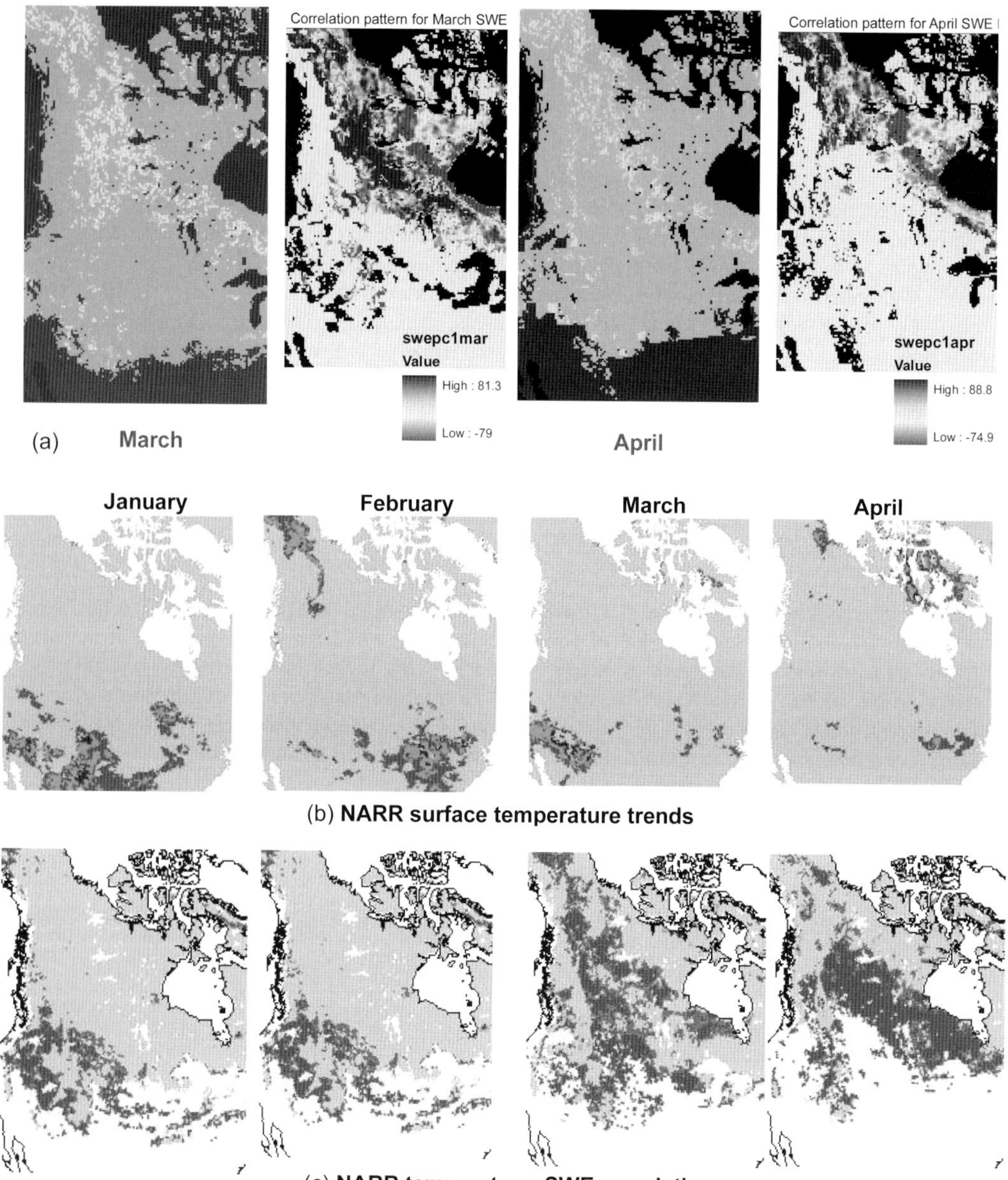

Fig. 1 (a) The spatial distributions of both decreasing [March and April] trends in SWE, the snow cover extent (grey colour) in NA for detected trends that are statistically significant at $\alpha/2 = 0.05$ (red), $\alpha/2 = 0.025$ (green) and $\alpha/2 = 0.01$ (blue), respectively. The PC1 patterns are also presented. (b) Monthly NARR surface temperature trends based on a 2-tailed significance level of 5% (red), 2.5% (green) and 1% (blue); and (c) NARR surface temperature–SWE correlation, where colour shadings are for 5% significance level (green for negative and light blue for positive correlations) and 1% significance level (red for negative and dark blue for positive correlations). (Note: printed in grayscale.)

climate anomalies have been shown to be teleconnected to the precipitation and streamflow of western Canada (e.g. Gobena & Gan, 2006). To reduce the dimensionality and capture any dominant signals in the SWE data, a PCA was conducted. Pearson correlation coefficients (ρ_p) of

Table 3 Pearson's Correlations (ρ_p) between PC1 to PC4 with the corresponding monthly PDO, one-season lag, seasonal PDO of OND, NDJ, DJF, and JFM.

ρ_p	Monthly PDO				OND		NDJ	DJF		JFM
Mon	Oct	Feb	Mar	Apr	Mar	Apr	Mar	Mar	Apr	Apr
PC1	–0.13	**0.44***	**0.43**	0.36	0.26	0.30	0.29	0.36	**0.39**	**0.44**
PC2	**0.38**	0.34	0.15	0.09	0.14	0.02	0.10	0.17	0.05	0.12
PC3	–0.06	–0.21	–0.36	0.36	**–0.57**	0.31	**–0.56**	**–0.52**	0.30	0.30
PC4	–0.14	–0.32	0.17	0.25	–0.04	**0.45**	–0.04	–0.01	**0.43**	0.34

Numbers in bold mean statistically significant at $\alpha/2 = 0.05$.

Table 4 Spearman's rank correlation (ρ_s) between PC1, PC2, and PC3 and the corresponding monthly PDO, one-season lag, seasonal PDO of OND, NDJ, DJF, and JFM.

ρ_s	Monthly PDO				OND		NDJ	DJF		JFM
Mon	Jan	Feb	Mar	Apr	Mar	Apr	Mar	Mar	Apr	Apr
PC1	–0.25	0.28	0.21	0.34	0.02	0.26	0.03	0.15	0.23	0.26
PC2	–0.25	0.25	0.13	0.10	0.17	0.09	0.21	0.18	0.08	0.09
PC3	–0.30	–0.24	–0.35	0.32	–0.53#	0.32	–0.54	–0.39	0.26	0.27

Table 3 show several statistically significant ρ_p between monthly PC1, PC2, PC3 and PC4 of NA's SWE and monthly or seasonal (OND to JFM) PDO. No significant ρ_p was found between PDO-JAS and SWE of October to December, which means that winter to early spring SWE are more affected by PDO than the autumn SWE. Table 3 shows more statistically significant ρ_p than statistically significant Spearman's statistics ρ_s (Table 4). Apparently, PDO exerts some influence on the overall snowpack SWE of NA during March, and possibly April. The climate of western NA have been related to interdecadal oscillations of PDO, with warm PDO being associated with warm, dry conditions and cool PDO being related to cool, wet conditions in western NA (Gobena & Gan, 2006). The PDO regime has also been shown to modulate winter temperature responses over most of western Canada during El Nino and La Nina events. By exerting an influence on the moisture and temperature regimes of NA, PDO is expected to affect its snowpack. Compared to PDO, the effect of PNA is relatively modest with significant ρ_s and ρ_p in February (PC2) and March (PC3 and PC4), while Nino3 and SOI exhibit less influence over the snow of NA. ENSO is teleconnected to the precipitation of western NA. The PC1 relationship is located mainly in the north (south) central to western regions of USA (Canada), while the PC3 relationship stretches from the mid-west to eastern USA and Atlantic Canada. The first four PCs of monthly SWE are also compared with these climate anomalies. The influence of PDO is mostly represented by PC1 for November, and February to April. Except for October and January, PC1 is mainly positive until about 1988 and is mainly negative thereafter, and it likely describes SWE variations of low-frequency. It seems PDO is more correlated to PC1 until 1988, after which PDO's influence became fuzzy. To better understand the spatial relationships between climate anomalies and SWE, some selected cases of correlation fields between PC1, PDO, PNA, and monthly SWE of individual pixels for February to April are examined. They show a wide range of negative and positive relationships that are mainly centred along northern Canada, above the Great Lakes on the east, to Alaska of USA on the west, and in the mid-west near the American Rockies. PDO and PNA had affected the snow packs of southern Canada and the USA since selected cases exhibit either statistically significant ρ_p or ρ_s (Tables 3 and 4).

7 SWE–TEMPERATURE AND SWE–PRECIPITATION RELATIONSHIPS

Higher air temperatures in the last three decades in NA could be a major contributor to the fairly wide spread reduction in its SWE. Similarly, possible changes to precipitation could be partly responsible. In lower latitude regions, not much change in the annual precipitation had been observed but less snowfall and more rainfall had been observed in some places. To find possible climatic factors leading to a general reduction in the snow packs of NA, the next step is to correlate changes of SWE to gridded precipitation and air temperature data to explore the possible relationships between temperature, precipitation and SWE. Trend analysis of both the gridded 2-m air temperature data of the University of Delaware (Willmott & Robeson, 1995) and that of the North American Regional Reanalysis (NARR) (Mesinger *et al.*, 2006) showed little agreement between areas of detected increasing temperature trends and that of decreasing SWE trends that are statistically significant based on passive microwave data. However, extensive areas of significant negative correlations between SWE and temperature exist both across the USA and Canada except in January, and the distribution of these areas of negative correlation closely follow the areas of the decreasing trends detected from the SWE data (Fig. 1(a)–(c)). Unlike SWE trends, only scattered significant precipitation trends are detected, which implies that extensive decreasing SWE trends detected were more likely caused by warming than by decreasing precipitation. Albeit climatic anomalies could contribute to part of the detected changes, it seems that extensive decreasing trends in SWE detected in Canada and parts of USA were caused more by increasing temperatures than by decreasing precipitation. The global-scale decline of snow and ice cover since 1980 (Lemke *et al.*, 2007), likely support our conclusion that warmer temperature, which we believe to be driven by a general warming of the atmosphere in response to rising greenhouse gases, caused the decline in SWE detected in NA. Relative to the 1906–1970 surface temperature of NA, Forster *et al.* (2007) found both observed and 58 cases of GCMs' modelled results showing drastic increase in NA's surface temperature in recent decades.

8 SUMMARY AND CONCLUSIONS

On the basis of the non-parametric Kendall's test applied to the SWE data of SMMR and SSM/I for NA, decreasing trends have been detected. About 30% of the detected decreasing trends in SWE for 1979–2004 are statistically significant at $\alpha/2 = 0.05$, which is about three times more than detected increasing trends in SWE that are statistically significant. Significant decreasing trends in SWE are more extensive in Canada than in the USA, where such decreasing trends are mainly found along the American Rockies. Because of uncertainties associated with SWE data retrieved from passive microwave data, results obtained for mountainous or forested areas of NA, the tundra in Arctic Canada with many frozen lakes or snowpacks that consist of depth hoar and wind slabs should be treated with caution. The overall mean trend magnitudes are about –0.4 to –0.5 mm/year, which means an overall reduction of snow depth of about 5 to 6 cm (assuming a snow density of about 200 to 250 g/cm^3). The PC1 of NA's SWE is found to be significantly correlated to the PDO, marginally correlated to the PNA pattern and to ENSO. The PDO for February to April was significantly correlated to the SWE of Canada, and the SWE PC1 scores and PDO indices show quite similar variation, such that the PC1 score and PDO are positive until about 1988, after which the values are mostly negative. From the detected significant increasing trends of gridded temperature data of NARR and the University of Delaware for NA, and their respective correlations with SWE data, it seems that the extensive decreasing trends in SWE detected in NA are more attributed to increasing temperatures than to decreasing precipitation.

REFERENCES

Armstrong, R., Brodzik, M. J. & Savoie, M. H. (2001) Recent Northern Hemisphere snow extent: a comparison of data derived from visible and microwave sensors. *Geophys. Res. Lett.* **23**(19), 3673–3676.

Chang, A. T. C., Foster, J. L. & Hall, D. K. (1987) Nimbus-7 SMMR derived global snow cover parameters. *Ann. Glaciology* **9**, 39–44.

Forster, P., Ramaswamy, V., Artaxo, P., Berntsen, T., Betts, R., Fahey, D. W., Haywood, J., Lean, J., Lowe, D. C., Myhre, G., Nganga, J., Prinn, R., Raga, G., Schulz, M. & Van Dorland, R. (2007) Changes in atmospheric constituents and in radiative forcing. In: *Climate Change 2007: The Physical Science Basis*. 4th Report of Intergovernmental Panel on Climate Change, Cambridge University Press, UK.

Barry, R. & Gan, T. Y. (2011) *Global Cryosphere – Past, Present and Future.* Cambridge University Press. 480 pp.

Gobena A. K. & Gan, T. Y. (2006) Low-frequency variability in southwestern Canadian streamflow: links to large-scale climate anomalies. *Int. J. Climatol.* **26**, 1843–1869, doi:10.1002/joc.1336.

Kendall, M. G. (1975) *Rank Correlation Methods*. Griffin, London, UK.

Lemke, P., Ren, J., Alley, R. B., Allison, I., Carrasco, J., Flato, G., Fujii, Y., Kaser, G., Mote, P., Thomas, R. H. & Zhang, T. (2007) Observations: changes in snow, ice and frozen ground. *In: Climate Change 2007: The Physical Science Basis*, 4th Report of Intergovernmental Panel on Climate Change. Cambridge University Press, UK.

Mann, H. B. (1945) Non-parametric tests against trend. *Econometrica* **13**, 245–259.

Mesinger, F., DiMego, G., Kalnay, E., Mitchell, K., Shafran, P. C., Ebisuzaki, W., Jovi, D., Woollen, J., Rogers, E., Berbery, E. H., Ek, M. B., Fan, Y., Grumbine, R., Higgins, W., Li, H., Lin, Y., Manikin, G., Parrish, D. & Shi, W. (2006) North American Regional Reanalysis. *BAMS* **87**, 343–360, DOI: 10.1175/BAMS-87-3-343.

NSIDC (2005) EASE-Grid version of BU-MODIS land cover characteristic, ftp://sidads.colorado.edu/pub/EASE/. Boulder, Colorado, USA: National Snow and Ice Data Center, *Digital Media*.

Sen, P. K. (1968) Estimates of the regression coefficient based on Kendall's tau. *J. Am. Sta. A* **63**, 1379–1389.

Willmott, C. J. & Robeson, S. M. (1995) Climatologically aided interpolation (CAI) of terrestrial air temperature. *Int. J. Climatol.* **15**(2), 221–229.

Temporal variation in acidity and ion concentration of snowmelt water in light and heavy snow years

YOSHIHIRO ASAOKA[1], YUKARI TAKEUCHI[2] & SO KAZAMA[1]

1 *Department of Civil Engineering, Graduate School of Engineering, Tohoku University, 6-6-06, Aramaki Aza Aoba, Sendai, Miyagi 980-8579, Japan*
asaoka@kaigan.civil.tohoku.ac.jp

2 *Tohkamachi Experimental Station, Forestry and Forest Products Research Institute, 614 Tatsh-Otsu, Tohkamachi, Niigata 9480113, Japan*

Abstract This paper describes the temporal variation in chemical components of snowfall and snowmelt in a temperate snowy area. We conducted snowfall and snowmelt water sampling and their water quality analysis in light and heavy snow years at the Tohkamachi experiment station, Japan. We compared the behaviour of acidity and ion concentration of snowmelt water in response to annual snow conditions. Our results show that the mean acidity of snowfall is slightly higher than that of snowmelt. More acidic melt water flows out of the snowpack into the ground when snowmelt is generated on the surface and meltwater reaches the bottom of the snowpack. Comparisons between the two years revealed that although the snowpack has higher capacity for storing chemical components with increase of snowdepth, the stored chemical components gradually flow out of the snowpack with melt water caused by the heat flux from soil.

Key words snowmelt; temperate snow area; annual snow condition; pH; electric conductivity; yellow sand

INTRODUCTION

Snowfall contains sulfur oxides and nitrogen oxides due to human activity or mineral dust generated by wind action on arid and semi-arid areas, present in the troposphere (Nishikawa *et al.*, 2000; Ooki & Uematsu, 2005). The chemical components, which are included in the snowfall, are stored in the snowpack during the winter season. The accumulated chemical components infiltrate through the snowpack with repeated water melting and freeze. Intensive chemical components may flow out of the snow bottom due to the lower freezing point and higher melting point of water contaminated by chemical components, relative to pure water (Colbeck, 1981; Gro & John, 2008). It has been suggested that this may lead to temporal acidification of river environments in cold climate regions (Johannessen & Henriksen, 1978; Jeffries *et al.*, 1979; Aga *et al.*, 2001).

The Japan Sea side of the main Japanese island is famous as a heavy snow area and belongs to the warm temperate zone. Generally, a cold dry winter monsoon blows from the Asian continent to Japan during winter. Heat and vapour are transported from the relatively warm sea surface to the lower atmospheric layer as it crosses the Japan Sea. This causes a decrease in the stability of the lower temperature layer, the occurrence of cumulo-nimbus, and brings heavy snowfall to the Japan Sea side during the intensification of typical winter synoptic patterns, which are characterized by high pressure over Siberia (Siberia high) and low pressure over the Pacific (Pacific low). Cold-air advection from the north is another factor that leads to heavy snow. High snowpack has the potential of storing chemical components. A high rate of melt is driven by the temperate climate during the snowmelt season (Asaoka *et al.*, 2002, 2007). Moreover, Asian dust occurs from arid regions in eastern Asia, such as the Taklamakan and Gobi deserts, in spring time, and the spring monsoon transports mineral dust over the North Pacific Ocean and Japanese mainland (Yabuki *et al.*, 2005).

Winter precipitation and snow depth in temperate snowy areas, such as the Japan Sea side of the Japanese main island, have greater variation than in the cold snow regions. In this study, we conducted snowfall and snowmelt water sampling and their water quality analysis in a light snow year and a heavy snow year in the temperate snowy area. We also compared the acidity and ion concentration of snowmelt water between these two years. There are few observed data in the temperate snowy area. Better understanding the characteristics of snowmelt acidity and chemical components is important for soil and river environment management.

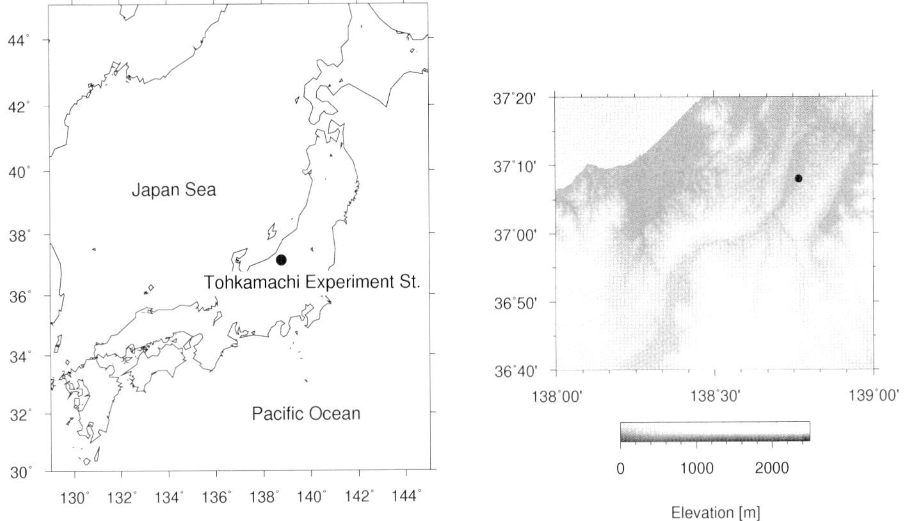

Fig. 1 Location of Tohkamachi Experimental Station, FFPRI.

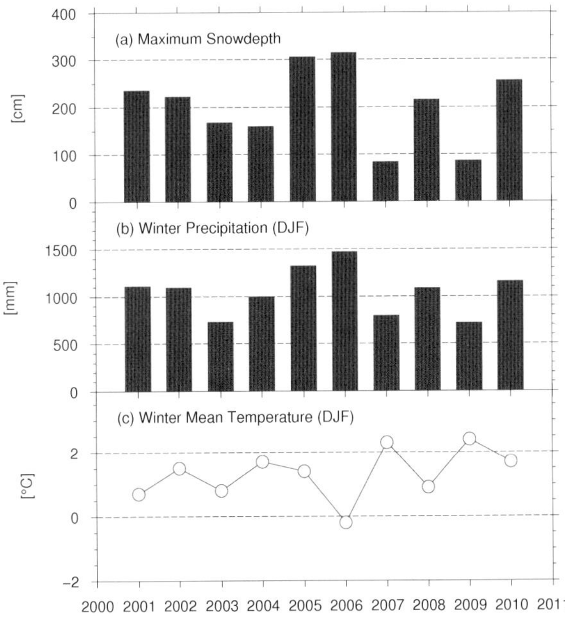

Fig. 2 Records of maximum snowdepth, winter precipitation and mean temperature from December to February, 2000–2011.

STUDY SITE AND METHODOLOGY

Snowmelt and precipitation sampling were carried out in a light snow year, December 2008 to March 2009, and a heavy snow year, December 2009 to March 2010, at the Tohkamachi Experimental Station, Forestry and Forest Products Research Institute (37°08′N, 138°46′E; 2000 m a.m.s.l., Fig. 1). Meteorological data such as temperature, precipitation, snow depth, etc. have been observed for more than 90 years at this site (Takeuchi *et al.*, 2008). Moreover, discharge

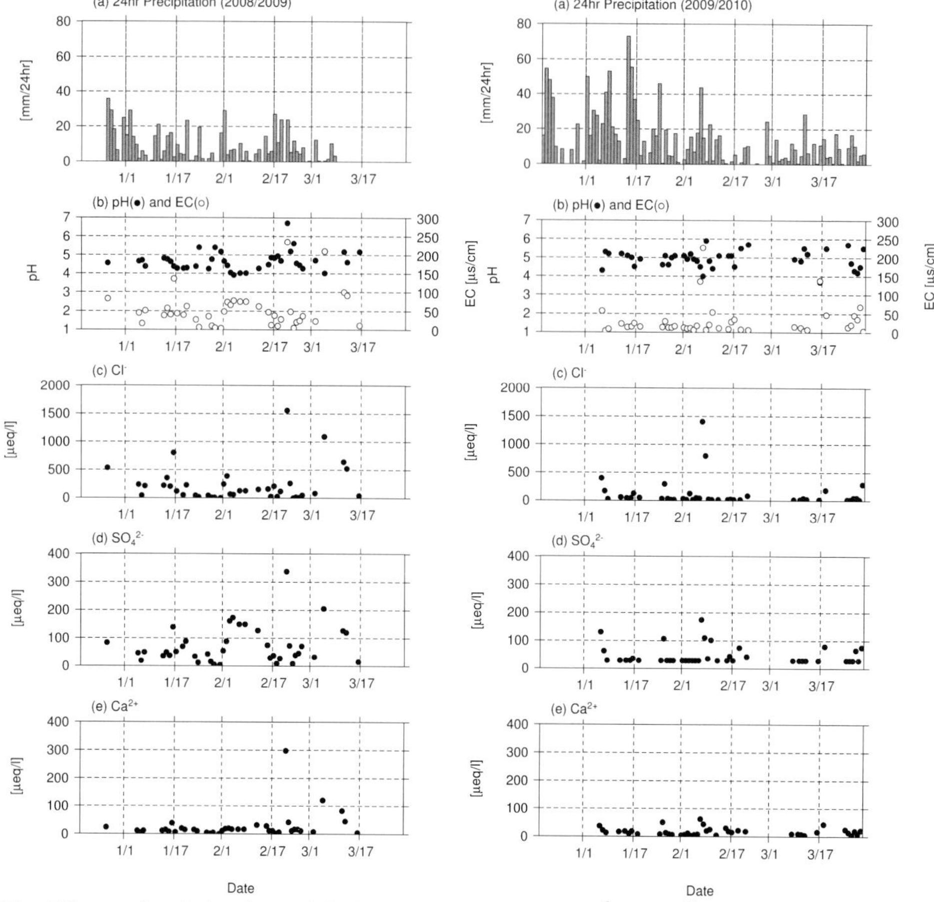

Fig. 3 Temporal variations in precipitation and its pH, EC, Cl^-, SO_4^{2-} and Ca^{2+}.

from the base of the snow and snow weight have been measured with a snow-lysimeter and metal wafer, respectively, for more than seven years (Takeuchi *et al.*, 2007). The study area is located on the Japan Sea side near the central mountains of the main island of Japan and belongs to a temperate climate zone. There are heavy snowfalls, and the annual maximum snow depth at Tohkamachi Experimental Station is more than 2 m on average. However, the monthly air temperature is higher than 0°C except in January, and rainfall sometimes occurs even in winter (Takeuchi *et al.*, 2009). It is reported that there is a distinct negative correlation between mean air temperature and total precipitation from December to February (Takeuchi *et al.*, 2008). In this study, we choose "a light snow year" and "a heavy snow year", December 2008–March 2009, and December 2009–March 2010, respectively. Annual maximum snow depth, mean air temperature, and precipitation during winter (December–February) for 2001 to 2010 at the Tohkamachi Experimental Station are shown in Fig. 2. The highest and lowest maximum snow depth over the 2001–2010 period were recorded in 2006 (314 cm) and in 2007 (84 cm), respectively, and its variability is very large, in spite of being a heavy snow area. The maximum snow depths in 2009 and 2010, when we conducted snowfall and snowmelt sampling, were 86 cm and 254 cm.

The snowfall samples, which were accumulated over about 24 hours on the plate, were collected at around 09:00 JST nearly every day. The plate was put on the snow surface for the next sampling after sampling the snowfall and removing the residue. With regard to snowmelt water sampling, we collected the discharge water out of the snow-lysimeter at around 17:00 JST nearly

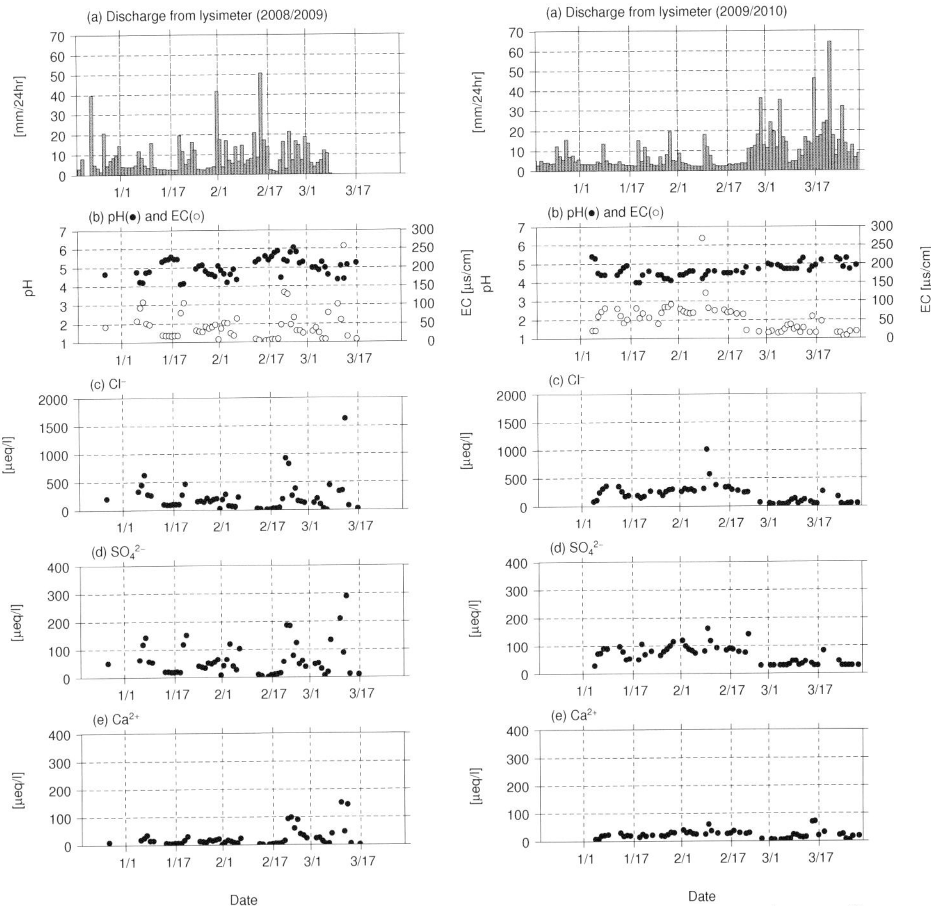

Fig. 4 Temporal variations in discharge from snow lysimeter and its pH, EC, Cl^-, SO_4^{2-} and Ca^{2+}.

every day. After filtering, for both types of sample, electric conductivity (EC) and pH were measured by the glass electrode method (DKK-TOA, CM-30R and HM-30R) and the respective concentration of the major dissolved components (Na^+, NH_4^+, K^+, Mg^{2+}, Ca^{2+}, Cl^-, NO_3^-, SO_4^{2-}) were measured by ion chromatography (DIONEX, ICS-1500).

RESULTS AND DISCUSSION

The temporal variations in precipitation and its chemical components (pH, EC, Cl^-, SO_4^{2-}, Ca^{2+}) in the light snow year (2009) and heavy snow year (2010) are shown in Fig. 3. Mean pH in snowfall in 2009 and 2010 is approximately 4.6 (SD: 0.4) and most of snowfall samples have a negative correlation between pH and EC. It is confirmed that pH and EC become lower and higher, respectively, with increasing SO_4^{2-}. However, high pH and EC snowfall were observed on 21 February 2009 with an increase in Ca^{2+} (Asaoka & Takeuchi, 2011). It is known that Ca^{2+} originates not only from the sea salt but also from soil particles such as yellow sand (Isawaka *et al.*, 1998). According to measured chemical components, snowfall nssCa^{2+} (non-sea salt Ca^{2+}) is 242.5 μeq/L and equivalent to approximately 80% of total Ca^{2+} (298.2 μeq/L). Calcium carbonate, which is one of the major components of Asian dust, has the ability to reduce high acidity. For this reason, high pH was observed in spite of high EC and SO_4^{2-}, and the slight increase of pH might be similarly caused by an increase of Ca^{2+} on 18 March 2010.

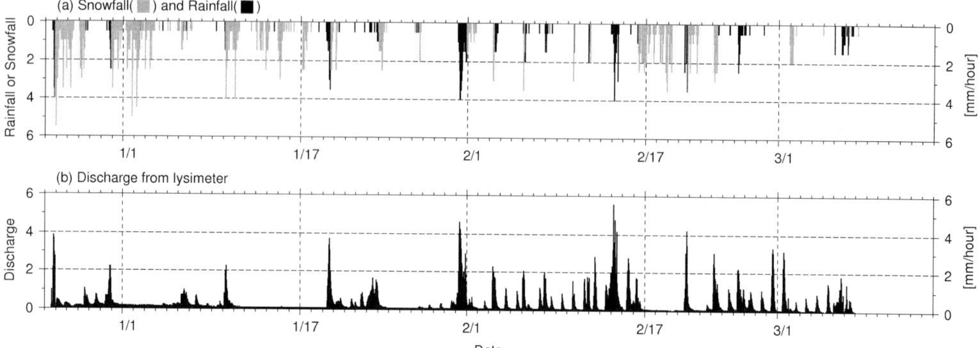

Fig. 5 Separated snowfall and rainfall and observed discharge during 2009 winter.

Temporal variations in discharge from the snow base by a snow-lysimeter and its chemical components (pH, EC, Cl^-, SO_4^{2-}, Ca^{2+}) in the light snow year (2009) and heavy snow year (2010) are shown in Fig. 4. The mean pH are 5.0 (2009) and 4.5 (2010), which are slightly higher than for mean snowfall. The temporal changes in EC and pH have the same tendency as those of the snowfall. Moreover, the pH and EC become lower and higher, respectively, at the time of abrupt snowmelt after the stable discharge, because the stable discharge is caused by the continuous heat flux from the soil, whereas the abrupt snowmelt is due to heat exchange between the snow surface and the atmosphere. If snowmelt water occurs on the snow surface and melt water reaches the snow bottom with abundant chemical components, then the pH increases. Nevertheless, if the surface melt water does not reach and freeze within the layer of snowpack because of high snowdepth, then the stored chemical components flow gradually out of the snowpack with the melt at the bottom due to heat flux from the ground. Therefore, the fluctuation of EC and pH in the light snow year is higher than in the heavy snow year. We separated snowfall and rainfall from observed precipitation by a discrimination method for precipitation using air temperature and humidity data (Yamazaki, 2001). It is noted that the precipitation was rainfall on 31 January and 14 February 2009, and rainfall flowed into the soil through the snowpack (Fig. 5). Therefore, the pH was stable and did not decline, in spite of the abrupt high discharge from the snow-lysimeter.

CONCLUDING REMARKS

We conducted snowfall and snowmelt water sampling and their water quality analysis in the temperate heavy snow area in both light and heavy snow years. There is little such field data in the temperate snowy area. In the main, the pH in snowfall and snowmelt becomes low with an increase in EC. However, it is suspected that snowfall which contains the yellow sand has higher pH in spite of its high EC due to the buffering action by calcium carbonate. More-acidic snowmelt water flows out of snowpack into the ground when meltwater occurs at the surface and reaches the snow bottom. Comparisons between the light and heavy snow years revealed that stored chemical components flow out of snowpack gradually with melt water caused by the heat flux from the soil below, although the snowpack has a high capacity for storing chemical components with the increase of snowdepth.

Acknowledgements This research was supported by JST/JICA, SATREPS (Science and Technology Research Partnership for Sustainable Development). We wish to thank Drs Yasoichi Endo and Shigeki Murakami, and Mr Shoji Niwano, Forestry and Forest Products Research Institute (FFPRI), for snowfall and snowmelt water sampling. Thanks are also extended to Dr Luminda Gunawardhana and Ms Youko Nakano, Tohoku University for their valuable assistance.

REFERENCES

Aga, H., Noguchi, I. & Sakata, K. (2001) Aquatic chemistry of a reservoir during the thaw season, *Water, Air, and Soil Pollution* **130**, 811–816.

Asaoka, Y., Kazama, S. & Sawamoto, M. (2002) The variation characteristics of snow water resources in a wide area and its geographical and climatic dependency. *J. Japan Soc. Hydrology and Water Resources* **15**, 279–289 (in Japanese with English abstract).

Asaoka, Y., Kominami, Y., Takeuchi, Y., Daimaru, H. & Tanaka, N. (2007) Estimation of snow water equivalent in wide area based on satellite remote sensing and regional characteristics of the snowmelt rate factor. *J. Japan Soc. Hydrology and Water Resources* **20**, 519–529 (in Japanese with English abstract).

Asaoka, Y. & Takeuchi, Y. (2011) Characteristics of acidity and major ion concentration of snowfall, snowpack and snowmelt water in warm snow area. *Ann. J. Hydraulic Engineering* **55**, 409–414 (in Japanese with English abstract).

Colbeck, S. C. (1981) A simulation of the enrichment of atmospheric pollutants in snow cover runoff. *Water Resour. Res.* **17**, 1383–1388.

Gro, L. & John, W. P. (2008) Ion enrichment of snowmelt runoff water caused by nasal ice formation, *Water Hydrol. Processes* **14**, 615–619.

Isawaka, Y., Yamamoto, M., Imasu, R. & Ono, A. (1988) Transport of Asia dust (KOSA) particles: importance of weak Kosa events on the geochemical cycle of soil particles. *Tellus* **40B**, 494–503.

Johannessen, M. & Heriksen, A. (1978) Chemistry of snow meltwater: concentration during melting. *Water Resour. Res.* **14**, 615–619.

Nishikawa, M., Hao, Q. & Morita, M. (2000) Preparation and evaluation of certified reference materials for Asian mineral dust. *Global Environ.Res.* **4**, 103–113.

Ooki, A. & Uemats, M. (2005) Chemical interactions between mineral dust and acid gases during Asian dust events. *J. Geophys. Res* **110**(D03201), doi:10.1029/2004JD004737.

Takeuchi, Y., Murakami, S. & Niwano, S. (2007) A comparative study of snowmelt measurements by the snow-lysimeter and the snow-weighing meter using the metal wafer. *Proc. Cold Region Technology Conference 23*, 156–260 (in Japanese).

Takeuchi, Y., Endo, Y. & Murakami, S. (2008) High Correlation between winter precipitation and air temperature in heavy-snowfall areas in Japan. *Ann. Glaciology* **49**, 7–10.

Takeuchi, Y., Niwano, S., Murakami, S., Yamanoi, K., Endo, Y. & Kominami, Y. (2008) Meteorological statistics during the past 90 years from 1918 to 2007 at Tohkamachi in Niigata Prefecture, Japan. *Bulletin of FFPRI* **7**, 187–244 (in Japanese with English abstract).

Takeuchi, Y., Endo, Y., Niyano, S. & Murakami, S. (2009) Data of meteorology and snow pit observations at Tohkamachi in Niigata Prefecture, Japan (VII) (2004-05 to 2008-09, five winter periods). *Bulletin of FFPRI* **8**, 227–277 (in Japanese with English abstract).

Yabuki, S., Mikami, M., Nakamura, Y., Kanayama, S. & Fengfu, F. U. (2005) The characteristics of atmospheric aerosol at Aksu, an Asian dust region of north-west China: A summary of observations over the three years from March 2001 to April 2004. *J. Meteorological Society of Japan* **83**, 45–72.

Yamazaki, T. (2001) A one- dimensional land surface model adaptable to intensely cold regions and its applications in eastern Siberia. *J. Meteorological Society of Japan* **79**, 1107–1118.

Modelling hydrological consequences of climate change in the permafrost region and assessment of their uncertainty

A. N. GELFAN

Water Problems Institute of the Russian Academy of Sciences, 3 Gubkina Str., 119333, Moscow, Russia
hydrowpi@aqua.laser.ru

Abstract A physically-based, distributed model of runoff generation in the permafrost regions is presented. The model describes processes of snow cover formation, taking into account blowing snow sublimation, snowmelt, freezing and thawing of the ground, water detention by a basin storage, infiltration, evaporation, overland, subsurface and channel flow. An important feature of the model is the detailed description of water and heat transfer within the active layer of soil during its seasonal thawing and freezing. A case study has been carried out for the Pravaya Hetta River basin (the catchment area is 1200 km^2) of Western Siberia within the Lower Ob River basin. The basin is located in tundra and forest-tundra vegetation zones. It has been shown that after precipitation, melt of ground ice is the second largest input to the basin water balance and accounts for about 70% of annual precipitation. Seasonal snow losses due to sublimation during blowing snow transport can reach almost 30% of the maximum snow accumulation. The model has been applied to assess the impact of climate change on hydrological processes in the permafrost basin. Uncertainty of the simulated hydrological consequences of climate change has been assessed by the multi-scenario approach. Simulated runoff response to the projected climate change varies significantly as a result of the uncertainty of the climate change scenario.

Key words permafrost hydrology; cold region modelling; climate change; uncertainty

INTRODUCTION

The permafrost region covers approximately one-quarter of the Earth's land surface, more than 60% of Russia and one-half of Canada. It constitutes a unique environment with an important role in the dynamics and evolution of the Earth system. Economic growth in the permafrost regions, the necessity of protection of the northern environment and maintenance of biodiversity, as well as increasing attention to global climate processes, make the investigations of permafrost hydrological processes, especially runoff generation, very important and urgent. It is well known that the permafrost region plays an important role in climate dynamics and represents potentially important sources and/or sinks of greenhouse gases. A changing climate could in turn lead to visible physical changes that could augment or retard global climate change and significantly impact ecosystems (Rouse *et al.*, 1997). There is increasing evidence that environmental change has reached an unprecedented degree. Many of these changes are related to the hydrological cycle and can result from both the direct and indirect impacts of human activities (Vörösmarty *et al.*, 2001).

The importance of the aforementioned issues requires advancing the current understanding of permafrost hydrology and developing, on this basis, adequate methods for their solution. Such an advance has to be associated with field and modelling studies. In recent decades, extensive experimental work in permafrost regions of different physiographic conditions has been carried out and summarized by Woo (1990), Quinton & Marsh (1999) and Carey *et al.* (2010). The accumulated knowledge of the main specific features of permafrost hydrology, which make it clearly distinguishable from the hydrology of temperate regions (Woo, 1990), has created a foundation for the development of process-based hydrological models representing these features, such as the models presented by Kuchment *et al.* (2000), Zhang *et al.* (2000), Quinton *et al.* (2004), Pomeroy *et al.* (2007), Dornes *et al.* (2008) and Zhang *et al.* (2008). Parameters of such models have clear physical meaning and can be related to measurable characteristics of river basins, such as topography, soil, vegetation, etc. Combined with, and resulting from the physical background of the models, this feature provides fresh opportunities for *a priori* assessment of the parameters and model validation in poorly gauged basins (Gelfan, 2005), which are typical for permafrost regions. In addition, these models allow one to obtain reasonable results in the case of environmental changes, particularly to assess hydrological consequences of the projected climate change.

The main objectives of this paper are to present the development of a physically-based, distributed hydrological model of permafrost watersheds located in the tundra vegetation zone, as well as applications of the model for assessing hydrological consequences of climate change and their uncertainty. In comparison with the model presented by Kuchment *et al.* (2000), more sophisticated modelling of heat and moisture transfer in the active layer has been developed in this study; it also uses the model of blowing snow transport (Essery *et al.*, 1999).

MODEL OF RUNOFF GENERATION: STRUCTURE, PARAMETER ASSESSMENT AND VALIDATION

The low-gradient Pravaya Hetta catchment is located in the north of Western Siberia, it drains west into the Nadym River (a tributary of the Ob River). The study area of approx. 1200 km^2 (up to the outlet in the Pangody town, 61°53′N, 147°43′E) is situated at the boundary of tundra and forest-tundra vegetation zones. Wetland and peatland occupy about 79% of the area covered by moss, lichen, and shrubs. The river basin is situated in the zone of continuous permafrost interrupted only by taliks under the river beds. The dominant soils are tundra podzols. On slopes of northern exposure, the average depth of the active layer is 0.2–0.8 m; on slopes of southern exposure and at the peatlands, the average depth of the active layer reaches 2.0 m. The study area has a sub-arctic continental climate, characterized by a large temperature range and low precipitation. Mean annual temperature is –7.8°C. Mean temperatures of January and July are –26.4°C and +15.4°C, respectively. Mean annual precipitation is 290 mm, of which about 100 mm falls as snow. The maximum depth of snow in the uplands is 10–20 cm, while in the lowlands it can exceed 40 cm and 70 cm in the areas covered by trees and shrubs, respectively. The majority of runoff results from snowmelt in the late spring and summer. Only minimal runoff occurs during the summer because of rainfall. Peak runoff discharges during snowmelt exceeded 300 m^3/s, while peak runoff events during summer were only about 30 m^3/s.

To calculate the characteristics of snow cover, a system of vertically-averaged equations of snow processes has been applied (Kuchment *et al.*, 2000). The system includes a description of temporal change of the snow depth, content of ice and liquid water, snow density, snowmelt, sublimation, refreezing melt water, and snow metamorphism. In order to simulate the effect of wind redistribution of snow, we applied the Simplified Blowing Snow Model, SBSM, presented in Essery *et al.* (1999). This simulation calculates downwind blowing snow transport and in-transit sublimation losses.

Vertical water and heat transfer associated with seasonal freezing and thawing of the active soil layer is described with the equations (Gelfan, 2006):

$$\frac{\partial W}{\partial t} = \frac{\partial}{\partial z}\left(D\frac{\partial \theta}{\partial z} + D_I\frac{\partial I}{\partial z} - K \right) \tag{1}$$

$$c_T\frac{\partial T}{\partial t} = \frac{\partial}{\partial z}\left(\lambda\frac{\partial T}{\partial z} \right) + \rho_w c_w\left(D\frac{\partial \theta}{\partial z} + D_I\frac{\partial I}{\partial z} - K \right)\frac{\partial T}{\partial z} + \rho_w \chi\frac{\partial W}{\partial t} \tag{2}$$

where W, θ and I are the total water content, liquid water content and ice content of soil, respectively, where $W = \theta + (\rho_i / \rho_w)I$; $K = K(\theta, I)$ is the hydraulic conductivity of soil; T is the soil temperature; λ represents the thermal conductivity of soil; $D = K(\partial\psi / \partial\theta)_I$ and $D_I = K(\partial\psi / \partial I)_\theta$; $\psi = \psi(\theta, I)$ is the capillary potential of soil; $c_T = c_{eff} + \rho_w\chi(\partial\theta / \partial T)$; c_{eff} is the effective heat capacity of soil $c_{eff} = \rho_g c_g (1 - P) + \rho_w c_w \theta + \rho_i c_i I$; ρ and c are the density and the specific heat capacity, respectively, where indexes w, i and g refer to water, ice and soil matrix, respectively; P is the soil porosity; χ is the latent heat of ice fusion.

The capillary potential $\psi = \psi(\theta, I)$ and the hydraulic conductivity $K = K(\theta, I)$ of the frozen soil are determined from the relationships (Gelfan, 2006) obtained by transformation of van Genuchten's formulas.

Melt water detention in the topographic depressions and peat mats is simulated under the assumption that free storage capacity is exponentially distributed over the basin area with the mean value DET_0. Under this assumption, cumulative water detention, DET_Σ, by the basin surface is calculated by:

$$DET_\Sigma = DET_0\left[1-\exp\left(-\frac{R_\Sigma}{DET_0}\right)\right] \tag{3}$$

where R_Σ is the snowmelt outflow from snowpack accumulated from the beginning of melt.

The rate of evaporation, E, from an unfrozen, snow-free soil is calculated as:

$$E = K_E d_a S_1 \tag{4}$$

where S_1 is the relative saturation of the upper soil layer; K_E is the empirical coefficient.

To simulate overland and subsurface flow over the Pravaya Hetta River basin, the catchment area was schematized as a series of rectangular reaches located along the main channel. Overland flow along each of the schematized reaches was described by the length-integrated kinematic wave equation written as:

$$L\frac{dh}{dt} = RL - q_l \tag{5}$$

$$h = \frac{m}{m+1}\left(\frac{q_l n_l}{i_l^{0.5}}\right)^{\frac{1}{m}} \tag{6}$$

where h is the average flow depth; m equals 5/3, L is the length of the reach; R is the rate of water inflow per unit length of the reach; q_l is the lateral overland inflow rate per unit length of the channel; i_l is the slope of the overland flow; n_l is the Manning's roughness coefficient.

It is also assumed that the horizontal movement of water in the active layer occurs only if the soil moisture content exceeds the soil field capacity θ_f. As a result, the subsurface flow is generated with the rate $R_g = \begin{cases}\frac{\partial}{\partial t}\left[(\theta-\theta_f)h_g\right], & \theta > \theta_f \\ 0, \text{ otherwise}\end{cases}$ where h_g is the depth of the subsurface flow.

Subsurface flow was described by the equations

$$(P-\theta_f)L\frac{dh_g}{dt} = R_g L - q_g \tag{12}$$

$$h_g = \frac{q_g}{2K_g i_l} \tag{13}$$

where q_g is the discharge; K_g is the coefficient of horizontal conductivity of soil.

To simulate channel flow, the one-dimensional kinematic wave equation is applied.

Calibration and validation of the model was carried out using the available daily hydro-meteorological measurement data for the period from 1 January 1981 to 31 December 1986. Initial snow depth and density values were assigned from snow survey measurements. Initial liquid water content in snowpack was assumed to equal zero. Because of the absence of the necessary measurements, initial soil heat and water content were assigned as constant on soil depth, equalled to the corresponding climatic values. Using these values, the initial ice content of soil was calculated by the model.

Soil porosity and density, as well as field capacity and horizontal hydraulic conductivity of the organic soil, were assigned *a priori* on the basis of the available measurements for the similar types of soil (Pavlov *et al.*, 1997; Sławiński *et al.*, 2004). The snow sub-model was calibrated using snow survey measurements as well as data of special observations on snow evaporation (Kuz'min, 1976). A comparison of the measured and calculated snow depth is shown in Fig. 1.

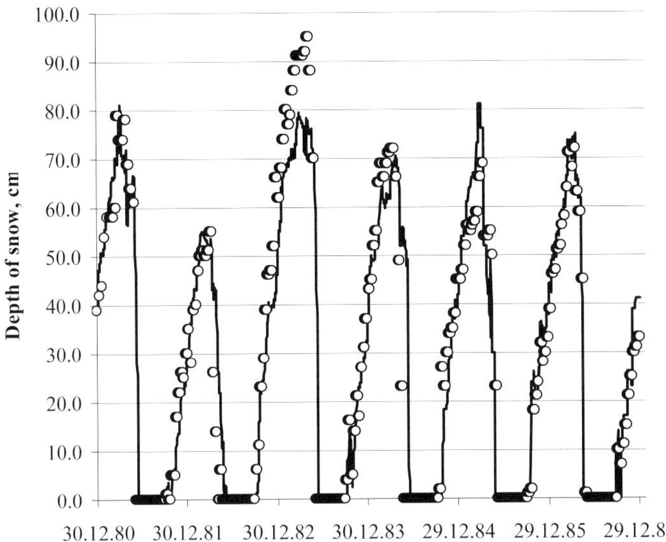

Fig. 1 Comparison of the observed (points) and calculated (line) depth of snow at the meteorological station Pangody.

The model showed that seasonal snow losses due to sublimation during blowing snow transport vary within 17–31 mm, depending on the meteorological conditions of the specific season, and reach 29% of the maximum snow accumulation; this result is close to that reported by Essery *et al.* (1999) for the shrub tundra.

Five parameters (saturated hydraulic conductivity of soil, the Manning's roughness coefficient of river channel, and the coefficients DET_0 and K_E in equations (3) and (4), respectively) were adjusted through calibration against observed hydrographs from 1 January 1981 to 31 December 1983. The model validation was carried out by comparison of the observed and simulated hydrographs for the period from 1 January 1984 to 31 December 1986. The simulated and the observed hydrographs for the six years of the calibration and validation periods are presented in Fig. 2. The Nash and Sutcliffe efficiency criterion for the discharge simulations is 0.74 for the calibration period and 0.73 for the validation period.

According to the simulation results, the largest floods in the basin are caused by overland snowmelt inflow into the channels. Most of the ice melt water formed within the thawing active layer of ground, as well as infiltrated rainwater, does not reach a channel and evaporates during a warm period. After precipitation, melt of ground ice is the second largest component of the basin water balance, and accounts for about 70% of annual precipitation.

UNCERTAINTY OF RUNOFF RESPONSE TO CLIMATE CHANGE

There are many estimates of hydrological consequences of the current and the projected climate change in the permafrost regions, particularly, for the Ob River basin (Yang *et al.*, 2004). These estimates contain significant uncertainty caused, most of all, by the diversity of the future climate scenario. An attempt is made below to account for this source of the uncertainty.

According to the recent IPCC report (*Climate Change*, 2007), annual temperature and precipitation in the Siberian north may increase by 1.3–4.7°C and 5–20%, respectively, depending on the chosen greenhouse gases emission scenario. In order to assess hydrological consequences of the projected climate change and their uncertainty, the following multi-scenario approach was applied (Fig. 3). At first, 1000 random combinations of annual changes of air temperature ΔT and

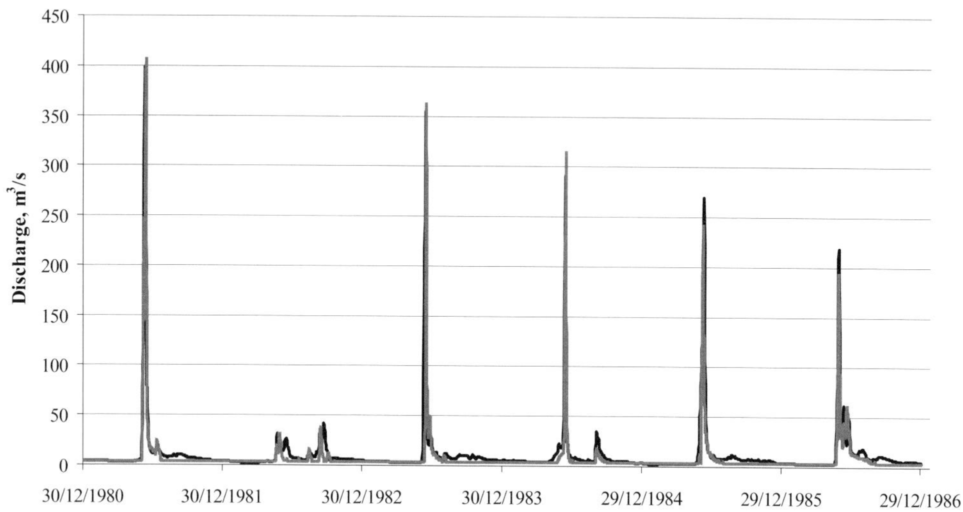

Fig. 2 Comparison of the observed (black line) and calculated (grey line) hydrographs at the gauge Pangody of the Pravaya Hetta River.

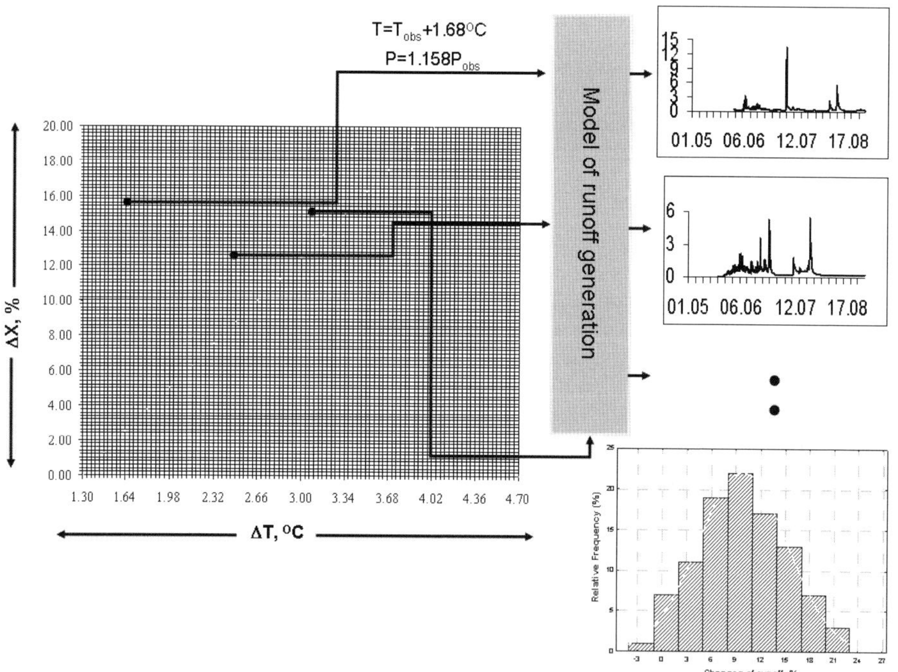

Fig. 3 Multi-scenario approach to assess uncertainty of hydrological consequences of the projected climate change.

precipitation ΔP were sampled from the intervals $\Delta T \in [1.3°, 4.7°]$ and $\Delta P \in [5\%, 20\%]$ by the Latin hypercube method. It was assumed that ΔT and ΔP are independent random variables, uniformly distributed within the respective interval, i.e. any combination of ΔT and ΔP was equally guessed. Then, the daily series of temperature and precipitation measured in the basin under consideration were changed according to the assigned values of ΔT and ΔP. As a result, 1000 synthetic series of these meteorological variables were obtained. Finally, the model was forced by these series and 1000 hydrographs were simulated reflecting uncertainty of the hydrological consequences of climate change.

Mean response of runoff to 1000 constructed scenario of climate change turned out to be insignificant: annual runoff and peak discharge increased by 5.6% and 6.7%, respectively. At the same time, variation of the runoff response is rather high. Changes of the annual runoff can vary within the 95%-confidence interval of [–1%, +16%]. The corresponding interval for the annual peak discharge is [–2%, +26%]. Decrease of runoff can occur under a large increase in air temperature along with a small change in precipitation. In this case, the small increase of precipitation cannot compensate for the increase of the basin evaporation.

Acknowledgements The author wishes to acknowledge the support of the Targeted Programme of the Presidium of RAS. This paper is a contribution to PUB, the Decade for Prediction of Ungauged Basins, International Association of Hydrological Sciences.

REFERENCES

Carey, S. K., Tetzlaff, D., Seibert, J., Soulsby, C., Buttle, J., Laudon, H., McDonnell, J., McGuire, K., Caissie, D., Shanley J., Kennedy, M., Devito, K. & Pomeroy. J. W. (2010) Inter-comparison of hydro-climatic regimes across northern catchments: synchronicity, resistance and resilience *Hydrol. Processes* **24**, 3591–3602.

Climate Change (2007) *Climate Change:The Physical Science Basis*, Summary for Policymakers. Contribution of Working Group I to the Fourth Assessment Report of the Intergovernmental Panel on Climate Change. Cambridge University Press.

Dornes, P. F., Tolson, B. A., Davison, B., Pietroniro, A., Pomeroy, J. W. & Marsh, P. (2008) Regionalisation of land surface hydrological model parameters in subarctic and arctic environments. *Phys. & Chemistry of the Earth* **33**, 1081–1089.

Essery, R., Li, L. & Pomeroy, J. W. (1999) Blowing snow fluxes over complex terrain. *Hydrol. Processes* **13**, 2423–2438.

Gelfan, A. N. (2005) Prediction of runoff in poorly gauged basins using a physically based model. In: *Prediction in Ungauged Basins: Approaches for Canada's Cold Regions* (ed. by C. Spence, J. Pomeroy & A. Pietroniro) (Proceedings of the Workshop), 101–118. Yellowknife, NWT, Canada.

Gelfan, A. N. (2006) Physically based model of heat and water transfer in frozen soil and its parametrization by basic soil data. In: *Predictions in Ungauged Basins: Promises and Progress* (ed. by M. Sivapalan *et al.*), 293–304. IAHS Publ. 303, IAHS Press, Wallingford, UK.

Kuchment, L. S, Gelfan, A. N. & Demidov, V. N. (2000) A distributed model of runoff generation in the permafrost regions. *J. Hydrol.* **240**, 1–22

Kuz'min, P. P. (1976) *Data on Snow Cover Evaporation Measurements*. State Hydrological Institute Publ., Leningrad. USSR (in Russian).

Pavlov, A. V., Dubrovin, V. A. & Haritonov, L. P. (1997) Experimental study of the ground thermal regime in the Arctic part of the Western Siberia. *Trans. VSEGINGEO*, 310–320 (in Russian).

Pomeroy, J. W., Gray, D. M., Brown, T., Hedstrom, N. R, Quinton, W. L., Granger, R. J. & Carey, S. K. (2007) The cold regions hydrological model, a platform for basing process representation and model structure on physical evidence. *Hydrol. Processes* **21**, 2650–2667.

Quinton, W. L., Carey, S. K. & Goeller, N. T. (2004) Snowmelt runoff from northern alpine tundra hillslopes: major processes and methods of simulation. *Hydrol. Earth System Sci.* **8**, 877–890.

Quinton, W. L. & Marsh, P (1999) A conceptual framework for runoff generation in a permafrost environment. *Hydrol. Processes* **13**, 2563–2581

Rouse, W. R, Douglas, M. S. V., Hecky, R. E., Hershey, A. E., Kling, G. W., Lesack, L., Marsh, P., McDonald, M., Nicholson, B. J., Roulet, N. T. & Smol, J. P. (1997) Effects of climate change on the freshwaters of Arctic and Subarctic North America. *Hydrol. Processes* **11**, 873–902.

Sławiński, C., Witkowska-Walczak, B. & Walczak, R. T. (2004) *Determination of Water Conductivity Coefficient of Soil Porous Media*. Publ. ALF-GRAF, Lublin, Poland.

Vörösmarty, C. J., Hinzman, L. D., Peterson, B. J., Bromwich, D. H., Hamilton, L. C., Morison, J., Romanovsky, V. E., Sturm, M. & Webb, R. S. (2001) *The Hydrologic Cycle and its Role in Arctic and Global EnvironmentalChange: A Rationale and Strategy for Synthesis Study*. Fairbanks, Alaska: Arctic Research Consortium of the US, USA

Woo, M.-K. (1990) Permafrost hydrology. In: *Northern Hydrology: Canadian Perspectives* (ed. by T. D. Prowse & S. L. Ommanney), NHRI Science Rep. no. 1., 63–76. National Hydrology Research Institute, Environment Canada, Saskatoon, Saskatchewan, Canada.

Yang, D., Ye, B. & Shiklomanov, A. (2004) Streamflow characteristics and changes over the Ob river watershed in Siberia. *J. Hydromet.* **5**, 69–84.

Zhang, Y., Carey, S. K. & Quinton, W. L. (2008) Evaluation of the algorithms and parameterizations for ground thawing and freezing simulation in permafrost regions. *J. Geophys. Res.* **113** (D17116), doi:10.1029/2007JD009343.

Zhang, Z., Kane, D. L. & Hinzman, L. D. (2000) Development and application of a spatially-distributed Arctic hydrological and thermal process model (ARHYTHM). *Hydrol. Processes* **14**, 1017–1044.

Permafrost loss and a new approach to the study of subarctic ecosystems in transition

WILLIAM L. QUINTON, LAURA E. CHASMER & RICHARD M. PETRONE
Cold Regions Research Centre, Wilfrid Laurier University, 75 University Ave W, Waterloo N2L 3C5, Canada
wquinton@wlu.ca

Abstract This study uses remote sensing to demonstrate the rate and spatial pattern of land-cover change resulting from permafrost loss in a subarctic region that typifies the southern boundary of permafrost. Permafrost occupied 0.70 km^2 of a 1.0 km^2 area in 1947, but by 2008 occupied only 0.43 km^2. This study also explains the need for an Earth Systems approach to properly examine the integrated mechanisms, interactions and feedbacks among physical, chemical and biological components of warming subarctic ecosystems.

Key words permafrost thaw; ecosystem change; subarctic; peatlands.

INTRODUCTION

Land-cover change induced by permafrost thaw introduces considerable uncertainty to the future availability of freshwater resources in northern Canada where permafrost controls water storage and drainage processes by limiting the amount of water infiltration to that which can be stored in the active layer, and by restricting hydrological interaction between near-surface supra-permafrost water and deep sub-permafrost groundwater. Along the wetland-dominated southern boundary of permafrost, permafrost typically occurs in the form of tree-covered plateaus that rise above the surrounding treeless and permafrost-free wetland terrain. As such, permafrost plateaus obstruct and redirect surface and near-surface drainage in the surrounding terrains. Thawing and subsidence of permafrost plateaus has led to increasing interconnectivity of drainage networks (Beilman & Robinson, 2003), and to changes in the local hydrological cycle due to changes in soil thermal and moisture regimes (Hayashi *et al.*, 2007), surface energy balances, and snow accumulation and melt rates and patterns (Wright *et al.*, 2009).

Northwestern North America is one of the most rapidly warming regions on Earth (Johannessen *et al.*, 2004). Climate warming and its environmental consequences are unlikely to proceed in an easily predictable manner due to complex responses and feedbacks. Permafrost thaw is one of the most important and dramatic manifestations of climate warming in Canada, and is strongly influenced by feedback processes (Jorgenson *et al.*, 2010). It also has the potential to alter other key aspects of ecosystems such as runoff and snow cover, forest composition, biodiversity and habitat for keystone species, surface–atmosphere interactions including greenhouse gas fluxes, forest fire regimes, and the quantity and quality flows to downstream ecosystems and the Arctic Ocean (Rowland *et al.*, 2010). While permafrost and ecosystem responses to warming occur in varying degrees throughout the North, the discontinuous permafrost zone of the subarctic is where the most dramatic permafrost thaw and landscape transformations are currently observed (Quinton *et al.*, 2011). Forecasted dramatic changes in temperature and moisture are expected to affect the processes governing the release of carbon dioxide and methane from the vast stores of carbon in northern peatlands (Lenton *et al.*, 2008). There are strong indications that the hydrology and hydrochemistry of the subarctic are changing as a result of permafrost thaw, yet little is known about the interactions and feedbacks among climate, water, biogeochemistry and the ecology of subarctic ecosystems. As a result, ecosystem consequences of warming and the impacts of thaw cannot be predicted with confidence (Frey *et al.*, 2007). Further, the implications of these feedbacks are not only prevalent within the subarctic region, but also to downstream aquatic and terrestrial ecosystems and the Arctic Ocean, the settled regions of southern Canada, due to downwind movement of water vapour in weather systems, and to global climate systems through changes to greenhouse gas fluxes, forest fire regimes and ground surface albedo.

The lack of knowledge on the mechanisms and rates of land-cover change induced by permafrost thaw, the impact on hydrological processes and resulting ecosystem impacts, and appropriate mitigation strategies, underscores the need for scientific research to provide the knowledge base needed for informed and sustainable water resource management. This article presents a selection of published and unpublished material for the purpose of: (1) demonstrating a clear example of the rate and spatial pattern of permafrost loss over the last ~60 years in the southern boundary of permafrost in the Northwest Territories, Canada, and (2) recommending an approach to examining subarctic ecosystem responses and feedbacks to permafrost thaw.

STUDY SITE

This study was conducted in the 152-km^2 Scotty Creek basin (61°18'N, 121°18'W; 285 m a.s.l.), which lies within the 140 000 km^2 Hay River Lowland of the Northwest Territories, Canada. Scotty Creek typifies the southern extent of permafrost in much of Canada where a high density of peatlands helps to preserve discontinuous permafrost due to the large thermal offset created by the dry, insulating peat covering the ground surface (Robinson & Moore, 2000; Smith & Riseborough, 2002). The peatlands in this region are dominated by a mosaic of permafrost plateaus, channel fens and ombrotrophic flat bogs (Quinton *et al.*, 2009). Unlike bogs and fens, the permafrost plateaus support a tree cover (*Picea mariana*) and are underlain by ~5–10 m thick permafrost (Burgess & Smith, 2000). Their crests rise 1 to 2 m above the surfaces of the surrounding bogs and fens (Robinson & Moore, 2000). Channel fens take the form of broad, 50 to >100 m wide channels. Unlike the floating peat mat characterising the surface of fens, flat bog surfaces are relatively fixed. Most flat bogs are small features that occur within permafrost plateaus, while others are relatively large, contain numerous permafrost plateaus, and are connected to channel fens.

METHODOLOGY

The forest cover distribution in the study area mirrors that of the permafrost, since the forest occurs only on the permafrost plateaus. This offers a unique opportunity to monitor permafrost thaw rates since such thaw transforms permafrost plateaus into treeless wetlands (i.e. bogs or fens), a change easily detected on the ground and from aerial and satellite imagery (Jorgenson *et al.*, 2001). Field observations at Scotty Creek (Quinton *et al.*, 2009) and elsewhere in the Hay River Lowland (Robinson & Moore, 2000) indicate that permafrost thaw leads to local inundation and water-logging, causing death of the tree cover within one or two years.

IKONOS multispectral satellite imagery (4-m resolution) from August 2000 and four sets of aerial photographs taken between 1947 and 2008 were obtained for Scotty Creek. The aerial photographs include high resolution (0.55-m to 1.22-m) historic black and white (visible) aerial photography (acquired July to September, 1947, 1970, 1977) and mosaiced near-infrared aerial photography (0.18-m resolution) acquired in August 2008 coincident with an airborne survey using light detection and ranging (LiDAR). The historical aerial photographs were ortho-rectified using the 2008 aerial photography and a LiDAR-derived digital elevation model (DEM) (resolution = 1 m) (Chasmer *et al.*, 2011). A 1 km × 1 km subset area was chosen, corresponding to the area covered by the LiDAR survey and also containing linear disturbances visible in all images (requirement of aerial triangulation and ortho-rectification processes). The three cover types (permafrost plateaus, wetlands (i.e. bogs and channel fens) and open water were classified based on land-cover spectral properties (Chasmer *et al.*, 2011). Using the time series of images, sequential maps were developed that show changes in the spatial distribution of the tree-covered area and, by proxy, the area underlain by permafrost. Errors in plateau edge detection/delineation range from between 8% and 12% (and up to 25% using IKONOS imagery). These errors are due to variable pixel resolutions between images, image geometry, and edge delineation at the boundary between forest and bog/fen. Due to the mismatch in pixel resolution, accuracy of aerial triangulation and ortho-rectification, ability to detect plateau features within aerial photographs, and forest/plateau edge delineation, the uncertainty in the delineation of the boundaries between different land covers results in error that ranges between 8% and 12% in the land cover area estimates from the air photos and up to 26% from the IKONOS imagery (Chasmer *et al.*, 2011). All linear disturbances were digitised using the IKONOS image to obtain an estimate of their total length within the Scotty Creek basin.

RESULTS AND DISCUSSION

Image analysis

Remote sensing analysis of the 1.0 km^2 subset area (Fig. 1) indicates that permafrost occupied 0.70 ± 0.06 km^2 in 1947 and decreased with time to 0.43 ± 0.03 km^2 by 2008, as evidenced by the expansion and merger of wetlands (i.e. bogs and fens) and the shrinkage and disappearance of the forest cover (i.e. permafrost plateaus). This loss rate of 38% loss (±8%) of the permafrost present in 1947 is similar in magnitude to that reported by Beilman & Robinson (2003) over the same period for other sites in the region. The 2008 imagery indicates 133 km of linear disturbance (i.e. winter roads and seismic cutlines) within the 152-km^2 basin, approximately five times greater (0.875 km^{-1}) than the basin's natural drainage density (0.161 km^{-1}) (Quinton *et al.*, 2003).

Need for an Earth-systems approach

For the past few decades, researchers have studied permafrost ecosystems in a compartmentalised fashion, focusing on individual components (e.g. water, terrestrial ecology) of the ecosystem. However, it has become clear that the integrated ecosystem response to climate change and anthropogenic disturbances in the subarctic cannot be addressed adequately by this approach. Instead, an Earth-systems approach is needed to properly examine the integrated mechanisms, interactions and feedbacks among physical, chemical and biological components of warming subarctic ecosystems. This would require the gathering of expertise from the diverse fields of hydrology, terrestrial ecology, biogeochemistry and wildlife ecology to address the common overarching goal of developing a comprehensive understanding of subarctic ecosystem responses and feedbacks to climate warming, and a new capacity to predict the rate and impact of permafrost loss in the subarctic. Such an approach will by necessity, bridge disciplinary boundaries and forge new collaborations.

Unlike the high-arctic, which was the major focus of the International Polar Year (IPY), there has been no large interdisciplinary research programme targeted to the subarctic, which has relatively thin, discontinuous and therefore sensitive permafrost. It is also the permafrost region with the highest population density, land disturbance and rate of resource development. This is also the region with extensive peatlands, and the deepest snow depth outside of the western cordillera, and as such represents an important runoff generation zone for freshwater flow to the Arctic Ocean. The co-occurrence of permafrost and extensive peatlands means that this region contains the largest soil carbon stocks in Canada, though this is not well quantified.

Wholesale ecosystem change in the subarctic will have wide-ranging effects from global climate forcing to regional socio-economic impacts as this region is at the forefront of mineral resource development and a northward expanding commercial forestry. An integrated, Earth-systems approach will reveal new insights on the effects of a warming global climate on biodiversity, trace element cycling and exposure/toxicity to living systems, and water resources and greenhouse gas fluxes in northern ecosystems. It will also specifically isolate the climate drivers of systems changes in ecology, hydrology and chemistry, and will reveal the existence of thresholds with respect to the structure and function of plant, microbial and animal species that are important to subarctic food webs, including species at risk such as woodland caribou.

IPY provided an excellent opportunity to intensify scientific research activity in the arctic and subarctic regions, resulting in new knowledge on permafrost ecosystems. For example, permafrost

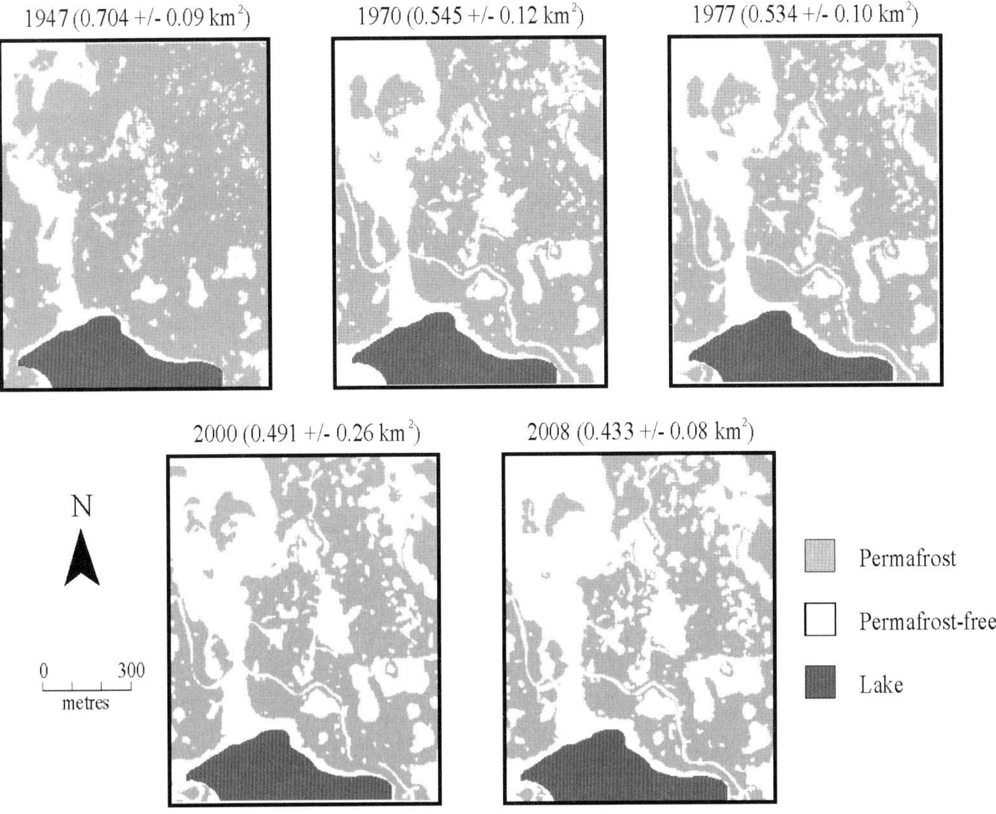

Fig. 1 The area of the 1-km^2 subset area of Scotty Creek, Northwest Territories, Canada underlain by permafrost in 1947, 1970, 1977, 2000 and 2008 (modified from Quinton *et al.*, 2011).

thaw in the southern Yukon is enhanced by shrub tundra expansion and height increases, which result in much deeper snowpacks and more rapid snowmelt that both work to further accelerate permafrost thaw (Quinton & Carey, 2008). In another example, Kemper & Macdonald (2009) indicated that plant communities in the low Arctic may not be in equilibrium with the present climate, but that disturbance associated with seismic exploration triggers a succession towards a new equilibrium. These and other individual research projects laid the foundation for the approach recommended in this report, in which the integrated ecosystem response is studied in a truly interdisciplinary manner.

REFERENCES

Beilman, D. W. & Robinson, S. D. (2003) Peatland permafrost thaw and landform type along a climate gradient. In: *Proceedings Eighth International Conference on Permafrost* (ed. by M. Phillips, S. M. Springman & L. U. Arenson) (Zurich, Switzerland), vol. 1, 61–65. A.A. Balkema.

Burgess, M. M. & Smith, S. L. (2000) Shallow ground temperatures. In: *The Physical Environment of the Mackenzie Valley, Northwest Territories: A Base Line for the Assessment of Environmental Change* (ed. by L. D. Dyke & G. R. Brooks), 89–103. Geological Survey of Canada Bulletin 547, Ottawa, Canada.

Chasmer, L., Hopkinson C. & Quinton, W. L. (2011) Quantifying errors in permafrost plateau change from optical data, Northwest Territories, Canada: 1947 to 2008. *Canadian J. Remote Sensing – CRSS Special Issue* **36**(2), S211–S223.

Frey, K. E., Siegel, D. I. & Smith, L. C. (2007) Geochemistry of west Siberian streams and their potential response to permafrost degradation. *Water Resour. Res.* **43**, W03406, doi:10.1029/2006WR004902.

Hayashi, M., Goeller, N. T., Quinton, W. L. & Wright, N. (2007) A simple heat conduction method for simulating the frost-table depth in hydrological models. *Hydrol. Processes* **21**, 2610–2622.

Johannessen, O. M., Bengtsson, L.. Miles, M. W., Kuzmina, S. I., Semenov, V. A., Alekseev, G. V., Nagurnyi, A. P., Zakharov, V. F., Bobylev, L. P., Pettersson, L. H., Hasselmann, K. & Cattle, A. P. (2004) Arctic climate change: observed and modelled temperature and sea-ice variability. *Tellus Series—Dynamic Meteorology and Oceanography* **56**(4), 328–341.

Jorgenson, M. T., Racine, C. H. Walters, J. C. & Osterkamp, T. E. (2001) Permafrost degradation and ecological changes associated with a warming climate in central Alaska. *Climatic Change* **48**, 551–579.

Jorgenson, M. T., Romanovsky, V., Harden, J., Shur, Y., O'Donnell, J., Schuur, E. A. G., Kanevskiy, M. & Marchenko, S. (2010) Resilience and vulnerability of permafrost to climate change. *Canadian J. Forest Research* **40**, 1219–1236, doi:10.1139/X10-060.

Kemper, J. T. & Macdonald, S. E. (2009) Directional change in upland tundra plant communities 20–30 years after seismic exploration in the Canadian low-arctic. *J. Vegetation Science* **20**, 557–567, doi: 10.1111/j.1654-1103.2009.01069.x.

Quinton, W. L. & Carey, S. K. (2008) Towards an energy-based runoff generation theory for tundra Landscapes. *Hydrol. Processes* **22**, 4649–4653. doi: 10.1002/hyp.7164.

Quinton, W. L., Hayashi, M. & Pietroniro A. (2003) Connectivity and storage functions of channel fens and flat bogs in northern basins. *Hydro.l Processes* **17**, 3665–3684.

Quinton, W. L., Hayashi, M. & Chasmer, L. (2009) Peatland hydrology of discontinuous permafrost in the Northwest Territories: overview and synthesis. *Canadian Water Resour. J.* **34**, 311–328.

Quinton, W. L., Hayashi, M. & Chasmer, L. (2011) Permafrost-thaw-induced land-cover change in the canadian subarctic: implications for water resources. *Hydrol. Processes* **25**, 152–158, doi: 10.1002/hyp.7894.

Robinson, S. D. & Moore, T. R. (2000) The influence of permafrost and fire upon carbon accumulation in High Boreal peatlands, Northwest Territories, Canada. *Arc. Antarc. Alp. Res.* **32**, 155–166.

Rowland, J. C., Jones, C. E., Altmann, G., Bryan, R., Crosby, B. T., Geernaert, G. L., Hinzman, L. D., Kane, D. L., Lawrence, D. M., Mancino, A., Marsh, P, McNamara, J. P., Romanovsky, V. E., Toniolo, H., Travis, B. J., Trochim, E., Wilson, C. J. (2010) Arctic Landscapes in transition: responses to thawing permafrost. *EOS, Trans. Am. Geophys. U.* **91**(26), 229–230, doi:10.1029/2010EO260001.

Smith, M. W. & Riseborough, D. W. (2002) Climate and the limits of permafrost: A zonal analysis. *Permafrost Periglacial Proc.* **13**, 1–15.

Wright, N., Hayashi, M., & Quinton, W. L. (2009) Spatial and temporal variations in active layer thawing and their implication on runoff generation in peat-covered permafrost terrain. *Water Resour. Res.* **45**, W05414.

Monte Carlo experiments for uncertainty investigation of glacier melt discharge predictions through surface energy balance analysis

FREDDY SORIA & SO KAZAMA

Civil Engineering Department, Tohoku University, Aoba Aramaki 6-6-06, PO Box 980-0871, Sendai, Japan

soria@kaigan.civil.tohoku.ac.jp

Abstract The spatial representativeness of point records is a concern in glacier discharge predictions. A Monte Carlo-based global sensitivity approach is used to investigate the predictive uncertainty in the net radiation (*Rn*) as the major component driving glacier melt in the Bolivian Andes. The *Rn* is inferred through the Surface and Energy Balance Algorithm, calibrated with point dry-season records monitored on a glacier's ablation area. High uncertainties are expected in the vicinity of the monitoring station (surface albedo (α) between 0.81 and 0.79, specific melt discharge (SMD) between 72 and 88 L s^{-1} km^{-2}); smaller uncertainties are expected on the glacier boundaries (α between 0.10 and 0.08, SMD between 128 and 143 L s^{-1} km^{-2}). Thus, with the incoming long wave radiation ($R_L\downarrow$) as the most sensitive model parameter, the spatial variability in α determines the spatial variability in the SMD predictive uncertainties.

Key words tropical Andes; sensitivity analysis; remote sensing

INTRODUCTION

The spatial representativeness of point records that arises from the complexity observed in natural systems, is a concern to mathematical modellers. In watershed numerical modelling (as a valid example), the classical "calibration" as a means to assure the adequacy of a given model has long been brought into question by several hydrological publications. These publications established our limitations in representing complex natural systems (e.g. Beven, 1993; Wagener *et al.*, 2004), emphasizing the need for the uncertainty reduction in predictions to be a primary aim. Thus, similar to watershed hydrology, the investigation of predictive uncertainty is a demanding topic in the prediction of glacier-melt discharge in remote areas, due to limited knowledge of the spatial distribution of surface processes resulting from the low spatial density of ground observations.

Our aim in this study is to investigate the uncertainty in the spatial representativeness of point data recorded on a glacier in the tropical Andes, through a sensitivity analysis approach inspired by the equifinality concept. The equifinality idea suggests that, given current levels of knowledge and measurement technologies, rather than a unique representation of a given system, the existence of a universe of behavioural models is likely. Beven (1993) formally introduces the equifinality idea in the hydrological literature. It became an inspiration for Wagener *et al.* (2004), who among others provide the basis for Tang *et al.* (2007) and Soria & Kazama (2011), whose ideas are applied in this study. The approach includes the application of remote sensing techniques as a tool to infer the spatial distribution of surface energy balance processes on glacial formations.

STUDY AREA AND DATA

The investigation is on the Zongo glacier in the Cordillera Real, situated in the tropical Andes in Bolivia. From a water resources engineering perspective, the melt from the ice caps of the Cordillera Real are worth studying because they provide freshwater for nearby ecosystems and for neighbouring urban settlements, La Paz and El Alto (approximate population of 1 000 000 people). The Zongo glacier (approx. 2 km^2 of ice cover, horizontal view) is a unique source of information for the study of tropical glaciers in the Andes of Bolivia.

The surface energy balance is investigated using point meteorological observations acquired on 26 July 2005 (at 5050 m a.m.s.l.) on the ablation zone of the Zongo glacier. The spatial distribution of the energy balance is inferred from the processing of a Landsat ETM+ 30-m horizontal resolution scene. The meteorological data are provided by the GLACIOCLIM (Les GLACIers, un Observatoire du CLIMat), and the Landsat scene was obtained from the US Geological Service (USGS) Earth Resources Observation and Science (EROS). The Landsat scene was acquired during the dry season (in the austral winter, 26 July 2005) at 10:30 h (local time). This scene was selected considering the small discrepancy with ground observations as observed in Soria & Kazama (2010). The calibration of the Landsat scene was carried out with the procedure suggested in Chander *et al.* (2009). The glacier-covered area was calculated from false colour composites (as in Soria & Kazama, 2009). The topographic information at 90-m horizontal resolution is from the Shuttle Radar Topography Mission Digital Elevation Model (SRTM DEM).

METHODS

The net radiation is assumed to be the major source of energy for glacier melt in the Cordillera Real. The instantaneous net radiation at the Landsat sensor acquisition time, *Rn*, is estimated through the Surface Energy Balance algorithm SEBAL (Bastiaanssen, 2000). The *Rn* is calibrated with the GLACIOCLIM point data. The uncertainty analysis is carried through a Monte Carlo variance-based global sensitivity analysis (Chan *et al.*, 2000).

Energy balance equation and simplifications

The energy available for glacier melt Q_M in W m^{-2} is investigated through equation (1), and the melt depth is calculated with equation (2) (Paterson, 1999):

$$Q_{\mathrm{M}} = Rs + R_{\mathrm{L,n}} + Q_{\mathrm{H}} + Q_{\mathrm{LE}} + Q_{\mathrm{G}} + Q_{\mathrm{P}} \tag{1}$$

$$M = \frac{Q_{\mathrm{M}} \cdot \Delta t}{D_{\mathrm{W}} \cdot \lambda_{\mathrm{f}}} \tag{2}$$

In equation (1), Rs is the net shortwave radiation, $R_{L,n}$ is the net longwave radiation, Q_H is the turbulent sensible heat, Q_{LE} is the turbulent latent heat flux, Q_G is the conductive-energy flux in the snow/ice or subsurface flux, and Q_P is the heat flux supplied by precipitation (Paterson, 1999). In equation (2), M (in m) is the melt depth for a time interval Δt due to Q_M in W m^{-2}, the density of water D_W is 1000 kg m^{-3}, and the latent heat of fusion of ice λ_f is 0.334 10^3 kJ kg^{-1}.

We assume that the calculations of Q_M are well represented by the net radiation *Rn* (i.e. the sum of Rs and $R_{L,n}$). The points below, (a) to (c), discuss such an assumption.

(a) The *Rs* and the incoming longwave radiation $R_L\downarrow$ are the most relevant sources of energy and can not be neglected. The *Rs* controls the variability of the energy balance during the melt season (which coincides with the glacier accumulation and the glacier ablation seasons), whereas the $R_L\downarrow$ is the main energy source for melting (Wagnon *et al.*, 1999).

(b) The turbulent fluxes Q_H and Q_{LE} tend to cancel each other during the melt season (the wet season) (Sicart *et al.*, 2008). This observation suggests that Q_H and Q_{LE} can be neglected for analysis conducted during the melt season. For the winter season (the dry season), the relevance of Q_H and the Q_{LE} is high due to the dry air at high elevations. However, for our analysis we assume that such error may not be large, given that average daily melt discharge rates during winter are notoriously low in comparison to melt season discharge rates (e.g. in the year 1999–2000, daily melt discharge in winter was about 10% of the peak melt discharge in the melt season (Sicart *et al.*, 2007).

(c) The Q_P is negligible because precipitation on the glacier always falls as snow (Wagnon *et al.*, 1999). The Q_G is negligible because it is excessively small in melt season; compared to *Rn* and turbulent fluxes, the Q_G is also small in winter (Wagnon *et al.*, 2009).

Other assumptions for glacier melt estimations

The equilibrium line is at approx. 5250 m a.m.s.l. (Sicart *et al.*, 2007). We assume that the runoff limit is at some distance above the equilibrium line (Rick, 2008). In the ablation zone, it is assumed that all melt contributes to runoff (Rick, 2008). In the accumulation zone, it is assumed that some melt is retained by refreezing of the percolated melt (Rick, 2008). We assume that Q_M is effective on the glacier portion below 5300 m a.m.s.l.. Above 5300 m a.m.s.l., we assume that melt refreezes before it reaches the glacier catchment outlet.

Estimation of the net radiation through remote sensing

The SEBAL inferences are carried out at the SRTM DEM resolution. The Rn in W m^{-2} is estimated using equation (3), where α is the dimensionless surface albedo, $R_L\downarrow$ is in W m^{-2}, and $R_L\uparrow$ in W m^{-2} is the outgoing longwave radiation. The ε_O is the dimensionless surface emissivity. The ε_O is 0.999 on snow surfaces (Morse *et al.*, 2000). For our analysis, ε_O is considered to be an uncertain parameter.

$$Rn = (1-\alpha)Rs + R_L\downarrow + R_L\uparrow - (1-\varepsilon_O)R_L\downarrow \quad (3)$$

The narrow band albedo (α_{TOA}) is transformed into α using equation (4) (Bastiaanssen, 2000), where ω is a dimensionless weighting coefficient, ρ is the dimensionless planetary reflectance at the top of the atmosphere, each Landsat band is denoted as Λ, α_p is the sun radiation reflected from the atmosphere, and τ_{SW} is the dimensionless atmospheric transmissivity of clear skies. The standardized mean solar exo-atmospheric spectral irradiance ESUN in W m^{-2} µm^{-1} is used to calculate ω (equation (5)) (Chander *et al.*, 2009). The α_p is assumed to be around 0.03 (Morse *et al.*, 2000). For our analysis, α_p is considered an uncertain parameter. The τ_{SW} is estimated with equation (6) (Bastiaanssen, 2000), where z is the SRTM DEM surface elevation in m a.m.s.l. Further uncertain topographic corrections are neglected (Riaño *et al.*, 2003).

$$\alpha = (\alpha_{TOA} - \alpha_p)\tau_{SW}^{-2} = [\Sigma(\omega_\Lambda \cdot \rho_\Lambda) - \alpha_p]\tau_{SW}^{-2} \quad (4)$$

$$\omega_\Lambda = \frac{\mathrm{ESUN}_\Lambda}{\Sigma(\mathrm{ESUN}_\Lambda)} \quad (5)$$

$$\tau_{SW} = 0.75 + 2 \cdot 10^{-5} \cdot z \quad (6)$$

Rs in W m^{-2} is estimated with equation (7) (Morse *et al.*, 2000), where the solar constant Gsc is 1367 W m^{-2}, $90 - \beta$ is the sun elevation angle, d^2 is the Earth–sun distance in astronomical units:

$$Rs = Gsc(90-\beta)\tau_{SW} \cdot d^{-2} \quad (7)$$

The $R_L\uparrow$ and $R_L\downarrow$ in W m^{-2} are estimated with the Stefan-Boltzmann Law using equations (8) and (9), respectively (Morse *et al.*, 2000), where Ts (in K) is the surface temperature, Ta (in K) is the absolute air temperature at the reference height, and the Stefan-Boltzmann constant σ_{SB} is 5.67 10^{-8} W m^2 K^{-4}. The Ts is estimated from the brightness temperature detected by the sensors Tb and ε_o using equation (10) (Chander *et al.*, 2009), where K1 and K2 in W m^{-2} sr µm^{-1} are Landsat calibration constants, and L_Λ in W m^{-2} sr µm^{-1} is the spectral radiance at the sensor's aperture. The ε_{eff} is the non-dimensional effective atmospheric emissivity (about 0.7 on snow and ice covered surfaces, Morse *et al.*, 2000). For our analysis, ε_{eff} is considered an uncertain parameter.

$$R_L\uparrow = \varepsilon_o \sigma_{SB} Ts^4 \quad (8)$$

$$R_L\downarrow = \varepsilon_{eff} \sigma_{SB} Ta^4 \quad (9)$$

$$Ts = \frac{1}{\varepsilon_o^{0.25}} Tb = \frac{1}{\varepsilon_o^{0.25}} \cdot \frac{\mathrm{K2}}{\ln(\mathrm{K1}/L_\Lambda + 1)} \quad (10)$$

Ta may be approximately equal to Ts on areas where most of the energy is spent on sublimation (e.g. Morse *et al.*, 2000). For our analysis in winter season, the latter mentioned assumption may be erroneous, because of which the spatial distribution of the Ta is inferred under the assumption that the temperature gradient between Ts and Ta at the observation site is constant along the entire surface of the glacier.

Uncertainty analysis, numerical experiments, and uncertain model parameters

Variance-based techniques have the advantage for interpreting the uncertainty contributions of mutual parameter interactions to the total output variance. In summary, the technique estimates the contributed variance of u model parameters (each denoted by sub indices i,j....k) to model output $Y = f(u_i, u_j, .., u_k)$. After theoretically decomposing the total variance of the model output $V(Y)$ into summands of decreasing dimensions, sensitivity indices measure the relevance of the parameter contribution to total variance (Chan *et al.*, 2000). Each term on the decomposition is computable by Monte Carlo integrations (see Chan *et al.*, 2000). In this analysis, the interpretation of the numerical experiments is carried through the total order index S_{Ti} at every grid cell. The S_{Ti} denotes the main effect of parameter u_i, as well as its interactions, and it is interpreted as the expected percentage of variance that remains if all parameters were known but u_i (Chan *et al.*, 2000). As an importance measure, S_{Ti} evaluates the importance of a parameter as the percentage of the output variance associated with it (Chan *et al.*, 2000). Equation (11) calculates the S_{Ti}, where V_{-i} denotes the influence on the variance of all the factors except u_i. For details on applications, the reader is referred to Tang *et al.* (2007) and Soria & Kazama (2011):

$$S_{Ti} = (1 - V_{-i}) \cdot [V(Y)]^{-1} \tag{11}$$

For our analysis we carried out 1024 experiments with a sample size of 128. The uncertain parameters are those that could be calibrated. Three model parameters were tested in the sensitivity analysis of the net radiation equation: ap on behalf of the α and Rs component, ε_O on behalf of the $RL\uparrow$ component, and ε_{eff} on behalf of the RL↓ component. The uncertainty bounds assumed for each uncertain parameter are the same for each grid cell. Table 1 summarizes the uncertain parameters and the corresponding uncertainty bounds. The Sobol quasi-random sequence for non-correlated parameters is used to generate the sample (Chan *et al.*, 2000), 1024 values of M are calculated, and the set of M values is interpreted through the S_{Ti} calculated for each grid cell.

Table 1 Uncertainty range for the model parameters.

Model component	Model parameter	Uncertainty range
α	α_{TOA}	None (non-calibratable parameter)
	ω_Λ	None (non-calibratable parameter)
	α_p	0.025 to 0.040 (calibratable parameter)
	z	None (non-calibratable parameter)
	τ_{SW}	None (non-calibratable parameter)
Rs	Gsc	None (non-calibratable parameter)
	$90 - \beta$	None (non-calibratable parameter)
	d^2	None (non-calibratable parameter)
$R_L\uparrow$	σ_{SB}	None (non-calibratable parameter)
	ε_O	0.900 to 0.999 (calibratable parameter)
Ts	Tb	None (non-calibratable parameter)
	K1 and K2	None (non-calibratable parameter)
	L_Λ	None (non-calibratable parameter)
	Ts	Function of ε_O
$R_L\downarrow$	ε_{eff}	0.6 to 1.0 (calibratable parameter)
	Ta	None (non-calibratable parameter)

α: surface albedo; α_{TOA}: albedo at the top of the atmosphere; ω_Λ: weighting coefficient for α_{TOA}; α_p: sun radiation reflected from the atmosphere; z: surface elevation; τ_{SW}: atmospheric transmissivity; Rs: incoming shortwave radiation; Gsc: solar constant; $90 - \beta$: sun elevation angle; d^2: Earth–sun distance; $R_L\uparrow$: outgoing longwave radiation; σ_{SB}: Stefan-Boltzmann constant; ε_O: surface emissivity; Ts: surface temperature; Tb: at-sensor brightness temperature; K1, K2: calibration constants; L_Λ: spectral radiance at the sensor's aperture; $R_L\downarrow$: incoming longwave radiation; ε_{eff}: effective atmospheric emissivity; Ta: absolute air temperature.

RESULTS

The uncertainty analysis summarized in Fig. 1 compares the S_{Ti}s calculated for ε_O and ε_{eff} and the α values. Changes in α_p show a very low sensitivity on the α component, because of which the corresponding results are not presented in Fig. 1. The sensitivity of ε_{eff} is higher compared to the sensitivity of the other two model parameters tested, which not only suggests the dominance of the $R_L\downarrow$ on the sensitivity of the net radiation calculations over the Zongo glacier during the winter season, but interestingly reveals the high predictive uncertainty expected on melt estimates over the ablation area around the region where the monitoring station is installed (marked as a white circle on the central region of the glacier in Fig. 1). However, the results show that in the region where the highest uncertainties in net radiation predictions are likely, low rates of specific melt discharge (SMD) should be expected because of the high surface albedo (the calculated SMD uncertainty range in the central zone of the ablation area is 72–88 L s^{-1} km^{-2}).

As the surface albedo decreases in the vicinity of the lateral moraines and the glacier terminus, it also decreases the predictive uncertainty of the net radiation, as well as the sensitivity of the net radiation component terms. Simultaneously, as the surface albedo values decrease, the rates of predicted SMD increase considerably; in consequence, it increases the relevance of a predictive uncertainty that is apparently smaller than the predictive uncertainty observed in the glacier zone with high surface albedo (the calculated SMD uncertainty range on the boundaries of the glacier is in the range 128–143 L s^{-1} km^{-2}).

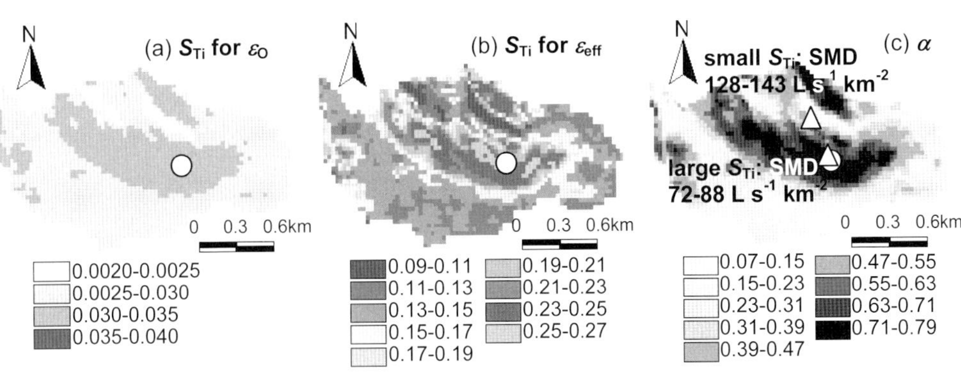

Fig. 1 Spatial distribution of the dimensionless total order index S_{Ti} on the Zongo glacier calculated for 26 July 2006 (10:30 h, local time). The figures correspond to the: (a) surface emissivity ε_O; (b) effective atmospheric emissivity ε_{eff}; and (c) spatial distribution of the surface albedo α for the Monte Carlo run with the highest calculated values. The white circles indicate the approximate location of the monitoring station; white triangles indicate the pixels where the smallest (small S_{Ti}) and the largest uncertainties (large S_{Ti}) are expected. Also shown are the ranges of specific melt discharge values (SMD) in L s^{-1} km^{-2} calculated on the latter mentioned locations.

In the melt season, the coincidence with the accumulation season should cause the glacier surface area having high albedo to grow; in consequence, the areal extent of the glacier where high melt predictive uncertainties are expected should also increase. However, considering that the melt rates on the region with high albedo are likely to be lower than the melt rates on the glacier boundaries, the overall uncertainty in melt predictions during the melt season should be expected to be smaller than the overall uncertainty expected in melt predictions during the dry-winter season over the glacier caps of the Cordillera Real.

CONCLUSIONS

The dominance of the albedo on the estimation of the spatial distribution of the net radiation is a known fact that has been demonstrated throughout our results. From the perspective of the net radiation investigation, the highest net radiation predictive uncertainties should be expected in glacier areal portions with higher surface albedo. Conversely, the numerical experiments described reveal that, when the aim is the calculation of glacier melt, a high predictive uncertainty should be expected over the glacier portions with low surface albedo, because those areal portions determine the areas with higher melt potential. For glacier melt estimations on the central part of the glacier with high surface albedo, the high predictive uncertainty may be attenuated by the low potential discharge associated with a high surface albedo. The conclusions also apply for assessing the installation of a monitoring network, because a more dense measuring network would be desired on the glacier areal portions that drive higher uncertainties in the predictions.

Acknowledgements The research was carried out by GRANDE project, supported by JST/JICA, SATREPS (Science and Technology Research Partnership for Sustainable Development), Japan. The meteorological data were provided by the French SO/ORE network GLACIOCLIM supported by IRD, the French Research Ministry, and local partners (IHH in Bolivia).

REFERENCES

Bastiaanssen, W. (2000) SEBAL-based sensible and latent heat fluxes in the irrigated Gediz Basin, Turkey. *J. Hydrol.* **229**, 87–100.

Beven, K. (1993) Prophecy, reality and uncertainty in distributed hydrological modelling. *Adv. Water Resour.* **16**, 41–51.

Chan, K., Tarantola, S., Saltelli, A. & Sobol, I. (2000) Variance-based methods. In: *Sensitivity Analysis* (ed. by A. Saltelli, K. Chan & E. Scott), 174–195. John Wiley, Chichester, UK.

Chander, G., Markhamb, B. & Dennis, H. (2009) Summary of current radiometric calibration coefficients for Landsat MSS, TM, ETM+, and EO-1 ALI sensors. *Remote Sens. Environ.* **113**(5), 893–903.

Morse, A., Allen, R., Tasumi, M., Kramber, W., Trezza, R. & Wright, J. (2000) Application of the SEBAL Methodology for estimating consumptive use of water and streamflow depletion. *Idaho Department of Water Resources – Final Report.*

Paterson, W. (1994) *The Physics of Glaciers*. Elsevier Science, Oxford, UK.

Riaño, D., Chuvieco, E., Salas, J. & Aguado, I. (2003) Assessment of different topographic corrections in Landsat TM data for mapping vegetation types. *IEEE Trans. Geosci. and Remote Sens.* **41**(5), 1056–1061.

Rick, U. (2008) Meltwater transport through firn in the accumulation zone of the Greenland ice sheet. PhD Thesis (abstract), University of Colorado, USA.

Sicart, J., Ribstein, P., Francou, B., Pouyaud, B. & Condom, T. (2007) Glacier mass balance of tropical Zongo glacier, Bolivia, comparing hydrological and glaciological methods. *Global and Planet. Change* **59**, 27–36.

Sicart, J., Hock, R. & Six, D. (2008) Glacier melt, air temperature, and energy balance in different climates: the Bolivian Tropics, the French Alps, and northern Sweden. *J. Geophys. Res.* **113**, D24113.

Soria, F. & Kazama, S. (2009) Evaluation of climate change effects on glacier area and vegetation using remote sensing imagery. *Proc. 7th Int. Symp. on Ecohydraulics ISE & 8th Hydroinformatics Int. Conf.*, Concepcion, Chile.

Soria, F. & Kazama, S. (2010) Potential impacts of climate change on the tropical Andes. *Annual J. Hydraul. Engng. JSCE.* (submitted).

Soria, F. & Kazama, S. (2011) Assessing streamflow source areas investigation through uncertainty evaluation of numerical experiments in small catchments. *Hydrol. Processes* (accepted).

Tang, Y., Reed, P., van Werkhoven, K. & Wagener, T. (2007) Advancing the identification and evaluation of distributed rainfall–runoff models using global sensitivity analysis. *Water Resour. Res.* **43,** W06415.

Wagener, T., Wheater, H. & Gupta, H. (2004) *Rainfall–Runoff Modelling in Gauged and Ungauged Catchments*. Imperial College Press, London, UK.

Wagnon, P., Ribstein, P., Kaser, G. & Berton, P. (1999) Energy balance and runoff seasonality of a Bolivian glacier. *Global and Planet. Change* **22**, 49–58.

3 Climate

Fluvial response to climate change: a case study of northern Russian rivers

SERGEY CHALOV & GALINA ERMAKOVA
Lomonosov Moscow State University, Faculty of Geography, 119991 Leninskie gory, 1, Moscow, Russia
srchalov@rambler.ru

Abstract The cold regions of North Eurasia include very sensitive fluvial systems. Rapid changes in climate are reported for these areas. The aim of this study is to propose a framework for fast climate-driven predictions of fluvial systems, and to apply it for rivers of the northern part of the East European Plain and West Siberian Plain. The general approach consists of integrating outputs from climate models into a hydrological model, and then driving a catchment and morphodynamic model using output from the hydrological model. Modelled by AOGCMs, future climate shifts are the drivers of significant changes in surface flow. Predictions of an up to 25% decrease in annual runoff by the middle of the 21st century enables us to forecast changes in sediment migration rates, stream energy and water-channel boundary interactions, changes in channel morphology and channel patterns shifts using corresponding physically-based equations. Whereas high dimensional models are still computationally too expensive for long-term morphological predictions, simple 1-D equations enable us to make assessments of channel system response. We tested a suit of 1-D models to estimate fluvial response to climate change for the middle of the 21st century of medium and large rivers draining the north of Russia. Comparison with regional predictions for other territories is the special task of the study.

Key words climate change; runoff calculations; fluvial systems; sediment load; channel patterns

INTRODUCTION

A fluvial system represents a combination of processes and forms expanding from hillslopes through rills and gullies to rivers. The indirect influence of climatic factors on fluvial systems appears in change of those factors that control fluvial processes and forms. A review of the literature provides insight into previous understanding of fluvial system evolution. The essence of this stability and its failure is given by Velikanov (1958): "*Interaction and mutual control between fluid flow and solid boundaries leads to certain combinations of stream hydraulic conditions and channel morphology*". The controlling factors described by Leopold *et al.* (1964) were width, depth, velocity, slope, discharge, sediment size, suspended sediment concentration, and channel roughness.

Climate change effects on fluvial processes should be considered through conceptual process-based modelling which aims to join together the spatial structuring, variability and scaling effects. Though a variety of studies have been done in fluvial geomorphology, the modelling capability in this area is still highly imperfect, even in terms of the general vision and understanding of the driving forces and links. Consideration of the joint impacts of climate change through controlling factors for the entire system emphasizes the nonlinear nature of fluvial response, and the possibly severe and synergistic effects that come from the combined direct effects of climate change. River runoff changes and other factors affect fluvial system evolution, such as rates of evapotranspiration, precipitation characteristics, plant distributions, sea level, glacier and permafrost melting, and human activities.

The modelling approach considered in this paper comprises use of climate model output which drives a hydrological model, and output of the hydrological model drives a catchment and morphodynamic model. The problem is tackled by applying a suite of 1-D models to the fluvial response to climate change. High-resolution models are computationally too expensive for long-term morphological predictions over extended river reaches. The main aim of this study is to develop a framework for fluvial processes, including climate-driven modelling and application of a 1-D approach in the rivers of the East European and West Siberian Plains. Based on a literature review and recent studies, we describe fluvial system response using governing 1-D equations. These equations were previously validated for the rivers of the past environments and proved their high adequacy. Validations of the models for the present or past time scales (testing against "real" data) is necessary, but beyond the scope of this short paper.

STUDY AREA

The study focuses on the rivers of two northern areas of the Russian Federation: the East European and West Siberian plains. The East European plain together with the Northern European Plain constitutes the European Plain. It is the largest mountain-free part of the European landscape. The East European plain rivers (Volga, Dnepr, Don, Neva, Severnaya Dvina and Pechora and their tributaries) are very important fluvial systems from both economic and ecological perspectives. The study area includes the watersheds of the Caspian Sea – Volga River and its tributaries (Kama, Oka, Vyatka, Sura), Azov and Black Seas – Don and Dnepr, White and Barents Seas – Onega, Mezen, Severnaya Dvina and Pechora. The West Siberian Plain is another extensive flat territory of Russia, which is represented mostly by the Ob watershed and its confluents (Irtysh, Tom, Chulym, Ket, Tavda, Tura, Sosva, Biya) and also some small watersheds of the northern rivers: Nadym, Pur and Taz. The flow of some rivers (Biya, Katun, Tom) is formed in mountain conditions, but most of the territory is a flatland. All the studied rivers are located in the northern part of Eurasia and flow into the Arctic seas, and are thus considered as cold region.

RUNOFF CALCULATIONS

Climate-driven changes in river flow will likely cause changes in the fluvial system. Its estimation is crucial for predicting sedimentation rates and morphology changes. Results from global atmospheric circulation models and their interaction with the ocean (atmospheric–ocean general circulation models, AOGCM) were used to assess surface runoff changes. The climate-driven hydrological model is based on the water balance equation:

$$\overline{Y} = \overline{P} - \overline{E} \tag{1}$$

where $\overline{Y}$, $\overline{P}$ and $\overline{E}$ are average values (mm) of river flow, precipitation and evapotranspiration, respectively. Evapotranspiration $\overline{E}$ is often evaluated using the equation proposed by V. S. Mezentsev (Kislov *et al.*, 2008):

$$\frac{E}{E_o} = \left\{1 + \left(\frac{E_o}{P}\right)^n\right\}^{-1/n} \tag{2}$$

where E_o is evapotranspiration, and n is the correction factor, which is usually applied as $n = 3.8$ for East European Plain, and $n = 4$ is appropriate for Siberia (West Siberian Plain) (Kislov *et al.*, 2008). Evapotranspiration E_o depends on the sum of positive monthly average temperatures T_o:

$$E_o = a\, T_o + b \tag{3}$$

where a and b are regional empirical coefficients (Kislov *et al.*, 2008). The approach was previously validated for the period of 1961–1989. Using the data for 21st century from the AOGCMs, runoff changes were evaluated for the middle of the 21st century (2050) and the end of the 21st century (2100). Calculated runoff values for the baseline (1961–1989 years) and predicted periods were compared using runoff changes rate: $K_y = Y_{predicted}/Y_{base}$.

The results (Figs 1 and 2) indicate a general reduction in average annual runoff in the southern part of the territory (south of around 55°N), and an increase in the northern part. Thus water flow will increase in high latitudes in the watersheds of the Severnaya Dvina, Mezen, Pechora, Nadym, Pur and Taz rivers. In the central part of the East European Plain, the model suggests a 10–15% decrease in annual runoff, while for southern watersheds (Dnepr, Don) the decrease attains 25% by the middle of the 21st century. These changes also occur across a large part of the West Siberian Plain. Water flow was found to reduce across central and southern parts of the Plain. This study

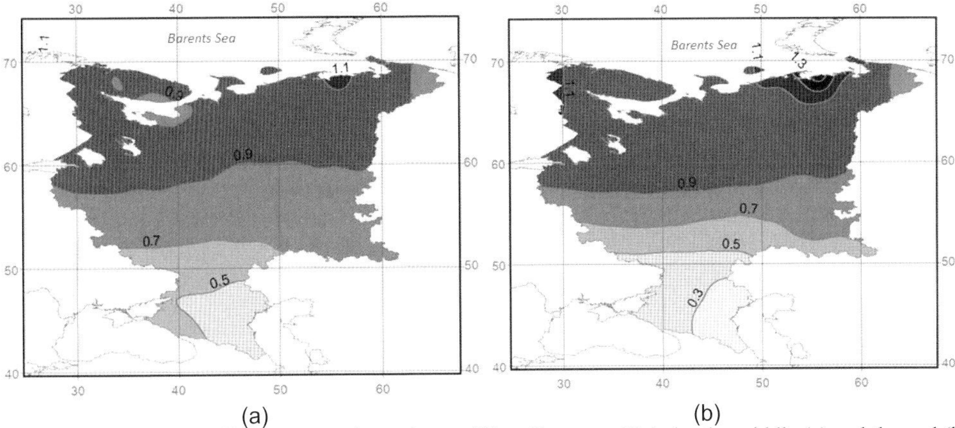

(a) (b)

Fig. 1 Relative runoff changes on the territory of East-European Plain by the middle (a) and the end (b) of the 21st century

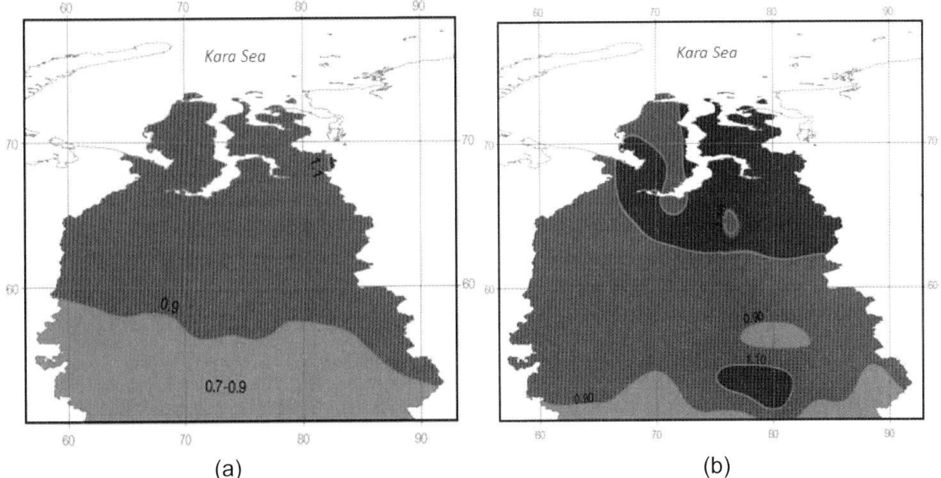

(a) (b)

Fig. 2 Relative runoff changes on the territory of West-Siberian Plain by the middle (a) and the end (b) of the 21st century.

attempts to quantify several sources of uncertainty, and to show that the effects of model uncertainty on the estimated change in runoff were generally small relative to the differences between scenarios and the assumed change in global temperature by 2050.

CATCHMENT PROCESSES AND SEDIMENT DELIVERY

Evidently, future catchment processes are strongly affected by changes in precipitation. Further changes in land-use management practices could represent the reaction of farmers to climate shifts in implementing different crops and land use. There are a few methods which can be used for the estimation of the rate of main driving forces – surface erosion. The well-known empirical Universal Soil Loss equation (Wischmeier *et al.*, 1971) is:

$$A = R_f K_f L_f S_f C_f P_f \tag{4}$$

where A is soil loss in tons acre^{-1} year^{-1}, R_f is rainfall factor, K_f is soil erodibility factor, L_f is slope-length factor, S_f is slope-steepness factor, C_f is cropping management factor, and P_f is the erosion control practice factor. A similar approach resulted in the Modified Universal Soil Loss Equation, MUSLE (Wischmeier *et al.*, 1971):

$$E = R_e K_e L_s S_g C_u P_c \tag{5}$$

where E is the mean of a couple of years of annual eroded soil mass from a unit of surface area, R_e is annual mean erosion from rainfall and runoff, K_e is coefficient of erodibility, L_s is nondimensional coefficient of slope length, S_g – nondimensional coefficient of slope gradient, C_u is nondimensional coefficient of type of cropping and land use, and P_c is a non-dimensional coefficient of erosion control practice. In the context of fluvial processes response to climatic change, all these parameters are regarded as changing variables, but R_e contributes most to changes in erosion (Kanatieva *et al.*, 2010). The main task for this work is to integrate the results of erosion modelling to the in-channel sediment data, which in the case of linear relationships, could be done using rate of sediment delivery and unit area discharge relationships (Dedkov & Mozjerin, 1984). Estimation of the amount of mobilized sediment that actually reaches the stream network depends on the delivery ratio *DR*, which determines the fraction of eroded sediment that is transferred from one location to another (i.e. field to field, or field to stream). It depends on the distance between the sediment source and channel, slope and roughness of the flow path, and availability of surface runoff. The latter provides background for climate-induced change assessments. The *DR* could be empirically estimated (Dickinson *et al.*, 1986) as:

$$DR = \lambda\left(\frac{H_c I^{1/2}}{n_r L}\right)^{\beta} \tag{6}$$

in which H_c is a hydrological coefficient expressing the ability of a certain area to generate surface runoff, I is the slope, L is distance between sediment source and channel, n_r is a roughness coefficient that depends on the type of land use, and λ and β are empirical parameters, being 9.53 and 0.79, respectively.

Where the data of sediment delivery is absent, the most useful method is connected with relationships with water discharge Q. The average relationship between river discharge and suspended sediment concentration is described by a sediment rating curve, which can be found through empirical statistical relationships (Syvitski *et al.*, 2000). The degree of $Q_s = f(Q)$ relationship is about 2–3. This leads to a 5% decrease in suspended sediment transport for the northern rivers (Severnaya Dvina, Onega, Mezen), and a 15–30% decrease for the central and southern rivers (Kama, Don, Irtysh).

IN-CHANNEL PROCESSES

In-channel changes appear in sediment load, bed-load particle stability, dune length and height, channel patterns (straight, meandering or braided), riverbank erosion, riverbed erosion and channel slope. The scale and intensity of these processes are influenced by a number of factors controlled by climate through water runoff changes.

Total sediment transport

The amount of total sediment load can be calculated through the Makkaveev (1955) equation:

$$R = A\, I^{n1} Q^{m} \tag{7}$$

where A is the erosive coefficient estimated for the gauged river basins (Makkaveev, 1955; Chalov & Shtankova, 2000), I is slope, Q is discharge, and n_1 and m are regional coefficients which depend on stream hydraulics (m usually equals 2 for plain rivers and 3 for mountain rivers; n_1 is about 1). Through the regional combination of m and A, the predictions were made. The results are shown in Table 1. For rivers where runoff decrease is expected, the total sediment load shows the same decreasing trend: for the Don by 40%, Kama 17% and Vyatka 22%. The Biya River demonstrates stable sediment transport rates ($R_{pr}/Ro = 1$) in the absence of significant flow changes.

Table 1 Total sediment transport changes for periods of the middle and the end of the 21st century.

River	Q_{pr}/Qo 2050	Q_{pr}/Qo 2100	A	m	R_{pr}/Ro 2050	R_{pr}/Ro 2100
Don	0.69	0.53	2.5	1.44	0.59	0.40
Kama	0.89	0.87	0.31	1.63	0.83	0.80
Vyatka	0.86	0.92	1.13	1.62	0.78	0.87
Biya	1.00	1.00		3.45	1.00	1.00

Ro – total load for the period (1961–1989), R_{pr} – total load for prediction period (2050 and 2100).

Bed load

Climate instability transforms the characteristics of bed load. A distinction can be drawn between transport in gravel-bed and sand-bed rivers. For alluvial sand-bed rivers, the best results to estimate sediment load are indicated through application of a structural approach (Alexeevskiy, 1998). The latter usually represents the moving of grains in groups such as ripples and thus forming the hierarchy of bed forms (riffles). Usually five types of ripples from small to large (Alekseevskiy, 1998) can be identified on rivers of different size. Their full height and shifting velocity depend on stream order N, in accordance with the equation:

$$(h_g)_i = \alpha N_{sh}^{\gamma} \tag{8}$$

where α and γ are the empirical coefficients, and i is bed form type. Stream order is calculated using (Scheidegger, 1966) $N_{sh} = \log_2 S + 1$, where S is the number of streams, the length of which is less than 10 km, located in the upstream of the drainage basin. Climate-induced changes of Q_o causes stream order transformation because of the logarithmic relationship between stream order and discharge: $N_{sh} = f(Q_o)$. Bed forms move along the stream with a certain velocity, C_g. The analysis of the bed form order of different type demonstrates the seasonal character of the relation between C_{gi} and stream order N_{sh}. During floods, bed forms move with the velocity:

$$C_g = \alpha_1 V_f N_{sh}^{\gamma_1} \tag{9}$$

where V_f is annual stream velocity during flood, and α_1 and γ_1 are empirical coefficients determined for each type of bed form (Alexeevsky, 1998).

The approach described was applied for the Chulim River (Ob basin), Severnaya Dvina and Oka rivers to estimate future changes in sediment transport. The relative changes in bed load were calculated using the Velikanov (1958) formula:

$$G_y = k \rho_n b_a \Sigma(h_i - C_i) \tag{10}$$

By the middle of the 21st century, due to flow changes, bed load will decrease by 1% for the Chulim and Severnaya Dvina, and 3% for the Oka. By the end of the century it seems to decrease by 2% for the Chulim and Severnaya Dvina, and by 5% for the Oka, in comparison with the present time.

Channel aggradation/degradation

Sediment delivery, sediment transport and transport capacity alterations enhance either sediment deposition or bed incision. Further changes in longitudinal profile occur indicating channel aggradation or degradation. Combination of the stream order–water discharge and the mass storage equations enables estimation of bed elevation change Δz (Alekseevskiy *et al.*, 2006):

$$\Delta z(x) = -\frac{1}{B\sigma_o}\frac{\Delta t}{\Delta x}\Delta[\alpha_2 e^{\beta_2 N} + \alpha_3 e^{\beta 3 N} \tag{11}$$

where B is channel width, σ_o is density of bed deposits (kg/m^3), α_2, α_3 and β_2, β_3 are regional coefficients, and N is stream order. Concerning $N = f(Q_o)$, the correlation between Δz and Q_o changes is determined and can serve as the basis of bed elevation shift analysis on the extensive river sections under different scenario of Q_o changes. For the rivers of the Severnaya Dvina watershed, a runoff decrease of 5% will lead to 0.1–0.3 m of channel aggradation. In the Don catchment a significant runoff decrease will induce 0.5–1.0 m of channel aggradation.

Riverbank migration

The instability of water regime transform intensity and probability of planform changes are important. The latter phenomenon is estimated through riverbank migration rates. Berkovitch (1992) claimed that riverbank erosion C_b (m year^{-1}) could be assessed as a function of annual water discharge Q:

$$C_b = K_1(Q^2 I/dH_b) \quad (12)$$

where I is channel slope; d is average sediment size, mm; H_b is the height of the washed bank relative to low water level, m; and K_1 is an empirical coefficient (m^3/s)$^{-1}$. Riverbank migration rates were estimated for the studied rivers. The results are shown in Table 2. Increase (5–15%) in the intensity of bank migration is expected for northern rivers (such as Sev. Dvina, Peshora, Neva, Taz, Pur, Nadym). Conversely bank migration will become slower on rivers of central southern areas, and reduce by from 10 to 50%.

Table 2 Rates of riverbank migration changes according to equation (12) and (13) by the middle and the end of 21st century.

River	C_{b2050}/C_{b1990}	C_{b2100}/C_{b1990}	BC_{x2050}/BC_{x1990}	BC_{x2100}/BC_{x1990}	River	C_{b2050}/C_{b1990}	C_{b2100}/C_{b1990}	BC_{x2050}/BC_{x1990}	BC_{x2100}/BC_{x1990}
Volga	0.86	0.81	0.94	0.92	Nadym	0.98	1.24	0.99	1.06
Kama	0.85	0.83	0.94	0.93	Pur	0.97	1.24	0.99	1.06
Dnepr	0.67	0.45	0.86	0.73	Taz	1.07	1.33	1.02	1.08
Don	0.59	0.40	0.82	0.71	Biya Katun	1.00	1.00	1.00	1.00
Sev.Dvina	1.04	1.10	0.96	1.01	Tobol	0.67	0.65	0.90	0.89
Pechora	1.04	1.12	0.99	1.04	Ishim	0.64	0.61	0.89	0.88
Chulym	0.95	0.84	0.99	0.95	Irtysh	0.82	0.83	0.95	0.95
Ket	0.99	0.94	1.00	0.98	Ob	0.90	0.91	0.97	0.98

C_b – bank migration rate according to equation (12) using the approach of Berkovich;
BC_x – bank migration according to equation (13), Bartley's approach.

Riverbank erosion (BC_x) can be estimated from bankfull discharge, which is determined by a process of basin-wide regionalization of the 1.58-year recurrence interval flow on the annual maximum time series (Bartley *et al.*, 2004). Because woody riparian vegetation is known to reduce bank erosion rates, a simple ramp function is introduced to exclude the proportion of bank with riparian vegetation (PR):

$$BC_x = 18(1 - PR)(Q_{1.58})^{0.6} L_x \quad (13)$$

Calculations for rivers of the East European and West Siberian Plain indicate (Table 2) that future changes mainly represent decreases of riverbank erosion. The most considerable changes are predicted for the Don and Dnepr rivers due to significant flow decrease (more than 25% by the end of the century). On most other rivers, these rates will be between 1% and 8%. There are also a few rivers with stable rates of bank erosion: the Ket, Taz, Biya and Katun. Although the results from both equations gave similar directions of change, the absolute values in a few cases differed by up to 20%. The Berkovitch equation (12) gave large values in bank migration, whereas according to Bartley's approach (13), only slight fluctuations in bank migration are predicted.

Channel patterns

The direction and intensity of bed deformations and bed form dynamics changes can lead to channel pattern shifts. Channel patterns recognition is typically based on the assumption that the transitions between the different channel types are threshold-governed processes. Conventional QI-type models predict the existence of either multi-thread or single-thread meandering and straight channel types using data on discharge Q and channel slope I. The more advanced model by Van den Berg (1995) takes into account the size of the sediment material. QI-type models can be used to predict channel pattern development. A probabilistic approach indicates that QI values (unit stream power) will still be in the transition zone around the critical threshold where two channels types may be expected.

Previous study on channel pattern prediction under future climatic conditions for 16 rivers of Severnaya Dvina and Pechora basin, using Van den Berg's model (Anisimov *et al.*, 2008) found potential transformations of the channel types, from single- to multi-thread, at 4 of the 16 selected locations in the next few decades, and at 5 locations by the middle of the 21st century. Hypothetical scenarios of 10%, 15%, 20% and 35% runoff increase were used. The results are confirmed by the recent investigations of several rivers (Alexeevsky *et al.*, 2006) where channel pattern changes were studied during 1960–2000 and were found to correspond with climate flow rate shifts. Thus for the small Protva River (Volga basin) the cumulative difference curve depicts a growth of flow rate during 1960–2000: $M = \frac{1}{n}\sum\frac{(K_i - 1)}{c_v} = 0.17$. For the period 1975–2000, M equals 0.75; this is the period of steady flow rate growth. A single braided reach transformed into a multi-thread channel with a few islands. Comparison of predicted and virtual channel pattern changes shows an absence of total transformation from a meandering channel to a braided or straight one.

COMPREHENSIVE ASSESSMENT OF FLUVIAL SYSTEM RESPONSE

The results obtained show the synergistic effects of the fluvial system's response to climatically-driven water discharge changes. Simulations for a given basin demonstrate that increase of surface runoff enhances surface erosion and sediment delivery, which is followed by sediment transport alterations. The latter contribute to distinctions in mass storage over channel reaches, caused in longitudinal profile evolution. Due to channel degradation decrease, a change from meandering and braided channel patterns to straight channels is usually observed. Otherwise, in the aggraded channel reaches, braided channels occur (Alekseevskiy, 1998). The meandering-braided threshold shows that climate decrease of Q_{max} (I = const) causes the transformation from the braided to meandering channels

The results of this investigation show the complex reaction of the fluvial system to climate changes. A number of regionally-based predictions of climate-driven changes of fluvial processes have been made recently (Asselman *et al.*, 2003; Gomez *et al.*, 2009; Verhaar *et al.*, 2010) to describe exact phenomena of the fluvial system (Table 3). It can be seen that in different regions of the world, flow and channel process fluctuations may be of different amount, and even of different trend. Asselman *et al.* (2003) suggested that sediment load transport of the River Rhine by 2100 could decrease by 13%. Morphological simulations for the 21st century of three tributaries of the Saint-Lawrence River (Verhaar *et al.*, 2010) predicted an overall increase in volumes of bed material that will reach the Saint-Lawrence River, as well as an effect on the longitudinal profile extending up to 10 km from the confluence with the Saint-Lawrence River. Decrease of the mean flow in the Waipaoa River (Gomez *et al.*, 2009) by an average of 13% in the 2030s, and 18% in the 2080s, will lead in the 2030s to the decline of bed load to 6.3±16.1 kt year^{-1}, further rising to 9.4±20.1 kt year^{-1} in the 2080s (after aggradation reduces the amount of accommodation space and modifies the long profile of the simulated river). The results depend mostly on the type of climate prediction used the given region, as this underlies the fluvial process model. Choosing the certain fluvial response approach also plays an important role in the resulting values, but the contribution is less significant.

Table 3 Overview of regional-scaled predictions of fluvial processes in 21st century.

Region	Time scale	Climate-driven water discharge scenario	Suspended load	Bed load	Bed elevation	Riverbank erosion	Authors
River Rhine	2100	Annual average discharge increase	Decrease by 13%				Asselman, *et al.*, 2003
Tributaries of the Saint-Lawrence	2010–2099	Reference base level 1 cm decrease		increase in average bed material delivery	Degradation		Verhaar *et al.*, 2010
Waipaoa River, New Zealand,	2010–2030s	Q_{mean} decrease by 13%	Either decline by 1 Mt/year or increase by 1.9±1.1 Mt/year.	Decrease 6.3±16.1 Kt/year	Aggrade by 0.31 m		Gomez *et al.*, 2009
	2010–2080s	Q_{mean} decrease by 18%		Increase 9.4±20.1 Kt/year	Aggrade by 0.85 m		
Northern Russian rivers (Onega, Pechora, Severnaya Dvina, Pur, Taz)	2050	Q_{mean} decrease by 3–5 %	Decrease	Decrease by 5–7%	10 cm aggradation	Increase by 2–3 %	Present study
	2100	Q_{mean} increase by 3–5 %	Increase	Increase by 5–15%	20 cm aggradation	Increase by 10 %	

CONCLUSION

There is evidence that the global climate is changing. Nowadays, hypotheses of global warming are largely applicable. Climate-induced changes of flow were assessed by a hydrological model which is able to recalculate meteorological data to runoff rates. AOGCMs modelling data and relationships between meteorological and hydrological parameters were used. Through direct and indirect climate influences, river runoff will decrease for the most part of northern Eurasia. The outputs of these assessments underlie the fluvial model, which was proposed for us through the review of existing approaches for predicting fluvial system behaviour. Further consideration of a variety of fluvial processes was made and the prediction models were tested. In a few cases, it was shown (e.g. bank erosion) that various models show similar directions of changes, but the absolute values of the changes could be different. It was found that 1-D assessments give physically-based results which could be easily joined to characterize the singular nature of the fluvial system. The suite of models could successfully cover the whole range of processes which govern the behaviour of the catchment–river system. The special task remaining for future work is to link channel pattern evolution and certain components of the fluvial system.

REFERENCES

Alekseevskiy, N. I. (1998) *River Sediment Production and Movement.* Izdatelstvo MSU (in Russian).

Alekseevskiy, N. I., Vlasov, B. N., Kononova, A. N., Sergeev, O. N. & Chalov, S. R. (2006) Water runoff and channel changes. *J. Water Industry* **2**, 80–99 (in Russian).

Anisimov, O., Vandenberghe, J., Lobanov, V. & Kondratiev, A. (2008) Predicting changes in alluvial channel patterns in North-European Russia under conditions of global warming. *J. Geomorphology* **98**, 262–274 (in Russian).

Asselman, N. E. M., Middelkoop, H. & Van Dijk, P. M. (2003) The impact of changes in climate and land use on soil erosion, transport and deposition of suspended sediment in the River Rhine. *Hydrol. Processes* **17**, 3225–3244.

Bartley, R., Ollet, J. & Henderson, A. (2004) A sediment budget for the Herbert River catchment, North Queenland, Australia. In: *Sediment Transfer Through the Fluvial System* (ed. by V. Golosov *et al.*). IAHS Publ. 288. IAHS Press, Wallingford, UK.

Berkovitch, K. M. (1992) *Channel Management.* Moscow, USSR (in Russian).

Chalov, R. S. & Shtankova, N. N. (2000) Sediment transport and channel types in Kama river watershed. In: *Issues of physical Geography and Geoecology of the Ural.* Perm University, Perm, Russia, 99–116 (in Russian).

Dedkov, A. P. & Mozjerin, V. I. (1984) *Erosion and Sediment Load on the Earth.* Kazanskiy University, USSR (in Russian).

Dickinson, W. T., Rudra, R. P. & Clark, D. J. (1986) A delivery ratio approach for seasonal transport of sediment. In: *Drainage Basin Sediment Delivery* (ed. by R. F. Hadley), 237–251. IAHS Publ. 159. IAHS Press, Wallingford, UK.

Gomez B., Cui, Y., Kettner, A. J., Peacock, D. H. & Syvitski, J. P. M. (2009) Simulating changes to the sediment transport regime of the Waipaoa River, New Zealand, driven by climate change in the twenty-first century. *J. Global Planet. Change* **67**, 153–166.

Kanatieva, N. P., Krasnov, S. F. & Litvin, L. F. (2010) Contemporary changes of climate factors and erosion in North Povolzhye. *Erosion and Fluvial Processes* **17**, 14–27 (in Russian).

Kislov, A. V., Evstigneev, V. M., Malkhazova, S. M., Sokolikhina, N. N., Surkova, G. V., Toropov, P. A., Chernyshev, A. V. & Chumachenko, A. M. (2008) *Forecast of Climate Induced Resources Provision on East-European Plain in the Conditions of XXI Century Warming*. Maks-press, Moscow, Russia (in Russian).

Leopold, L. B., Wolman, M. G. & Miller, J. P. (1964) *Fluvial Processes in Geomorphology*. W. H. Freeman and Co., San Francisco, USA.

Makkaveev, N. I. (1955) *River Channel and Basin Erosion*. Izdatelstvo AN USSR (in Russian).

Scheidegger, A. E. (1966) Effects of map scale on streams orders. *Hydrol. Sci. J.* **11**(3), 56–61.

Syvitski, J. P. M., Morehead, M. D., Bahr, D. B. & Mulder, T. (2000) Estimating fluvial sediment transport: the rating parameters. *J. Water Resour. Res.* **36**, 2747–2760.

Van den Berg, J. H. (1995) Prediction of alluvial channel pattern of perennial rivers. *J. Geomorphology* **12**, 259–270.

Velikanov, M. A. (1958) *Channel Process*. Gosfizmatizdat, Moscow, USSR (in Russian).

Verhaar, P. M., Biron, P. M., Ferguson, R. I. & Hoey, T. B. (2010) Numerical modelling of climate change impacts on Saint-Lawrence River tributaries. *Earth Surf. Processes Landf.* **35**, **10**, 1184-1198.

Wischmeier, W. H., Johnson, C. B. & Cross, B. V. (1971) A soil erodibility nomograph for farmland and construction sites. *J. Soil Water Conserv.* **26**, 189–193.

Local understanding of hydro-climatic changes in Mongolia

S. R. FASSNACHT[1], T. SUKH[1], M. FERNANDEZ-GIMENEZ[2], B. BATBUYAN[3], N. B. H. VENABLE[1], M. LAITURI[1] & G. ADYABADAM[4]

1 *Watershed Science Program, Warner College of Natural Resources, Colorado State University, Fort Collins, Colorado 80523-1472 USA*
srf@cnr.colostate.edu

2 *Rangeland Science Program, Warner College of Natural Resources, Colorado State University, Fort Collins, Colorado 80523-1472 USA*

3 *Institute of Geography, Ulaanbaatar, Mongolia*

4 *Institute of Meteorology and Hydrology, Ulaanbaatar, Mongolia*

Abstract Air temperatures in semi-arid regions have increased more over the past few decades than those in many other parts of the world. Mongolia has an arid/semi-arid climate where large portions of the population are herders whose livelihood depends upon limited water resources. This paper combines local knowledge and understanding of recent changes in water availability in streams, springs and wells, with an analysis of climatic and hydrological change from meteorological station data to illustrate the degree of change among Mongolian water resources. We find that herders' perceptions of hydro-climatic change are very similar to the results of the station-based analysis. Additionally, since station data are spatially limited, local knowledge can emphasize smaller-scale variability in changes to climate and hydrology. For this paper, we focus on a site in the Khangai Mountains and another in the Gobi desert-steppe, both in Central Mongolia.

Key words Mongolia; perceptions of hydro-climatic change; local knowledge; climate change

INTRODUCTION

Due to their intimate relationships with land and water resources, farmers and herders possess local knowledge or memory of changes in climate and hydrology. In semi-arid regions where water resources are limited, herders and farmers rely on their understanding of local water sources for cropping and to water livestock. Local, traditional, and/or indigenous knowledge has proved very useful for identifying the degree and effects of climate change (e.g. the special issue of *Climatic Change* on "Indigenous Peoples' Knowledge of Climate and Weather" edited by Green & Raygorodetsky, 2010), and has been observed and quantified in several studies (e.g. Berkes & Folke, 2002).

With a changing climate, the possibility of reduced water resources (e.g. Milly *et al.*, 2008) has an impact on those reliant on these resources. In Mongolia, ecological changes due to climate variation have been observed (Yu *et al.*, 2003; Angerer *et al.*, 2008), in particular alteration in vegetative patterns related to proximity to water sources (Fernandez-Gimenez & Allen-Diaz, 2001), including the effects of grazing gradients within different climate/ecological zones. However, the linkage between hydrological change and climate change has yet to be illustrated for Mongolia.

Recent work has been successful in using local knowledge to analyse climate change for Mongolia at smaller scales than attainable using downscaled climate models, especially for climate extremes (Marin, 2010). Change in other semi-arid regions, such as drought in Kenya, has been identified using local knowledge (Ifejika Speranza *et al.*, 2010). This paper assesses both climatic and hydrological changes in Central Mongolia and compares them to the observations and perceptions of herders in the region.

MONGOLIAN CLIMATE AND CHANGE

Global air temperatures have increased by 1.34°C per century over the past 50 years, with greater increases in many semi-arid regions (IPCC, 2007). The Mongolian climate is characterized by a long cold winter, dry and hot summer, low precipitation, large temperature fluctuations, and a relatively large number of sunny days (on an average, 260 days per year). Across the country, meteorological records show an extreme minimum temperature of –52.9°C in January and an extreme maximum temperature of 43.1°C in July (Batima & Dagvadorj, 2000). On average, the annual air temperature has increased by approx. 1.6°C in the past 60 years (2.6°C/century) with warming starting in the 1970s and intensifying at the end of the 1980s (Batima & Dagvardorj, 2000). Temperature increases have been more prominent in the winter months with temperature warming of 3.6°C (6°C/century) since the 1940s (Batima & Dagvardorj, 2000). A recent investigation showed that temperatures have seen a significant increase at all locations, with increases in Central Mongolia from 2 to 4°C/century (Jamiyansharav, 2010). These increases have been more noticeable in the mountain areas of western Mongolia, in the Gobi Desert, and the steppe areas, with warming of up to 6°C/century (Jamiyansharav, 2010). Northern areas, such as Lake Hövsgöl, have also warmed more than the average, with maximum and minimum increases of 1.8 (4.5°C/century) and 1.95 (4.9°C/century) degrees in the past 40 years, respectively (Nandintsetseg *et al.*, 2007). These temperature increases have been linked to the accelerated degradation of permafrost in the past 15 years (Sharkhuu *et al.*, 2007).

The average annual precipitation in Mongolia is highest in the northern regions and decreases southward; it ranges from 50 mm in the Gobi Desert to 400 mm in the northern parts, with more than 60% occurring during the summer months. Batima & Dagvardorj (2000) saw no changes in total annual recorded precipitation when combining stations across Mongolia, while Jamiyansharav (2010) reported increased precipitation across central Mongolia and decreased precipitation in drier areas such as the Gobi Desert. Changes in precipitation were less statistically significant than temperature increases. Changes in indices of precipitation extremes (taken from Nicholls & Murray, 1999) have been observed (e.g. Nandintsetseg *et al.*, 2007). However, these changes in precipitation extremes are less distinctive than changes in temperature.

Previous studies on climate change in Mongolia have used one station per aimag (province). For example, Jamiyansharav (2010) used 17 stations to investigate national trends. Over the area of Mongolia (1.6×10^6 km^2), this yields about one station per 100 000 km^2. Other areas of the world with extreme climates, such as the Arctic region, had almost an order of magnitude more hydrological monitoring sites (Shiklomanov *et al.*, 2002). Therefore, the use of local knowledge can provide an expanded spatial coverage, or as suggested by Marin (2010), address changes at the regional or local scale.

To date, Mongolian hydrological data have not been analysed for changing trends. In other parts of the world, these have been defined by changes in annual runoff volume, annual peak flow, the timing of peak flow, and the timing or centroid of the runoff volume (Fassnacht, 2006). For example, Stewart *et al.* (2005) observed an earlier peak flow in most of the US Pacific Northwest, associated with a change in the timing of peak snow accumulation.

STUDY SITES

This work examines potential climate change, in terms of the variability of precipitation, temperature, and hydrology for Jinst soum (county) of Bayankhongor aimag and Ikhtamir soum of Arkhangai aimag (Fig. 1). For the meteorological analysis, the Horiult and Bayankhongor stations were used to represent Jinst, and the Erdenemandal and Tsetserleg stations were used to represent Ikhtamir. The Horiult station had 25 complete years of record for daily maximum, minimum and average temperature and precipitation data, while the other three stations had at least 45 complete years of record (Table 1).

Daily streamflow was recorded at the following four stations in the area that had at least 25 complete years of record: Tuin River at Bayankhongor and Bogd (Jinst), Khoit Tamir River at Ikhtamir and Hanui River at Erdenemandal (Ikhtamir) (Table 2). The Tuin River originates from the south slope of the Khangai Mountains and flows to Orog Lake, and is part of the internal drainage basin of Central Asia. Orog Lake has dried up in recent years, but filled again in 2010 due to late season snows. The river runs for 243 km and has a basin area of 9410 km^2. Summer floods dominate, with 60% of the Tuin River's annual runoff coming from summer rainfall. The summer floods usually begin in early July, reaching their peaks in late July through early August.

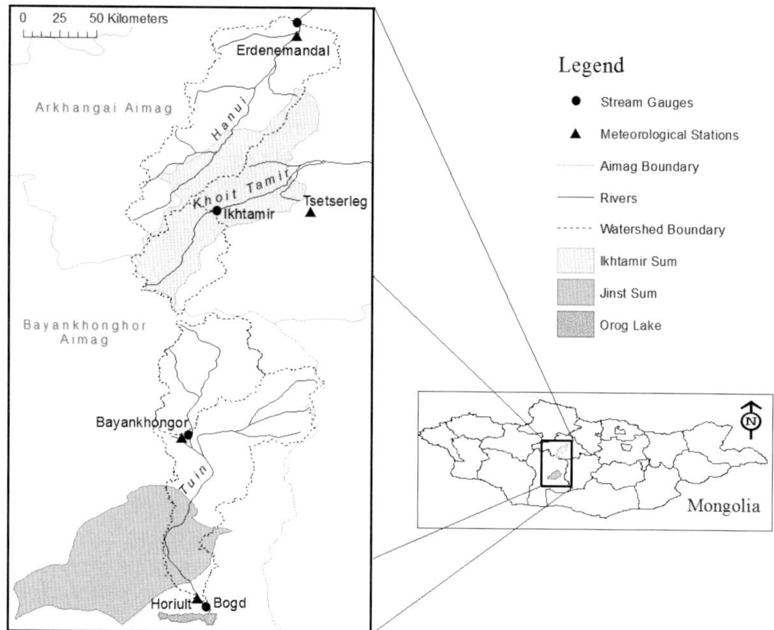

Fig. 1 Location of the Jinst and Ikhtamir soum study sites within Mongolia.

Table 1 Annual trend summary for Mongolian case study meteorological stations from daily data. Statistical significance is given as + for >90%, * for >95%, ** for >99%, and *** for >99.9% significant.

	Temperature (°C): Average	Maximum	Minimum	Precipitation: Annual (mm)	Days
	Horiult (25 years: 1971–2008)				
Mann-Kendall Z-score	2.6	1.2	3.0	0.44	3.1
Statistical significance	**		**		**
Sen's slope (per century)	4.3	1.9	4.7	23	50
	Bayankhongor (46 years: 1963–2008)				
Mann-Kendall Z-score	4.7	2.8	5.2	–1.3	–0.18
Statistical significance	***	**	***		
Sen's slope (per century)	5.0	3.5	5.3	–94	0
	Erdenemandal (45 years: 1964–2008)				
Mann-Kendall Z-score	4.9	3.4	5.9	–2.2	–1.9
Statistical significance	***	***	***	*	+
Sen's slope (per century)	5.4	4.1	7.3	–190	–25
	Tsetserleg (48 years: 1961–2008)				
Mann-Kendall Z-score	4.8	3.9	5.0	–1.8	–0.66
Statistical significance	***	***	***	+	
Sen's slope (per century)	4.4	4.4	4.2	–120	–6.0

The Hanui and Khoit Tamir rivers originate from the north slope of the Khangai Range and flow mostly through high mountainous areas. These rivers belong to the Arctic Ocean drainage basin. The Khoit Tamir River is a tributary of the Orkhon River and the Hanui River is a tributary of the Selenge River. Streamflow, temperature, and precipitation data were obtained from the Mongolian Institute of Meteorology and Hydrology (http://www.icc.mn/Meteoins/index.html). All data underwent a series of quality control evaluations implemented by the Institute.

Table 2 Trend summary for Mongolian case study hydrometric stations from daily data. Statistical significance is given as + for >90%, * for >95%, ** for >99%, and *** for >99.9% significant.

	Annual average discharge	Annual maximum daily discharge	Date of peak flow
	Tuin River at Bayankhongor (28 years: 1976–2008)		
Mann-Kendall Z-score	–2.0	–0.85	–0.73
statistical significance	*		
Sen's slope (per century)	–7.4	–30	–36
	Tuin River at Bogd (32 years: 1971–2008)		
Mann-Kendall Z-score	–0.86	0.61	–1.1
statistical significance			
Sen's slope (per century)	–2.4	18	–64
	Khoit Tamir River at Ikhtamir (26 years: 1976–2005)		
Mann-Kendall Z-score	–4.4	–3.6	0.93
statistical significance	***	***	
Sen's slope (per century)	–41	–310	63
	Hanui River at Erdenemandal (25 years: 1976–2005)		
Mann-Kendall Z-score	–4.3	–4.4	–0.72
statistical significance	***	***	
Sen's slope (per century)	–25	–170	–48

METHODS

The climate change analysis was performed using the annual average daily maximum, average, and minimum temperatures, as well as the annual total amount of precipitation and the number of precipitation days per year. Since most of the precipitation occurred seasonally as rainfall, there was less analytical emphasis on precipitation as snow. Zhang *et al.* (2004) corrected the daily precipitation due to biases, with the largest being undercatch of solid precipitation due to wind effect on gauge catch. However, wind speed data were not available for the entire period of record. A standard undercatch ratio, such as the gauge efficiency for solid and liquid precipitation of 50% and 90% used by Knowles *et al.* (2006), could not be applied since the phase of precipitation was not recorded. Therefore the precipitation data used in this study were gauge measurements without bias corrections.

Annual averages were computed from daily data when less than 15 days of record were deemed missing or of poor quality. No further discrimination was needed since periods of missing data were continuous and a month or more in length. Missing data were more common for the streamflow records than the meteorological records.

The statistical significance of the annual trend was computed using the Mann-Kendall test and the rate of change was determined from the Sen's slope (Gilbert, 1987). For the Mann-Kendall test, the data are ordered into sequential time series, with missing years of data allowed. For each time series record starting at the first year, all subsequent years are computed to determine whether an increase or decrease is observed. The total number of increases between pairs are subtracted from the total number of decreases and converted into a probability using the number of points in the time series (Gilbert, 1987). This probability is equivalent to the z-score. This non-parametric test is not biased by outliers or missing data, i.e. years with no average value. The Sen's slope is subsequently computed as the median slope (50th percentile) computed from the slopes between all data pairs. These methods are routinely used for climatic change analysis (e.g. IPCC, 2007) and hydrological change analysis (e.g. Burn *et al.*, 2010).

To determine local perceptions of climate and hydrological changes, in the summer of 2010 we conducted a short closed-end survey of 17 herders in Jinst and 20 herders in Ikhtamir, followed by a more open-ended discussion. Herder households were randomly selected in each area. Herder's ages ranged from 30 to 78 and each had at least 14 years of experience in the field of livestock husbandry. Survey questions focused on herders' perception of climatic change, including precipitation, questions related to temperature, and hydrological changes related to streamflow and river characteristics, and the state of springs and wells. It was deemed that questions very specific to temperature change were difficult to assess, thus questions were asked about changes to seasonal conditions, including their length and timing. The survey period of focus compared the current status of hydro-meteorological conditions with that which the herders perceived to exist when they were in their 20s. Responses to the closed-end survey questions were summarized and the frequencies of responses were computed. Qualitative discussion results were recorded and summarized. The narratives of the survey were analysed using a technique similar to that of Auerbach & Silverston (2003).

RESULTS

Temperature trends

Increasing trends in the annual mean average, maximum and minimum temperatures (Fig. 2) are statistically significant at all stations, except for the maximum temperature at Horiult (Table 1). The average temperatures are increasing by 4.3 to 5.4 °C per century, while the maxima are increasing by 3.5 to 4.4 °C per century when significant, i.e. not Horiult. As observed in many semi-arid regions, average minima are increasing the most at Horiult (4.7°C/century), Bayankhongor (5.3°C/century), and Erdenmandal (7.3°C/century). At Tsetserleg, the annual averages and maximums are increasing by 4.4°C per century, and 4.2°C for the annual minimum.

Precipitation trends

There has been a decrease in the annual precipitation from 1961 to 2008 at the Erdenemandal station (Fig. 3) computed to be 186 mm per century, and a less significant decrease in the

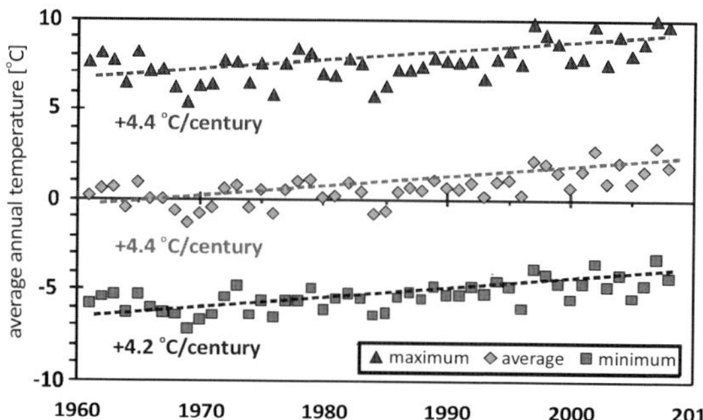

Fig. 2 Annual time series of average daily minimum, daily average and daily maximum air temperature at Tsetserleg.

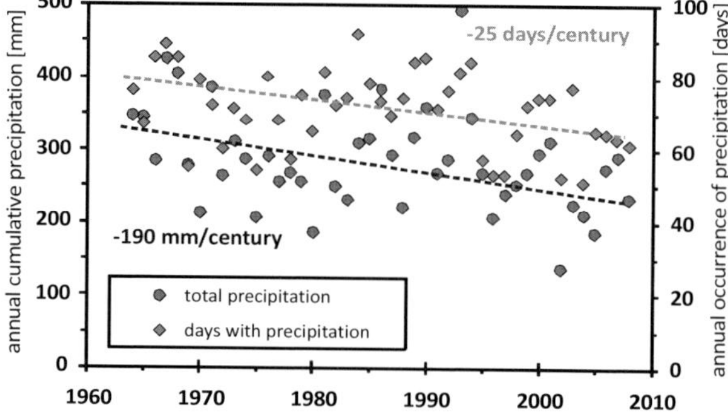

Fig. 3 Time series of annual cumulative precipitation and occurrence of precipitation at Erdenemandal.

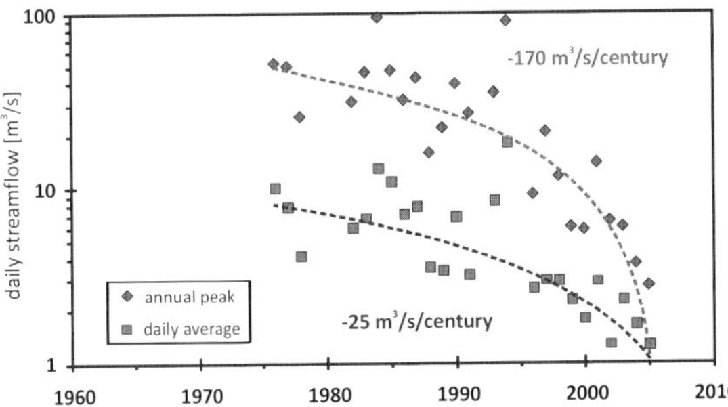

Fig. 4 Time series of annual daily maximum and average daily streamflow on the Hanui River at Erdenemandal.

occurrence of precipitation by almost 25 days per year per century (Table 1). The decrease in annual precipitation at Tsetserleg of 122 mm per century is less statistically significant. The increase in the number of days with precipitation of 50 days per year per century at Horiult was more statistically significant than other changes in precipitation.

Streamflow trends

The Hanui River at Erdenemandal (Fig. 4) and the Khoit Tamir River at Ikhtamir have seen a highly statistically significant decrease in the average annual streamflow of 24.7 and 40.7 m^3/s per century, respectively, and in the annual maximum daily discharge of 166 and 314 m^3/s per century, respectively ($p < 0.001$ for both quantities and locations, as seen in Table 2). The Tuin River at Bayankhongor saw a less significant ($p < 0.05$) decrease in annual streamflow of 7.44 m^3/s per century.

Herder observations

The survey questions were quantitative and addressed changes in precipitation and changes related to temperature, i.e. snowmelt (Table 3), as well as changes in streamflow and spring characteristics (Table 4). The supplemental notes provided a qualitative assessment of all these aspects (Table 5).

Table 3 Percentage change in the amount of precipitation observed by local pastoral herders. Questions were asked on changes in the amount rainfall and snowfall, rainfall intensity and the timing of snowmelt.

Location	Large decrease / much less intense / much earlier	Small decrease/ less intense / earlier	No change	Small increase/ more intense / later	Large increase / much more intense / much later
	The amount of rainfall has …?				
Jinst	71	29	0	0	0
Ikhtamir	100	0	0	0	0
	The amount of snow has ...?				
Jinst	24	29	35	6	6
Ikhtamir	55	15	25	5	0
	The rains have become ...?				
Jinst	0	6	29	65	0
Ikhtamir	0	15	5	80	0
	The timing of is snowmelt is ... ?				
Jinst	0	18	41	29	12
Ikhtamir	0	30	25	40	0

Table 4 Percentage changes in hydrological variability observed by local pastoral herders.

Location	Have you observed any changes in amount of water magnitude and flow in river basin?			YES	NO
Jinst				100	0
Ikhtamir				100	0
	Large decrease / much earlier	Small decrease /earlier	No change	Small increase /later	Large increase / much later
	Has the volume of water in the river …?				
Jinst	73	27	0	0	0
Ikhtamir	100	0	0	0	0
	Has the timing of peak flow become …?				
Jinst	0	0	40	53	7
Ikhtamir	5	35	35	20	5
	Have you observed any river / spring dry up over time?				
	Slight	Moderate		Severe	Complete
Jinst	14 / 0	43 / 11		29 / 11	14 / 78
Ikhtamir	0 / 10	20 / 5		50 / 15	30 / 80

While there were no direct questions related to changes in temperature, herders mentioned specific changes in seasonal extremes, such as hotter summer days and cooler summer nights, and that winters were warming (Table 5). From the seasonal change questions, there was more consistency at Ikhtamir than at Jinst, but some general trends were observed. All herders perceived that summer started later and all but 12% at Jinst thought that summer was shorter. The start of winter was different for the two sites, with 47% perceiving a later start at Jinst (35% earlier and 18% no change) while 75% at Ikhtamir perceived an earlier start to winter. A majority thought that spring started earlier (59% Jinst and 75% Ikhtamir). However, while most thought that autumn started earlier in Ikhtamir (90%), all in Jinst thought there was no change (59%) or it started later (41%).

Most herders perceived that the amount of rainfall has decreased greatly (71% in Jinst and 100% in Ikhtamir) or decreased somewhat (29% in Jinst). While a majority saw a large or small decrease (Table 3), 35% in Jinst and 25% in Ikhtamir saw no change and several herders perceived increases. The rainfall was seen to be more intense (65% at Jinst and 80% in Ikhtamir).

Snowmelt was perceived to start earlier now by some, and later by others. Changes were due in part to these warmer temperatures. Herders commented that the warmer winters have made it more difficult to distinguish between the winter and spring seasons. In Mongolia, much more precipitation occurs in the summer related to the Indian Monsoon. Precipitation has been observed to decrease, including shorter rainfall events, less infiltration, and less snowfall. Despite more intense rainfall events, discharge in the rivers has decreased with less depth and decreased velocity. Springs have been seen to dry up and the water level in wells has gone down.

DISCUSSION

Climate conditions of the mountain regions of Mongolia are changing, as indicated by a long-term warming trend during the last 45 years (Batima & Dagvadorj, 2000; Jamiyansharav, 2010; Tables 1 and 5). Change in climate variability and extremes of weather events have received increased attention in the last few years. Understanding changes in climate variability and extremes is made difficult by interactions between the changes in the mean and variability (Meehl *et al.*, 2000).

Mongolia's livestock are raised in open pastures that directly depend on favourable climate conditions year round. Extreme natural events such as drought and severe winter weather are serious events in Mongolia that cause great damage, not only to the livestock sector, but also to the national economy (Batima *et al.*, 2005).

Table 5 Summary of herder observations and perceptions of climate and hydrology at Jinst and Ikhtamir.

Variable	Jinst	Ikhtamir
Change in climate		
Temperature	Herders in this region have the same perceptions as those in Ikhtamir soum. They have observed climate and weather changes over the years, with winters warming since the 1980s.	Climate and weather have changed over the last 20 years. The number of extremely hot days in summer has increased, while the nights and mornings are getting colder. Herders have stated that it is more difficult to distinguish spring from winter, implying that winters are getting warmer and springs are getting colder. Since the 1980s winters have got warmer, and have been similar to spring, while in the 1970s herders had never observed snowmelt in the winter. In the past few decades, they have observed snowmelt earlier in the winter due to warmer winter temperatures.
Precipitation	Herders identified that from the 1970s through the 1980s, the occurrence of rainfall with low intensity increased and lasted for 1 to 3 days with high infilt-ration. Recently, precipitation comes about a month later at a higher intensity lasting much less time, 3 to 4 hours, with less infiltration. Precipitation has become more patchy. The amount of snowfall has decreased in the past few years, except in 2000 and 2002 which had much more snowfall and resulted in the death of a number of livestock.	Summer precipitation amounts have decreased considerably in recent years. Some herders mentioned that when they were children, the rainfall duration was longer, lasting 3 or 4 days. All of their clothes would get wet and they did not have any dry clothes to wear. Rainfall is now considered rare, with the duration of rainfall becoming shorter with high intensity, low infiltration, and high runoff. The amount of snowfall has decreased in the past few years, with 20 to 30 cm falling 20 years ago.
Change in hydrology		
Streamflow	The water level in the Tuin River has severely decreased and been cut off in some places, especially near Jinst soum. In the 1970s and 1980s the water level was much deeper with more rainfall.	The water level in the Ikhtamir River has decreased every year. Herders mentioned that 20 years ago water in the Ikhtamir River was deeper with stronger stream velocities, and horses and cars could not cross through the river. Currently, the river flows have decreased and rivers are much narrower. However, water levels have been higher this year due to more snowfall and rainfall occurring earlier. The Hanui River has dried up and been cut off in some places. This year, the river has started to flow again.
Springs and well	Most of the springs and lakes used by herders have dried up in the last few years. Water levels in wells have decreased since 2008. Some of the herders have been using wells due to water shortages.	Herders state that most springs have dried up in the last few years. They used to access springs at their winter and spring sites, some of which were very large and flowed year round. Most of these have dried up completely. Due to these water shortages, herders have started using wells in the last few years, and the well water levels are decreasing every year, which is attributed to drought and reduced rainfall.

Any analysis of climate change needs to examine the detailed characteristics of the changes. Specifically, since humans and the environment often respond to extremes, it is important to determine the variability or trends in a range of extreme values, rather than just mean conditions. The herder observations correspond well with the computed trends in climate and hydrology. As temperatures have increased significantly (more so in Ikhtamir than Jinst), precipitation amounts have decreased, which corresponds to a decrease in streamflow, in particular the average annual streamflow and the annual peak discharge. At Erdenemandal, the number of days with precipitation has decreased, while at Horiult is has increased significantly.

Beyond the decreases in precipitation, the decline in streamflow has likely been influenced by hydrological processes, such as increased evaporation, due to increased temperatures. There could also be a change in the phase of precipitation, with more rainfall occurring than snowfall, resulting in less snowpack or earlier onset of melting. These changes could have significant impacts (in terms of flooding, drought, water availability, etc.) on the local and regional environment.

The long-term average streamflow of the Tuin River has not changed significantly, while the herders have observed a depletion of water resources in the area. This is interesting because the drainage point of the Tuin River, Orog Lake, has been completely dry for a number of years. This may be due to a seasonal depletion of water resources from a change of river water regime. The infiltration of water into soil has been seen to decrease and has been attributed by some locals to pasture overgrazing, soil compaction, forest logging activity impacts, fires and insect outbreaks. Consequently, direct surface runoff may have increased, contributing to river flow while precipitation has decreased, resulting in lowered water tables.

The livelihood of Mongolian pastoral herders is directly associated to a large extent with weather extremes, forcing them to observe and record both quantitative and qualitative characteristics of a large number of climatic variables (Marin, 2010). The amount of snowfall has varied in the Jinst area over the past 11 years, resulting in several dzud events, the last of which occurred in 2010, killing a large number of livestock in the area.

When herders were asked about observed changes in precipitation, they all responded that changes had occurred in the past few decades. They stated that the duration of rainfall has become shorter, and that it lasted only a few hours, with a high intensity and high amount of runoff (Table 4). As a result of shortened precipitation durations and reduced rainfall amounts, most of the young herders had no experience with raincoats, while Russian military rain gear was regularly used by herders 20 years ago.

All the temperatures in the study areas were noted as increasing at a rate greater than the global average. Assuming precipitation in later years was less likely to fall as snow, and since gauge efficiency for rain is higher than for snow, it follows that actual precipitation amounts, (with wind bias held constant) would increase due to less undercatch (e.g. Knowles *et al.*, 2006). However, at Erdemendal and Tsetserleg, where a significant change in precipitation occurred (5% and 10%, respectively), the annual amounts decreased by 186 and 122 mm per century, respectively. At Horiult, the number of days with precipitation increased (50 days per century at a 1% significance level), but this should not be influenced by gauge undercatch or related biases.

CONCLUSIONS

This study examined meteorological trends in environmentally sensitive areas in the western mountain and steppe regions of Mongolia. The climate of the western region of Mongolia is warming. Warming is most pronounced in the high mountainous areas and their valleys. The average, maximum and minimum temperatures have significantly increased by at least 2°C between 1961 and 2008. These temperature changes corresponded to warming conditions observed by the herders.

In Ikhtamir, annual precipitation amounts and to a lesser extent the number of days with precipitation have decreased, corresponding with a decrease in the average and peak streamflow. These trends have been observed by herders, as well as the drying up of springs and decreased water levels in wells. In Jinst, the only observed precipitation change is an increase in the number of days at Horiult, but there are no measured overall streamflow changes. Despite the instrument measured results, the herders have seen less water in the rivers and springs (and wells) of the area.

Acknowledgements This research was supported by the Center for Collaborative Conservation at Colorado State University, the World Bank (award 7155627 entitled *Understanding Resilience in Mongolian Pastoral Social-Ecological Systems: Adapting to Disaster Before, During and After the 2010 Dzud*), and the National Science Foundation Dynamics of Coupled Natural and Human Systems (CNH) Program (award BCS-1011801 entitled *Does Community-Based Rangeland Ecosystem Management Increase Coupled Systems' Resilience to Climate Change in Mongolia?*).

REFERENCES

Angerer, J., Han, G., Fujisaki, I. & Havstad, K. (2008) Climate change and ecosystems of Asia with emphasis on Inner Mongolia and Mongolia. *Rangelands* **30**(3), 46–51.

Auerbach, C. F., & Silverstein, L. B. (2003) *Qualitative Data: An Introduction to Coding and Analysis*. New York University Press: New York, New York, USA. 203pp.

Batima, P. & Dagvadorj, D. (eds) (2000) *Climate Change and Its Impacts in Mongolia*. JEMR Publishing, Mongolia, 227pp.

Batima P., Natsagdorj, L., Gombluudev, P. & Erdenetsetseg, B. (2005) *Observed Climate Change in Mongolia*. Assessments of Impacts and Adaptations to Climate Change (AIACC) Working Paper No.12.

Berkes, F., & Folke, C. (2002) Back to the future: ecosystem dynamics and local knowledge. Chapter 5 in: *Panarchy: Understanding Transformations in Human and Natural Systems* (ed. by L. H. Gunderson & C. S. Holling), 121–146. Island Press, Washington DC, USA.

Burn, D. H., Sharif, M. & Zhang, K. (2010) Detection of trends in hydrological extremes for Canadian watersheds. *Hydrol. Processes* **24**(13), 1781–1790, doi: 10.1002/hyp.7625.

Fassnacht, S. R. (2006) Upper versus Lower Colorado River sub-basin streamflow: characteristics, runoff estimation and model simulation. *Hydrol. Processes* **20**, 2187–2205, doi: 10.1002/hyp.6202.

Fernandez-Gimenez, M. E., & Allen-Diaz, B. (2001) Vegetation change along gradients from water sources in three grazed Mongolian ecosystems. *Plant Ecology* **157**, 101–118.

Green, D. & Raygorodetsky, G. (2010) Indigenous knowledge of a changing climate. *Climatic Change* **100**(2), 239–242.

Gilbert, R. O. (1987) *Statistical Methods for Environmental Pollution Monitoring*. John Wiley & Sons, New York, 320pp.

Ifejika Speranza, C. I., Kiteme, B., Ambenje, P., Wiesmann, U. & Makali, S. (2010) Indigenous knowledge related to climate variability and change: insights from droughts in semi-arid areas of former Makueni District, Kenya. *Climatic Change* **100**(2), 295–315.

IPCC (2007) *Climate Change 2007: The Physical Science Basis*. Working Group I Report (WGI): Intergovernmental Panel on Climate Change, available at http://www.ipcc.ch.

Jamiyansharav, K. (2010) Long-term analysis and appropriate metrics of climate change in Mongolia. PhD Dissertation, Graduate Degree Program in Ecology, Colorado State University, Fort Collins, Colorado, USA.

Knowles, N., Dettinger, M. D. & Cayan, D. R. (2006) Trends in snowfall *versus* rainfall in the Western United States. *J. Climate* **19**, 4545–4559.

Marin, A. (2010) Rider under storms: Contributions of nomadic herder's observations to analyzing climate change in Mongolia. *Global Environmental Change* **20**, 162–176.

Meehl, G. A., T. Karl, T., Easterling, D. R., Changnon, S., Pielke, R Jr, Changnon, D.,. Evans, J., Groisman, P. Ya, Knutson, T. R., Kunkel, K. E., Mearns, L. O., Parmesan, C., Pulwarty, R., Root, T., Sylves, R. T., Whetton, P. & Zwiers, F. (2000) An introduction to trends in extreme weather and climate events: Observations, socioeconomic impacts, terrestrial ecological impacts & model projections. *Bull. Am. Met. Soc.* **81**, 413–416.

Milly, P. C. D., Betancourt, J., Falkenmark, M., Hirsch, R. M., Kundzewicz, Z. W., Lettenmaier, D. P. & Stouffer, R. J. (2008) Stationarity is dead: whither water management? *Science* **319**, 573–574, doi:10.1126/science.1151915.

Nandintsetseg, B., Greene, J. S. & Goulden, C. E. (2007) Trends in extreme daily precipitation and temperature near lake Hövsgöl, Mongolia. *Int. J. Climatology* **27**(3), 341–347, doi: 10.1002/joc.1404.

Nicholls N, & Murray B. (1999) Workshop on indices and indicators for climate extremes: Asheville, NC, USA, 3–6 June 1997. Breakout Group B: Precipitation. *Climatic Change* **42**, 23–29.

Sharkhuu, A., Sharkhuu, N., Etzelmüller, B., Flo Heggem, E. S., Nelson, F. E., Shiklomanov, N. I., Goulden, C. E. & Brown, J. (2007) Permafrost monitoring in the Hovsgol mountain region, Mongolia. *J. Geophys. Res.* **112**, F02S06, doi:10.1029/2006JF000543.

Shiklomanov A.I., Lammers, R. B.& Vorosmarty, C. J. (2002) Widespread decline in hydrological monitoring threatens pan-Arctic research. *EOS* **83**, 13–17.

Stewart, I.T., Cayan, D. R. & Dettinger, M. D. (2005) Changes toward earlier streamflow timing across western North America. *J. Climate* **18**(8), 1136–1155.

Yu, F., Price, K. P., Ellis, J. & Shi, P. (2003) Response of seasonal vegetation development to climatic variations in eastern central Asia. *Remote Sensing Environ.* **87**(1), 42–54.

Zhang, Y, Ohata, T., Yang, D. & Davaa, G. (2004) Bias correction of daily precipitation measurements for Mongolia. *Hydrol. Processes* **18**(16), 2991–3005.

Water cycle changes during the past 50 years over the Tibetan Plateau: review and synthesis

YINSHENG ZHANG[1] & Y. GUO[2]

1 *Key Lab. of Tibetan Environment Changes and Land Surface Processes (TEL), Institute of Tibetan Plateau Research, CAS, China*
yszhang@itpcas.ac.cn

2 *National Climate Centre, China Meteorological Administration, China*

Abstract The evidence for water cycle changes during the past 50 years on the Tibetan Plateau (TP) is synthesised by analyses of the meteorological observations and reanalysis data, and review of relevant studies. Robust warming has been evident, and decreasing wind speed has led to a weak atmospheric forcing. Snow depth decreased and the active layer depth increased in the permafrost region. In response to these changes, evapotranspiration slightly increased due to a wetter ground surface. Inhomogeneous changes in precipitation result in uncertainties regarding trends in river discharge over the regions and basins.

Key words water cycle; Tibetan Plateau; climate change

1 INTRODUCTION

In response to global warming, a critical question in hydrology is: *if the climate warms in the future, will there be an intensification of the water cycle? and, if so, what evidence is currently available to address this problem?* Empirical evidence for ongoing intensification of the water cycle would provide additional support for the theoretical framework that links intensification to warming (Huntington, 2005). To understand changes in the hydrological environment, it is necessary not only to precisely investigate the process at a local scale, but also to review long-term and large-scale data to obtain background knowledge.

The Tibetan Plateau (TP), the largest geomorphologic unit (Fig. 1) on the Eurasian continent, and highest plateau in the world, contains an "Asian water tower"; a large amount of water is stored in the world's highest and largest plateau. It is vital to understand whether supplies to these water resources have experienced any changes during recent global warming. Climate warming on the plateau during recent decades has been suggested by meteorological observations and ice core records. The climatic warming seems more evident on the plateau than the global mean trend during the past 50 years (Yao & Zhu, 2006).

There are five hydrological basins on the TP: Yangtze River, Yellow River, Southwest, Northwest, and the Plateau closed basin. More than 70 metrological observatories and some hydrological stations are located on the TP. These data mostly represent natural processes because human water use is negligible and can thus be ignored in these gauged catchments. These data have been widely used in estimations of variations in runoff and evaporation. Various satellite and reanalysis data provide useful information of the water cycle, such as precipitation and soil moisture. This work briefly reviews the current state of science regarding historical trends in hydrological variables, including precipitation, runoff, water vapour, soil moisture, evapotranspiration and snow cover. Data are often incomplete in spatial and temporal domains, and regional analyses are variable and sometimes contradictory; however, most studies do indicate an ongoing change of the water cycle.

2 TEMPERATURE CHANGE

Recent analysis of air temperatures (Liu & Chen, 2000) shows that the major portion of the plateau has experienced statistically significant warming since the mid-1950s. Temperature anomalies during 1960–2009 averaged for the different basins on the TP revealed a robust and synchronous warming over space and time (Fig. 2). The linear rates of annual mean temperature increases for

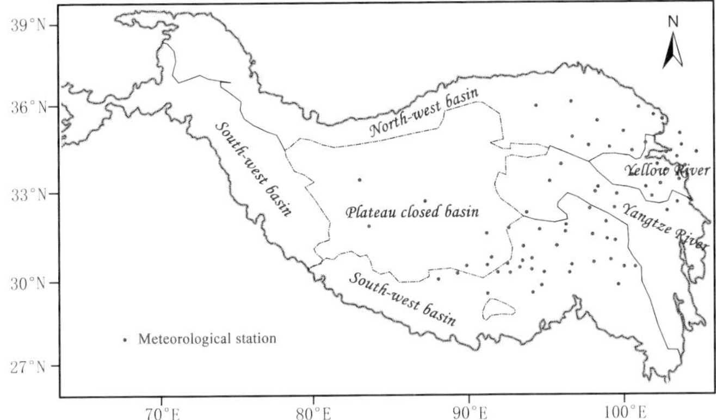

Fig. 1 Distribution of hydrological basins and meteorological stations on the Tibetan Plateau (TP).

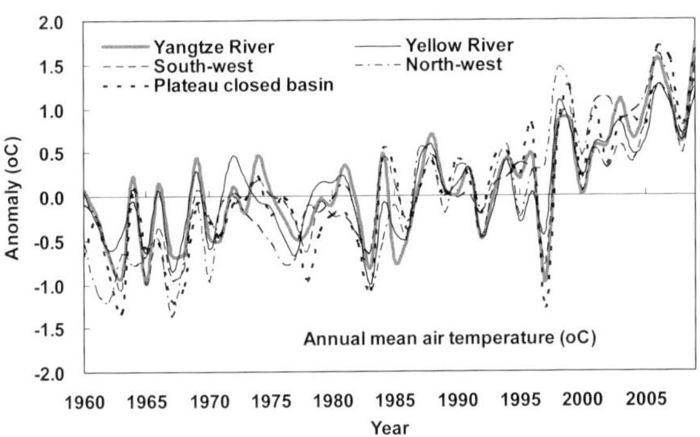

Fig. 2 Variation of air temperature anomalies during 1960–2009, averaged for different basins on the TP.

the different basins during 1960–2009 were about 0.30, 0.24, 0.28, 0.45 and 0.38 °C/decade for the Yangtze River, Yellow River, Southwest, Northwest and Plateau closed basin, respectively. These changes exceed the warming rates for the Northern Hemisphere and the same latitudinal zone for the same period (IPCC, 2007). Such robust warming may lead to significant changes in the water cycle in this cryosphere dominant region (Yao & Zhu, 2006).

3 CHANGES IN WATER VAPOUR AND PRECIPITATION

Recent satellite measurements indicate trends towards increasing water vapour in the tropical atmosphere during 1993–1999 that were consistent with model predictions (Minschwaner & Dessler, 2004). Table 1 summarizes the decadal changes of surface vapour pressure and precipitation during 1960–2009 on the TP. Surface vapour pressure, an indicator of atmospheric water content, synchronously increased among the five basins since 1990, but with different amplitudes.

On a global basis, precipitation over land increased by about 2% over the period 1900–1998 (Dai *et al.*, 1997; Hulme *et al.*, 1998). Regional variation in precipitation changes was highly significant. For example, zonally averaged precipitation increased by 7–12% between 30°N and 85°N, and by 2% for the regions between 0°S–55°S. Precipitation has decreased substantially in some regions (IPCC, 2007). Precipitation over the TP has different changes relative to atmospheric

Table 1 Decadal mean anomalies of vapour pressure and precipitation for the five basins on the TP.

	Mean	Yangtze	Yellow	Southwest	Northwest	Plateau closed basin
Vapour pressure (hPa)						
1960–69	–0.06	–0.08	–0.04	–0.17	0.07	–0.04
1970–79	–0.09	–0.10	–0.04	–0.13	–0.08	–0.10
1980–89	–0.08	–0.03	–0.05	–0.12	–0.05	–0.09
1990–99	**0.10**	**0.09**	**0.06**	**0.12**	**0.10**	**0.13**
2000–09	**0.14**	**0.09**	**0.06**	**0.15**	**0.19**	**0.30**
Precipitation (mm)						
1960–69	–11.2	–0.4	–5.9	–18.5	–12.6	–23.6
1970–79	–8.2	–6.8	–2.2	–14.5	–6.2	–11.6
1980–89	**6.9**	**28.9**	**10.5**	–6.3	**13.2**	**5.8**
1990–99	–2.4	–20.2	–6.0	**6.9**	–6.8	–1.3
2000–09	**16.6**	–0.4	**22.5**	**17.5**	**14.3**	**42.6**

Bold type notes positive anomalies

water content (Table 2). Regional precipitation mainly experienced positive anomalies during the decades 1980–89 and 2000–09, with the exceptions of the Southwest basin and Yangtze River basin. In the period of 1990–99, precipitation decreased in most of the basins, but increase in the Southwest basin. The changes in precipitation, clearly, were rather inhomogeneous both in time and space.

4 CHANGING IN WIND, SUNSHINE AND CLOUDINESS

Decreases in pan evaporation (P-EVA) have been observed on the TP (Fig. 3(a)), similar to the dominant trend over the world (Peterson *et al.*, 1995). Such decreases seem inconsistent with the observed trends of increasing temperature and precipitation, resulting in an "evaporation paradox" (Brutsaert & Parlange, 1998). Several analyses have suggested, however, that decreasing P-EVA is consistent with increasing surface warming and an acceleration of the water cycle (Roderick & Farquhar, 2004). Various mechanisms have been suggested to explain the apparent paradox, e.g. it has been suggested that decreases in solar irradiance, resulting from increasing cloud coverage and aerosol concentrations, and decreases in diurnal temperature range, would cause decreases in pan evaporation (Peterson *et al.*, 1995; Roderick & Farquhar, 2002). From our results, however, decreases in P-EVA may be primarily due to significant decreases in wind speed (WS) (Fig. 3(b)) and in sunshine duration (Fig. 3(c)), although the increase in lower cloudiness shows a complex trend (Fig. 3(d)).

5 CHANGES IN SNOW COVER AND ACTIVE LAYER DEPTH

It is difficult to evaluate the hydrological impact of snow cover changes on the TP, due to lack of supports to process studies. Table 2 summarizes decadal changes of annual maximum snow depth, which is considered to be the most important index of the influence of snow cover on hydrological processes. In the period of 1980–1989, the entire TP showed positive anomalies of maximum snow depth, implying more melting water penetrating the soil or forming runoff directly in the next spring. Decrease in the maximum snow depth was evident since 1990, as demonstrated by Qin *et al.* (2006).

Significant shrinkage of the permafrost has been reported on the TP (Yao & Zhu, 2006). Compared to other changes, active layer depth in the permafrost region is more important to hydrological processes, because it alters the hydraulic features of the surface layer. Table 2 summarizes decadal changes in the active layer depth. The positive anomalies reveal a significant thickening of the active layer, which may be due to climate warming.

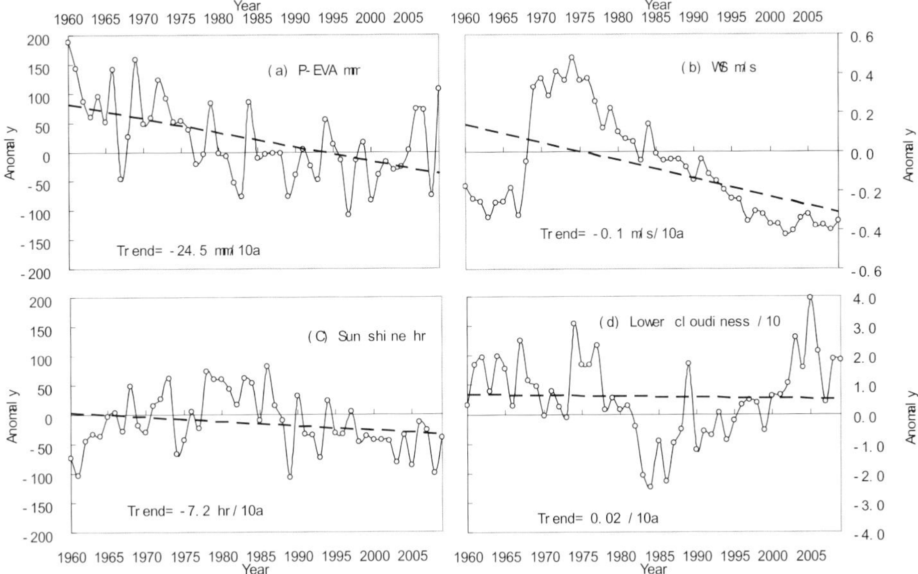

Fig. 3 Meteorological anomalies during 1960–2009 on the TP, (a) pan evaporation (P-EVA), (b) wind speed (WS), (c) sunshine duration, and (d) lower cloudiness.

Table 2 Decadal mean anomalies of maximum snow depth and active layer depth for the five basins on the Tibetan Plateau.

	Mean	Yangtze	Yellow	Southwest	Northwest	Plateau closed basin
Max depth of snow (cm)						
1960–69	–0.7	–0.7	–0.9	–1.3	–0.2	**1.3**
1970–79	–0.1	0.1	–0.1	–0.3	–0.1	**0.1**
1980–89	**0.5**	**0.8**	**0.4**	**0.7**	**0.2**	**0.3**
1990–99	–0.1	–0.7	–0.3	0.0	**0.1**	–0.3
2000–09	–1.0	–1.3	–0.8	–1.1	–0.5	–1.1
Active layer depth (cm)						
1960–69	–6.7	–0.6	–6.7	–14.8	**2.3**	**16.6**
1970–79	–2.1	–3.6	–0.5	–1.8	–4	–9.8
1980–89	–3.6	–2.6	–5.6	–3.7	–1.2	–6.1
1990–99	**3.5**	**3.7**	**4.9**	**3.0**	**2.9**	**7.8**
2000–09	**12.5**	**14.2**	**7.5**	**12.4**	**12.6**	**6.8**

6 CHANGES IN SOIL MOISTURE, EVAPOTRANSPIRATION AND RUNOFF

Due to lack of observations, we use ERA-40 reanalysis data to elucidate changes in surface water storage and evapotranspiration (ET). Figure 4 plots both annual ET and summer soil moisture in the 0–10 cm layer during 1960–2000. The averaged curves imply slight increases both in ET and soil moisture in the 0–10 cm layer, with linear increases of 1.0 mm and 0.6% per decade, respectively. Some previous reports (e.g. Brutsaert & Parlange, 1998), concluded that increasing water vapour concentration results in a warming-induced increase in ET. For the cryosphere, an increasing ET would be caused by increases in surface soil moisture, such as shown in Fig 4.

Table 3 shows discharge trends for the three TP basins, cited from Cao *et al.* (2006) that includes the Mann-Kendall tests. A positive (negative) value of Zc indicates an ascending (descending) trend. Results suggest a decreasing discharge trend in the Yangtze and Yellow River basins during 1956–2000. However, an increasing trend in discharge was observed for the Southwest basin.

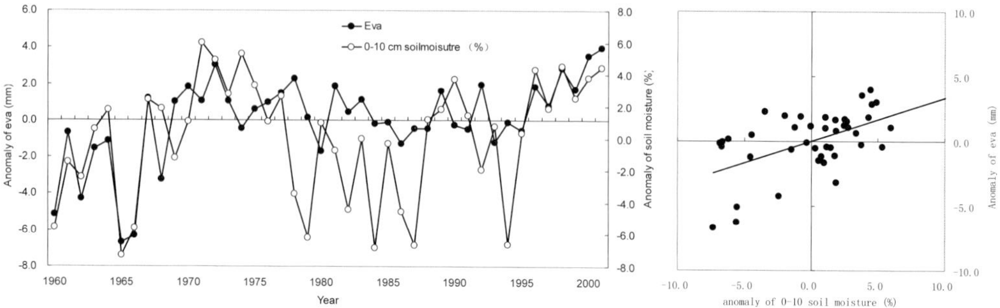

Fig. 4 Anomalies of ERA-40 annual ET (eva) and summer soil moisture in the 0–10 cm layer during 1960-2000 (left) and their correlation (right).

Table 3 Trends in mean annual discharge and Mann-Kendall tests (after Cao *et al.*, 2006).

Basin	First year	Last year	Series length	Test *Zc*
Yangtze River	1956	2000	45	–0.01
Yellow River	1956	2000	45	–0.30
Southwest	1956	2000	45	0.47

7 SUMMARY

Figure 5 summarizes the results of this analysis as follows:

1 Robust and synchronous warming over space and time was a primary characteristic of climatic change over the TP.
2 Significant decreases in wind speed and sun shine, and also an increase in lower cloudiness caused weakening of ground surface potential evaporation, as indicated by a decrease in pan evaporation.
3 Precipitation has slightly increased since 1990, but with spatial inhomogeneity.
4 Due to global and regional warming, maximum snow depth decreased and active layer depth increased.
5 Although pan evaporation tended to decrease, ET still slightly increased in response to increasing surface soil moisture.
6 There remains substantial uncertainty with regard to trends in river discharge.

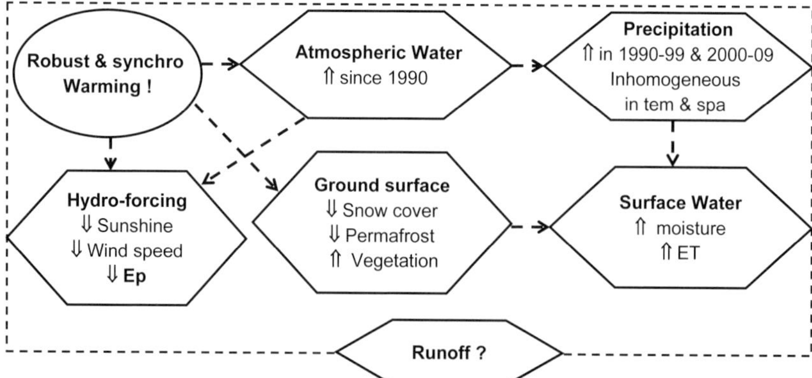

Fig. 5 Schematic changes of hydrological variables on the TP during 1960–2009, ⇑ & ⇓ for increasing and decreasing trends, respectively. Ep-pan evaporation, ET-evapotranspiration.

It is necessary to mention that we did not account for changes in glacier mass. The glaciers in this region are shrinking and that may significantly affect river discharge and water resources over the TP (Yao & Zhu, 2006).

Acknowledgements This work was supported by the Major State Basic Research Development Program of China (973 Program) under Grant no. 2010CB951701, the Natural Science Foundation of China (no. 595 40775045 and no. 41071042), the Innovation Project of Chinese Academy of Sciences (KZCX2-YW-BR-22).

REFERENCES

Brutsaert, W. & Parlange, M. B. (1998) Hydrologic cycle explains the evaporation paradox. *Nature* **396**, 30.

Cao, J., *et al.* (2006) River discharge changes in the Qinghai-Tibet Plateau. *Chinese Science Bull.* **51**(5), 594–600.

Dai, A., Fung, I. Y. & DelGenio, A. D. (1997) Surface observed global land precipitation variations during 1900–1988. *J. Climate*. **10**, 2943–2962.

Hulme, M., Osborn, T. J. & Johns, T. C. (1998) Precipitation sensitivity to global warming: comparisons of observations with HadCM2 simulations. *Geophys. Res. Lett.* **25**, 3379–3382.

Huntington. T. G. (2006) Sugar and syrup evidence for intensification of the global water cycle: Review and synthesis. *J. Hydrol.* **319**, 83–95.

IPCC (2007) Climate Change 2007: *The Scientific Basis*. Cambridge University Press, UK.

Liu, X. & Chen, B. (2000) Climatic warming in the Tibetan Plateau during recent decades. *Int. J. Climato*l. **20**(14), 1729–1742.

Minschwaner, K., & Dessler, A. E. (2004) Water vapour feedback in the tropical upper troposphere: model results and observations. *J. Climate* **17**, 1272–1282.

Qin, D., *et al.* (2006) Snow cover distribution, variability, and response to climate change in western China. *J. Climate* **19**, 1820–1833.

Peterson, A. T., *et al.* (1995) Evaporation losing its strength. *Natur*e **377**, 687–688.

Roderick, M. L. & Farquhar, G. D. (2004) Changes in Australian pan evaporation from 1970 to 2002. *Int. J. Climatol.* **24**, 1077–1090.

Yao, T. D. & Zhu, L. P. (2006) The response of environmental changes on Tibetan Plateau to global changes and adaptation strategy. *Adv. Earth Sci.* **21**, 459–464.

Spatial–temporal variation of temperature over China during 1961–2009

SUN LE-QIANG, HAO ZHEN-CHUN & WANG JIAHU

State Key Laboratory of Hydrology-Water Resources and Hydraulic Engineering, Hohai University, Nanjing, 210098, China

hzchun@hhu.edu.cn

Abstract The spatio-temporal variation of temperature is one of the basic signals for climate change. Analysis of its detailed distribution is useful for humans to adapt the ongoing and coming climate change. In this study, monthly mean temperature during 1961–2009 in China was used and processed by the classical Mann Kendall (MK) test. A Significant Year was defined as: (a) the time of break point for the temperature series, or (b) the time of 95% confidence level for the temperature series with monotonic trend. The rate of temperature changes before and after the Significant Year, and the trend magnitude were discussed. Our analysis shows: (a) all four annual regional average temperatures over China were decreasing before the 1970s, slightly or significantly; (b) the Tibetan Plateau and southwest Yunnan were the most significant warming areas during 1961–2009; and (c) the warming in northern China is much more significant than in the south, and the east coastal area was getting warmer more rapidly than the neighbouring interior.

Key words average temperature; Mann-Kendall test; spatial and temporal distribution; China

1 INTRODUCTION

Temperature is the most direct indicator of climate change. The change of temperature influences hydrological factors, such as streamflow and evapotranspiration. Hence, it is vital to analyse the spatial–temporal variation of temperature when assessing the change of local water resources (Hao *et al.*, 2006, 2007). There are many studies of temperature changes over China. Tu (1984) and Chen *et al.* (2010) report that, in China, except for the Northeast and northern Xinjiang, the basic pattern of temperature change over the past century is in accordance with that of the Northern Hemisphere. That is, a warm period from the 1900s until approximately 1945, followed by a cold period until the 1970s, and then another warm period. Yu (2000) shows that temperature change in China is not as significant as for the North Hemisphere. Ding & Dai (1994) obtained similar results, and they also undertook specific exploration into the temperature changes over Southeast China. Ban *et al.* (2006) did a similar analysis for Southwest China. Ding & Dai (1994) also discussed the influence of urbanization to the observed temperature changes. Shi *et al.* (1994) analysed the main characteristics of regional meteorological elements, including air temperature and precipitation, for the different decades in the last century. Chen *et al.* (1998) argue that temperature increases in China mainly happened over the regions north of 35°N, with Xinjiang and north Heilongjiang as the most significant warming areas. The areas between 23°N and 35°N and east of 100°E, however, were actually cooling during the past half century (Chen *et al.*, 1998). Qin (2005) agreed with Chen's results and reported that North China is indeed the main warming area in the past 50 years; furthermore, he calculated the increasing rate of temperature and pointed out that in the northern part of the Northeast, Inner Mongolia, as well as some western basins, the temperature increase rate was as high as 0.8°C per 10 years.

Most of the studies above focused on the overall trend of the temperature and do not contain detailed information. This may lead to a misunderstanding of regional climate trends. This paper aims to provide a more detailed and comprehensive depiction of the spatial–temporal variation of monthly temperature in China during 1961–2009. We use the newest temperature data provided by the China Meteorological Administration CMA (http://cdc.cma.gov.cn/) from more than 700 stations over China.

2 METHODOLOGIES

The linear trend assessment is important for time series analysis. Hess reviewed the method for linear trend detection (Hess *et al.*, 2001). The classical Mann Kendall test was used to examine the possible trends and abrupt change points in the average monthly temperature time series. This method was first mentioned by H. B Mann in 1945 (Mann, 1945), and since M. G Kendall's improvement (Kendall & Gibbons, 1962) it has become one of the most widely-used methods for trend detection in time series analysis (WMO, 1988).

In this method, the null hypothesis H_0 for the test is that there is no significant trend in the time series consisting of n independent random samples that all comply with a normal distribution. The alternative hypothesis test is double-tailed. The statistic S is given by:

$$S = \sum_{k=1}^{n-1} \sum_{j=k+1}^{n} \text{sgn}(x_j - x_k) \tag{1}$$

where:

$$\text{sgn}(x_j - x_k) = \begin{cases} +1, (x_j - x_k) > 0 \\ 0, (x_j - x_k) = 0 \\ -1, (x_j - x_k) < 0 \end{cases} \tag{2}$$

S is normally distributed with:

$$E(s) = 0 \tag{3}$$

$$V(s) = \frac{n(n-1)(2n+5)}{18} \tag{4}$$

When $n > 10$, the test statistic Z of the standardized normal distribution can be calculated as:

$$Z = \begin{cases} \dfrac{S-1}{\sqrt{Var(S)}}, S > 0 \\ 0, S = 0 \\ \dfrac{S+1}{\sqrt{Var(S)}}, S < 0 \end{cases} \tag{5}$$

A positive value of Z indicates an increasing trend and *vice versa*. Given a significance level α, if $Z \geq Z_{1-\alpha/2}$ then the null hypothesis will be rejected; that is to say, there exists a significant increasing (decreasing) trend in the target time series. If this method is used to detect abrupt change(s), the trend existed in the time series will be represented by another series UF_k rather than Z (Fu *et al.*, 1992); UF_k is calculated as follows:

$$S_k = \sum_{i=1}^{k} \sum_{j=1}^{i-1} \alpha_{ij} \quad (k = 1,2, \ldots, n) \tag{6}$$

where:

$$\alpha_{ij} = \begin{cases} 1, x_i > x_j \\ 0, x_i > x_j \end{cases} \tag{7}$$

then:

$$UF_k = \frac{|S - E(S_k)|}{\sqrt{Var(S_k)}} \quad (k = 1,2, \ldots, n) \tag{8}$$

where $E(S_k) = k(k+1)/4$ and $Var(S_k) = k(k-1)(2k+5)/72$.

In order to qualify the trend magnitude, the Theil-Sen approach (TSA) was employed (Kumar *et al.*, 2009). The TSA slope β is given by:

$$\beta = Median\left[\frac{(X_j - X_i)}{(j-i)}\right] \quad \text{where } i < j < n \tag{9}$$

A positive β indicates the increasing rate of the trend, and a minus value indicates the decreasing rate (Xu *et al.*, 2002). The unit of β here is °C/year.

The Mann-Kendall method can also be applied in change-point detection. However, it is not effective enough if there are more than one change-points in the time series (Fu *et al.*, 1992). An important work in change point detection is to set a confidence level and make sure the detected change-point can surpass this level. The most-used confidence levels are 90% and 95%. Temperature is relatively steady compared with other meteorological elements in long time periods, and considering the time span we chose, it is unlikely to be able to determine a significant change-point that is highly reliable (Jiang *et al.*, 2004). While it is believed the trend of the observed temperature time series is not monotone (Chen *et al.*, 1998; Jiang *et al.*, 2004; Wang *et al.*, 2009), and different parts of China may have different patterns in the trend of temperature evolution. This makes it necessary to use alternative statistics to indicate any change in the monotone series.

Calculation results show that there are only two types of trends here: slight decrease then increase, or continuous increase. We arrive at such conclusions not only in terms of the Four Geographic areas of China, but also from the statistical results of all the 730 stations. For the first situation, we call it a "mixed" trend, while the second type is called "monotone" or "monotone-like". If a mixed type of trend exists, we determine the time when the shift occurs; if it is monotone or monotone-like, we establish the time when the trend becomes significant. Here we define the time when the UF_k series first surpasses the confidence level as Significant Year (SY) to represent the transfer time of different trends or the significant trend happens. For example, the SY of regional monthly average temperature of Qinghai-Tibet is round about 1968.

3 *Z* SPATIAL DISTRIBUTION

The *Z* values of the Mann-Kendall test for 730 stations seasonal and annual temperature series of for 1961–2009 were calculated and interpolated via the ordinary kriging method. The results are shown in Fig. 1.

In spring season (MAM), most of China has a warming process. The most significantly warming areas include most parts of Eastern China and Northern China and the southeastern part of Northwestern China. The warming trends in northern Xinjiang, most of Southwest and Southern China do not surpass the confidence level of 95%. However, the results show that all other areas have a significant warming trend in spring with a confidence level of 95%.

Compared to spring, there are fewer areas with significant warming trend in summer (JJA). Heilongjiang, Jilin, Inner Mongolia, Beijing, Tianjin, most of Northwest China, Qinghai-Tibet Plateau, southwest Yunnan as well as parts of the Southeast coastal areas have significant warming trends. All other places, however, can not pass the confidence test of 95%. An interesting finding is the significantly cooling area with its centre in the border area of the central provinces Henan, Shannxi and Hubei.

Autumn (SON) temperatures increased significantly (SON). Unlike summer, most parts of China were warming in autumn over the last 50 years, and most of this warming process is reliable. Among those warming areas, the northeast Qinghai-Tibet Plateau is the most significant place. Only quite small parts, such as some sections of mid-South China and the Southwest can not pass the confidence test, and so very few parts were cooling and they can almost be ignored.

Similar to autumn, winter (DJF) was getting warmer nation-wide. Monthly average winter temperatures in the North coastal area, Qinghai-Tibet Plateau and west Yunnan show the most significant increase. Parts of Southwestern China like Sichuan, Chongqing, Guizhou and Guangxi can not surpass the confidence test, and some were even cooling.

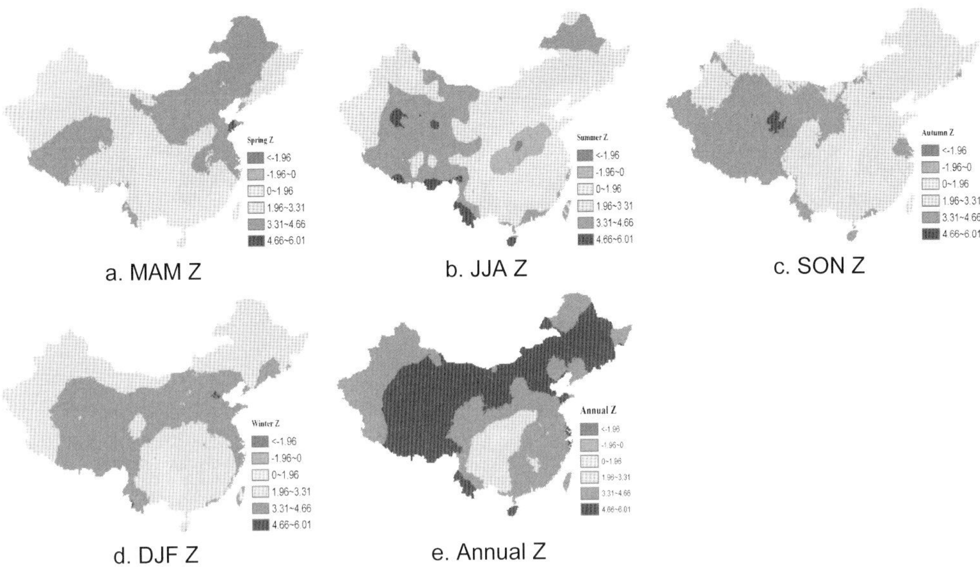

Fig. 1 *Z* value spatial distribution for annual and seasonal temperature over China.

The *Z* distribution for annual temperature shows that most of China was warming and the results are significant at a higher confidence level than for any single season; this is especially obvious for the Qinghai-Tibet Plateau, mid-east of the Northwest, the Northeast and southwest Yunan. A *Z* value greater than 1.96 can be regarded as significant at a confidence level of 95%. *Z* values for those regions are commonly higher than 5, thus it is almost certain that these regions were getting warmer in terms of annual temperature between 1961 and 2009. In the Yangtze River Basin, Wuhan and its surrounding area have significant warming trends, while neighbour areas, Chongqing and east Sichuan province, are the only places with a cooling trend.

It is a common pattern that Northern China and the coastal areas have more significant warming trends than Southern China and its neighbouring inland area. There is an overall increasing trend in both seasonal and annual temperatures over China in the past 50 years. The warming magnitude and span of annual temperature are much greater than seasonal temperatures. Northern China has warmed more rapidly than Southern China. Average temperatures in Qinghai-Tibet Plateau and southwest Yunnan have the most significant increases at both annual and seasonal time scales.

4 SY DETERMINATION

It is useful to determine the SY values of stations and to analyse their temporal and spatial distribution. For regional mean temperatures, we calculated the average temperature UF_k series for the four geographical regions. As shown in Fig. 2, SY of the South and Qinghai-Tibet occurs in 1970 or so, which means their trend is mixed with a decrease and an increase; neither should be neglected. The North and Northwest regions have similar trends, though we see decreases in the first several years of the records. The trends are insignificant to bring about a SY. The real SY happened after 1990, when the increasing trend is significant. The statistics of all 730 stations are listed in Table 1.

It is clear that there are two main peaks in the SY distribution, and the year of 1981 divides them into these two groups. Judging from the station and regional Mann-Kendall test results, the group before 1981 likely indicates a significant decreasing trend in the 1960s or 1970s, and the group after 1981 possibly indicates a lasting increasing trend without significant decrease since 1961.

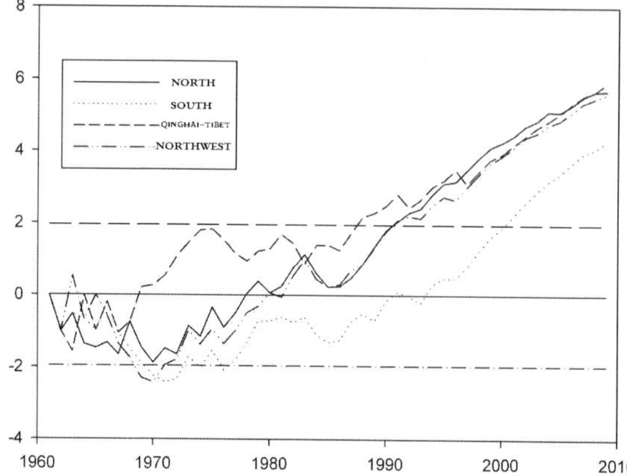

Fig. 2. The M-K test results of the annual temperature.

Table 1 Statistical characteristics of annual and seasonal SY.

	Annual	MAM	JJA	SON	DJF
MEAN	**1981**	**1993**	**1982**	**1988**	**1988**
SEM	0.519	0.495	0.556	0.536	0.497
MEDIAN	**1977**	**1998**	**1979**	**1991**	**1991**
MODE	**1964**	**2001**	**1964**	**1971**	**1964**
SD	14.012	13.361	15.011	14.47	13.432
VAR	196.341	178.511	25.321	209.386	180.422
SKW	0.26	–0.977	0.362	–0.154	–0.622
SES	0.09	0.09	0.09	0.09	0.09

SEM is Stand Error Mean; SD is Standard Deviation; VAR is Variance; SKW is Skewness; SES is Standard Error of Skewness.

Specifically, the distributions of annual and JJA (summer) Z are comparatively uniform and alike, even though two SY values of annual average make up more than 10% of all the annual SY values, respectively. The Mean and Median value of annual and summer SY values are quite alike as well (Table 1). Compared to annual and JJA, the other three seasons have more SYs than later than 1981. This verifies the last conclusion that unlike the summer and annual temperatures, spring, autumn and winter temperatures increased steadily over the past 50 years.

The spatial distribution of annual SY values in Fig. 4 indicates some cluster features: SY values in the Yangtze River and Yellow River basins are commonly earlier than 1981; Northeast, Northwest, Southwest and far Southern China have most SY values after 1981. The middle and eastern part of China, therefore, did not become warmer in the last 50 years. Instead, they have significant cooling trends prior to 1981.

5 MAGNITUDE DISTRIBUTION

The TSA slopes of annual temperatures, i.e. β to indicate the trend magnitudes, were calculated for three periods: 1961–1981, 1982–2009, and 1961–2009, respectively. Figure 5 presents the results of β in °C/year.

The β values are all positive during 1961–2009 over China. Most of the maximum values are found in the northern areas of the Northeast, mid-east part of Inner-Mongolia, Beijing, Tianjing, northeastern Qinghai-Tibet Plateau, and the north of Xinjiang. β values in these areas are as great as 0.40–0.50 °C/10 years, making them the most significantly warming areas in China during

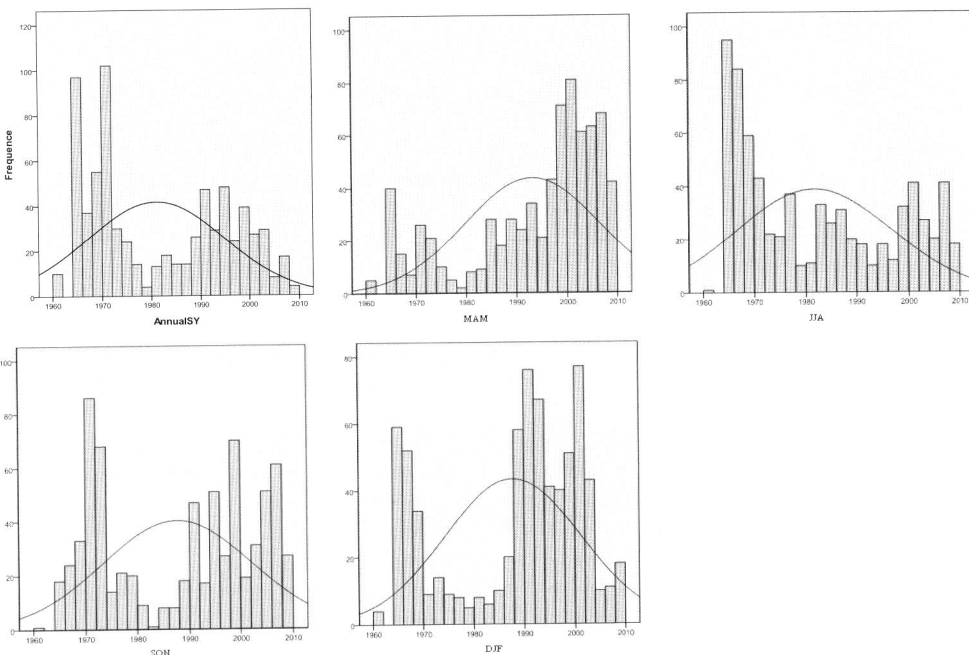

Fig. 3 Frequency distribution of the Significant Years of the annual and seasonal temperatures. The vertical axis (frequency) indicates the number of stations that have specific SY.

Fig. 4 Spatial distribution of annual SY (crosses stand for SY prior to 1981, circles stand for SY after 1981; in the base map, the shading distinguishes Northern China, Southern China, Northwest China and Qinghai-Tibet).

1961–2009. On the contrary, in the southwestern areas, such as Sichuan, Guizhou, Guangxi and Chongqing, the β values are no more than 0.14°C/10 years, although they are all positive. The coastal areas, such as Jiangsu, Zhejing and Guangdong, have greater β values than the neighbouring inland region, like Anhui and Hunan. Whether this has anything to do with urbanization is still uncertain and requires more research.

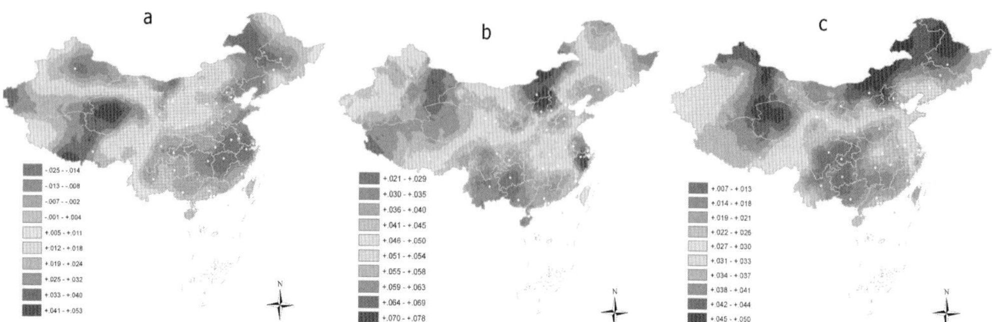

Fig. 5 Annual temperature changes, β in °C/a: (a) 1961–1981; (b) 1982–2009; and (c) 1961–2009.

The values are negative over a large percentage of China, but the Qinghai-Tibet Plateau might be warming at a rate as fast as 0.41–0.53°C/10-years. The major cooling areas are along the middle and lower reaches of the Yangtze River and the Northeast, where the decreasing rates are as great as 0.14–0.25 °C/10year. Temperatures all around China rebounded strongly during after 1981. In particular, increasing rates in parts of Zhejiang, Jiangsu, North China (especially the adjacent areas of Shanxi, Hebei and Inner Mongolia) are astonishing, 0.69–0.78°C/10 year. Even the least significant warming area, seen in parts of Guangxi and Guizhou, have rates greater than 0.21°C /10 year (see Fig. 5).

6 CONCLUSIONS

In the past 49 years, temperature in China has increased extensively. The temporal and spatial distributions of monthly and seasonal temperature trends are different. Comparatively, summer (JJA) has the least increase, while winter (DJF) has the largest increase. For yearly temperature, different patterns of seasonal temperatures may affect each other; a very significant decrease in one season may be offset by a significant increase in another season.

Average temperature over the Qinghai-Tibet Plateau (the Asian water tower), has increased in the past 49 years. This should draw attention when considering the possible consequences of snow and ice melt and glacier retreat to the local and regional energy and water cycle, as well as ecological security. North China has warmed faster than South China, if this trend continues in future, it would be wise to rethink the infrastructure building criteria for drought adaptation and water supply insurance in North China. Similarly, the Southeast Coastal areas have become warmer than their adjacent inland regions. Whether this can be explained by urbanization deserves more discussion.

The reason for setting a Significant Year (SY) is to describe the complex spatial and temporary distribution of different trends over time more effectively, and help to determine a proper divide if the trend is not continuous or changes over time. With the use of SY description, we successively determined the temporary similarity of annual and summer temperature variation. It is also an important reference to choose 1981 as the break point of trend calculation and periods.

Acknowledgements The financial support from the National Natural Science Foundation of China (40830639, 50879016), and the National Basic Research Program of China (2010CB951101) is gratefully acknowledged. The authors would like to extend their appreciation to the Special Fund of the State Key Laboratory of Hydrology-Water Resources and Hydraulic Engineering (1069-50985512).

REFERENCES

Ban Jun-mei, Miao Qi-long & Li Xiong (2006) Analysis of characteristics of temperature variations in Southwest China in recent 50 years. *Resources and Environment in the Yangtze Basin* **15**(003), 346–351.

Chen Longxun, Zhu Wenqin, Wang Wen, Zhou Xiuji & Li Weiliang (1998) Studies of climate change in China in recent 45 years. *Acta Meteorologica Sinica* **56**(003), 257–271.

Chen Quan-liang *et al.* (2010) Regional air temperature change of China since 1950s. *J. Anhui Agricultural Sciences* **7**, 3571–3574.

Dahe Qin, Yihui D, Jilan Su, *et al.* (2005) Assessment of climate and environment changes in China: climate and environment changes in China and their projection. *Advances in Climate Change Research* **1**(1), 1–10.

Ding Yihui & Dai Xiaosu (1994)Temperature variation in China during the last 100 years. *Meteorological Monthly* **20**(012), 19–26.

Fu Congbin & Wang Qiang (1992) The definition and detection of the abrupt climatic change. *Chinese J. Atmospheric Sciences* **16**(004), 482–493.

Hao Zhenchun, Li Li, Wang Jiahu, *et al.* (2007) Impact of climate change on surface water resources. *Earth Science (J. China University of Geosciences)* **32**(003), 425–432.

Hao Zhen-chun, Wang Jia-hu; Li Li, Wang Zhen-hua & Wang Ling (2006) Impact of climate change on runoff in source region of Yellow River. *J. Glaciology and Geocryology* **28**(001), 1–7.

Hess, A., Iyer, H. & Malm, W. (2001) Linear trend analysis: a comparison of methods. *Atmos. Environ.* **35**(30), 5211–5222.

Jiang Tianhan (2004) Detrended fluctuation analysis of monthly average temperature over China during the recent 100 years. *Scientia Meteorologica Sinica* **24**(002), 199–204.

Kendall, M. G. & Gibbons, J. D. (1962) *Rank Correlation Methods*. Griffin, London, UK.

Kumar, S., Merwade, V., Kam, J., *et al.* (2009) Stream-flow trends in Indiana: Effects of long term persistence, precipitation and subsurface drains. *J. Hydrol.* **374**(1-2), 171–183.

Mann, H. B. (1945) Nonparametric tests against trend. *Econometric* **13**(3), 245–259.

Shi Neng, Chen Jiaqi & Tu Qipu. (1995) 4-phase climate change features in the last 100 years over China. *Acta Meteorologica Sinica* **53**(004), 431–439.

Tu Qipu (1984) Trend and periodicity of temperature change in China during the past hundred years. *J. Nanjing Institute of Meteorology* **7**(2), 151–161.

Wang Haijun, Zhang Bo, Zhao Chuanyan, *et al.* (2009) The spatio-temporal characteristics of temperature change in recent 57 years in northern China. *Progress in Geography* **28**(4), 643–650.

WMO (1988) Analyzing long time series of hydrological data with respect to climate variability.

Xu, Z. X., Takeuchi, K. & Ishidaira, H. (2002) Long-term trends of annual temperature and precipitation time series in Japan. *J. Hydroscience and Hydraulic Engineering* **20**(2), 11–26.

Yu Jinhua, Ding Yuguo & Jiang Zhihong (2000) Singular spectrum analysis of temperature variations in China during the recent 100 years. *J. Nanjing Institute of Meteorology* **23**(004), 586–593.

Evaluation of IPCC AR4 global climate model simulation over the Yangtze River Basin

QIN JU[1], HAO ZHEN-CHUN[1], WANG LU[2], JIANG WEI-JUAN[3] & LU CHENG-YANG[4]

1 *State Key Laboratory of Hydrology-Water Resources and Hydraulic Engineering, Hohai University, Nanjing 210098, China*
super_juqin@126.com

2 *Delft University of Technology, Delft, The Netherlands*

3 *Ningbo Hongtai Hydraulic Information and Technology Co. Ltd, Ningbo, 315016, China*

4 *Yellow River Riverhead Institute, Yellow River Conservancy Commission, Lanzhou 730000, China*

Abstract The Fourth Assessment Report of the Intergovernmental Panel on Climate Change (IPCC AR4) presents 22 global climate models. This paper discusses the accuracy of the models in different temporal and spatial scales and evaluates their performances in simulating the temperature and precipitation over the Yangtze River Basin in China. The results indicate that the models are capable of simulating past climate. However, several climate models underestimate surface air temperatures and overestimate precipitation. Performances vary greatly among the models. Most models need to be improved since only a few produce correct seasonal cycles of climate. The results of scenarios analysis show differences among the models. The predicted tendencies of climate change, indicating the increase of temperature and precipitation in some regions, are consistent among the models. The results also show that the temperature and precipitation increase under different scenarios. The increase in temperature for the A2 scenario is the highest while the increase for the B1 scenario is the lowest. Eight models, that is: BCCR_BCM2.0, CCCMA_CGCM3.1, CNRM_CM3, GFDL_CM2.1, UKMO_HadCM3, MRI_CGCM2.3.2, NCAR_CCSM3 and NCAR_PCM, are able to precisely represent the characteristics of annual temperature and precipitation variations over the Yangtze River Basin. They have been selected to aid forecasting trends in water resources under future climate changes.

Key words IPCC AR4; simulation evaluation; Yangtze River Basin; temperature; precipitation

1 INTRODUCTION

According to the predictions of the *Fourth Assessment Report* by the Intergovernmental Panel on Climate Change (IPCC), average global temperatures could rise by 1.1–6.4°C by the end of the 21st century (IPCC, 2007). Climate change will affect the global and regional hydrological cycle, altering the spatial and temporal distribution of water cycle elements, such as precipitation, evaporation, runoff and soil moisture, causing re-allocation of water resources over time and space, thus increasing the probability of a variety of hydrological extremes.

At present, two methods are generally adopted to evaluate the impact of climate change on hydrology and water resources (Zhang & Wang, 2007). First are incremental scenarios, that is, to make use of different combinations of the assumed temperature and precipitation changes to form hypothetical scenarios of the future climate change. For example, Nash & Gleick (1991) adopt a conceptual hydrological model to study the influence on basin annual runoff for a temperature increase by 2°C and precipitation increase or decrease by 10–20%. Hao & Su (2000) used the improved distributed Xin'anjiang hydrological model assuming various temperature and precipitation changes to analyse the sensitivity of the Huaihe Basin runoff to climate changes. Nemec & Schaake (1982), Wang *et al.* (2000), Jia *et al.* (2008), Feng *et al.* (2006) and Zhu & Zhang (2005) also used similar methods for their research in different basins and regions.

The second method is the coupling of GCM predictions and the hydrological model. A good example is the work of Ozkul (2009), who combined the climate change scenarios of the IPCC AR4 report, with a water balance model, and forecast that surface water resources under future climate changes will reduce by 20% in 2030, and by 35% and 50% by 2050 and 2100, respectively. Nijssen *et al.* (2001) adopted scenarios of four climate models coupled with the VIC model to predict hydrological cycle changes in 2025 and 2045 for nine river basins. Hao *et al.* (2006) constructed a large-scale distributed hydrological model, considering the impact of melting snow and frozen soil, coupled with climate model predictions to evaluate water cycle and water resource changes due to climate change over the Yellow River source region. Chiew & McMahon (2002), Wetherald & Manabe (2002), Ellis *et al.* (2008), Edwin (2007) and Christensen & Lettenmaier (2007) have also done similar research. Most of the future water resources evaluation by coupling GCMs and hydrological models are based on a single or a few GCM outputs. Because different models have different simulation abilities, a single GCM has greater uncertainty in the output. There is no sufficient reason for selecting multiple GCMs, even though differences between GCMs have been considered and their statistical characteristics also have a certain stability and reliability. Therefore the principles and methods of selecting GCMs become extremely important.

The Yangtze River Basin, having large water resources, is the most important region in China. Its drainage area is 180×10^4 km^2 and it plays a significant role in the water security, energy security, the South-to-North Water Transfer Project implementation and future economic development of China. Therefore, understanding the evolution and future changes in water resources in the Yangtze River Basin under climate changes will provide a basis for water resources management and planning. In this paper, simulations by the 22 GCMs, released in the Fourth Assessment of IPCC Data Distribution Center, are evaluated. GCMs covering the Yangtze River have been selected for this work. Temperature and precipitation forecasts of these GCMs during 2010–2099 under three typical emission scenarios: A1B, A2 and B1, combined with one hydrological model, have been used to forecast future water resources changes.

2 SELECTION OF THE GLOBAL CIRCULATION MODELS

2.1 Data selection

Monthly temperature and precipitation data of the 22 global climate models in the contemporary climate (20C3M) conditions of the reference period 1961–1990 (IPCC AR4, 2007) have been selected for this study. We compare them with observations of 118 major meteorological stations in the Yangtze River Basin for the same period. Models with relatively better simulations of precipitation and temperature have been selected. Hao *et al.* (2010) introduced the global climate models in the Yangtze River in detail and their performance. Figure 1 is the distribution of the 118 weather stations (dots) and the two hydrological stations (small triangles) at Yichang and Datong. Table 1 presents the basic information of the 22 GCM models, and the grid boundary and numbers in the Yangtze River. More detailed information of the models can be found at http://www.pcmdi.llnl.gov/ipcc/about-ipcc.

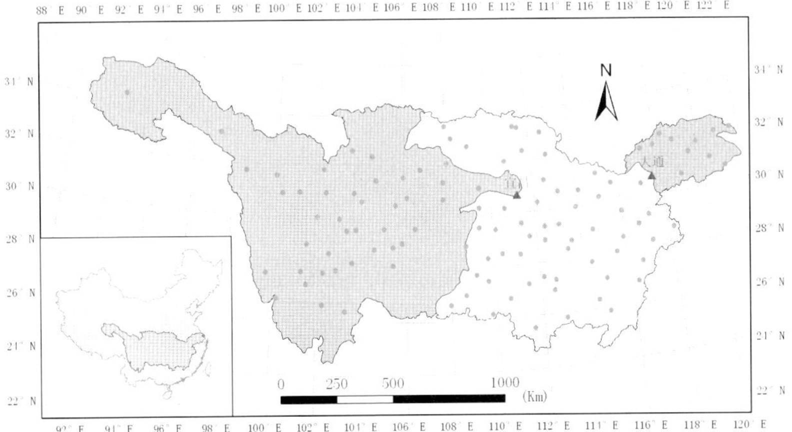

Fig. 1 The Yangtze River Basin and distribution the weather and hydrological stations. The dots represent 118 weather stations and the two triangles represent the two hydrological stations.

Table 1 Information of the 22 climate models of IPCC AR4 and boundary configuration for the Yangtze River Basin.

Number	Model	Country	Grid resolution /°	Longitude range /°	Latitude range /°	Grid number
1	BCCR_BCM2.0	Norway	2.81×2.79	87.18–118.12	20.93–37.67	35
2	CCCMA_CGCM3.1 T47(med-res)	Canada	3.75×3.71	86.25–120	20.41–38.96	22
3	CNRM_CM3	France	2.81×2.79	87.18–118.12	20.93–37.67	35
4	CSIRO_Mk3.0	Australia	1.88×1.87	88.12–120	23.31–38.23	60
5	MIUB_ECHO-G	Germany	3.75×3.71	86.25–120	20.41–38.96	22
6	LASG_FGOALS-g1.0	China	2.81×2.79	87.18–118.12	20.93–37.67	35
7	GFDL_CM2.0	USA	2.50×2.00	88.75–118.75	23–37	46
8	GFDL_CM2.1	USA	2.50×2.00	88.75–118.75	23–37	46
9	GISS_AOM	USA	4.00×3.00	86–118	22.5–37.5	21
10	GISS_E-H	USA	5.00×4.00	87.5–117.5	22–38	18
11	GISS_E-R	USA	5.00×4.00	87.5–117.5	22–38	18
12	UKMO_HadCM3	UK	3.75×2.50	86.25–120	22.5–37.5	26
13	UKMO_HadGEM1	UK	1.88×1.25	88.125–120	23.75–37.5	83
14	INM_CM3.0	Russia	5.00×4.00	85–115	20–40	18
15	INGV_SXG 2005	Italy	1.13×1.12	88.87–120.37	22.99–36.44	146
16	IPSL_CM4	France	3.75×2.54	86.25–120	21.55–39.29	29
17	NIES_MIROC3.2 hires	Japan	1.13×1.12	88.87–120.37	22.99–36.44	139
18	NIES_MIROC3.2 medres	Japan	2.81×2.79	87.18–118.12	20.93–37.67	35
19	MPI-M_ECHAM5-OM	Germany	1.88×1.87	88.12–120	23.31–38.23	60
20	MRI_CGCM2.3.2	Japan	2.81×2.79	87.18–118.12	20.93–37.67	35
21	NCAR_CCSM3	USA	1.41×1.40	88.59–120.93	23.11–37.12	99
22	NCAR_PCM	USA	2.81×2.79	87.18–118.12	20.92–37.67	35

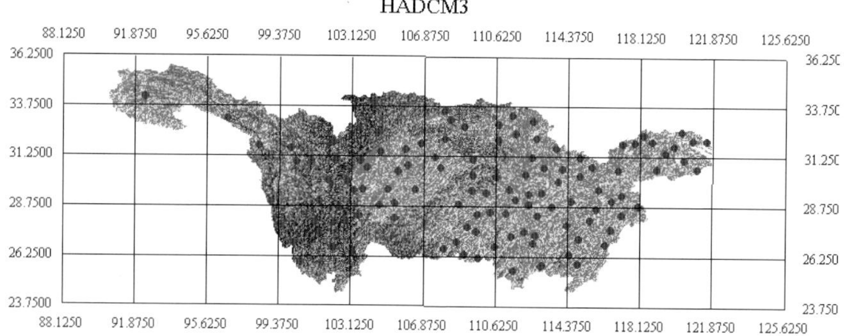

Fig. 2 Grid patterns of the UKMO_HadCM3 Model on Yangtze River Basin, dots present the 118 weather stations.

2.2 Evaluation of the model performance

According to the original mesh, monthly temperature and precipitation data of the GCMs has been calculated, using the geometric average method for the reference period 1961–1990. Relative error, absolute error, correlation coefficient, and coefficient of determination (Ju, 2009) have been chosen as the performance evaluation index. The average temperature of the Yangtze River Basin in the reference period is 15°C, and the average precipitation is 1186.2 mm. The simulated temperature and precipitation are compared with the observation values at the 118 weather stations. The results show that all model simulations underestimate temperature (Table 2). Each model's correlation coefficients are above 0.9 and their deterministic coefficients are low.

Table 2 Comparison between simulated temperature and precipitation with station observations: temp = temperature, Ppt = precipitation (after Wetherald & Manabe, 2002).

No.	Model	Analogue value Temp (°C)	Analogue value Ppt (mm)	Absolute error of Temp	Relative error of Ppt	Correlation coefficient Temp	Correlation coefficient Ppt	Coefficient of determination Temp	Coefficient of determination Ppt
1	BCCR_BCM2.0	8.5	1324.8	–6.5	11.7	0.986	0.836	0.256	0.609
2	CCCMA_CGCM3.1 T47 (med-res)	10.2	1135.3	–4.8	–4.3	0.979	0.813	0.559	0.591
3	CNRM_CM3	9.2	1289.5	–5.7	8.7	0.978	0.823	0.387	0.534
4	CSIRO_Mk3.0	7.4	850.6	–7.5	–28.3	0.976	0.836	–0.041	0.468
5	MIUB_ECHO-G	–8.0	463.4	–22.9	–60.9	0.983	0.799	–8.926	–0.671
6	LASG_FGOALS-g1.0	–6.0	585.8	–21.0	–50.6	0.975	0.557	–8.002	–0.544
7	GFDL_CM2.0	6.9	1048.7	–8.0	–11.6	0.975	0.854	–0.159	0.675
8	GFDL_CM2.1	8.3	1089.9	–6.7	–8.1	0.972	0.852	0.176	0.627
9	GISS_AOM	–6.6	591.5	–21.6	–50.1	0.978	0.827	–7.857	–0.205
10	GISS_E-H	–7.4	315.9	–22.3	–73.4	0.975	0.695	–9.051	–1.312
11	GISS_E-R	–8.0	286.1	–23.0	–75.9	0.968	0.645	–9.38	–1.462
12	UKMO_HadCM3	10.0	1241.2	–5.0	4.6	0.981	0.873	0.54	0.714
13	UKMO_HadGEM1	7.9	1443.8	–7.1	21.7	0.986	0.827	0.076	0.29
14	INM_CM3.0	–4.7	558.5	–19.7	–52.9	0.967	0.75	–6.57	–0.396
15	INGV_SXG 2005	–3.6	583.0	–18.6	–50.9	0.977	0.734	–6.001	–0.367
16	IPSL_CM4	–10.1	476.8	–25.1	–59.8	0.975	0.808	–11.214	–0.601
17	NIES_MIROC3.2 hires	–5.9	475.1	–20.9	–59.9	0.978	0.751	–7.705	–0.613
18	NIES_MIROC3.2 medres	–6.8	525.1	–21.8	–55.7	0.980	0.788	–8.354	–0.395
19	MPI-M_ECHAM5-OM	–6.5	598.9	–21.4	–49.5	0.979	0.814	–7.708	–0.146
20	MRI_CGCM2.3.2	10.9	1106.6	–4.0	–6.7	0.981	0.853	0.676	0.712
21	NCAR_CCSM3	9.4	961.0	–5.6	–19.0	0.986	0.847	0.432	0.605
22	NCAR_PCM	9.5	1233.8	–5.5	4.0	0.980	0.867	0.436	0.735

Comparisons of simulations of average temperature changes (1961–1990) in the Yangtze River basin with the observed temperature (Fig. 3(a)) indicate that the annual cycle or seasonal cycle of all models can be divided into two groups. One group with positive coefficients of determination is consistent with the measured data (Table 2). The other group with negative coefficients of determination are greatly different (with a large variation). Comparisons between simulated precipitation and the observations also show larger coefficients of determination.

The coefficients of determination of both temperature and precipitation greater than 0 are selected as the standard in choosing the model. Nine models, such as BCCR_BCM2.0, have been chosen. From Table 1, it can be seen that in these nine models, only UKMO_HadGEM1 model's relative error of precipitation simulation is higher than 20%. The relative errors of the other models are all less than 20%. The correlation coefficients are all above 0.8. Therefore, the UKMO_HadGEM1 model will be excluded. The temperature and precipitation forecasts for 2010–2099 of the following 8 climate models, i.e. BCCR_BCM2.0, CCCMA_CGCM3.1 T47(med-res), CNRM_CM3, GFDL_CM2.1, UKMO_HadCM3, MRI_CGCM2.3.2, NCAR_CCSM3 and NCAR_PCM, under three typical emission scenarios are chosen, in order to evaluate streamflow changes at the Yichang and Datong gauges in the Yangtze River Basin.

3 PREDICTION OF TEMPERATURE AND PRECIPITATION

Assessment of the eight models for annual precipitation and temperature of the Yangtze River Basin in the 21st century are shown in Table 3. Most models indicate increases, except the GFDL_CM2.1. The GFDL_CM2.1 predicts precipitation decreases in for A1B and B2 scenarios. In the A2 scenario, the proportion of increase is highest, followed by the A1B, while the B1 scenario has the smallest increase, and the UKMO_HadCM3 has the maximum increase. All the models show increase in temperature under the three scenarios. In the A1B scenario, the average temperature will increase by 3.2°C/100 year, which is slightly below the A2. The B1 scenario has

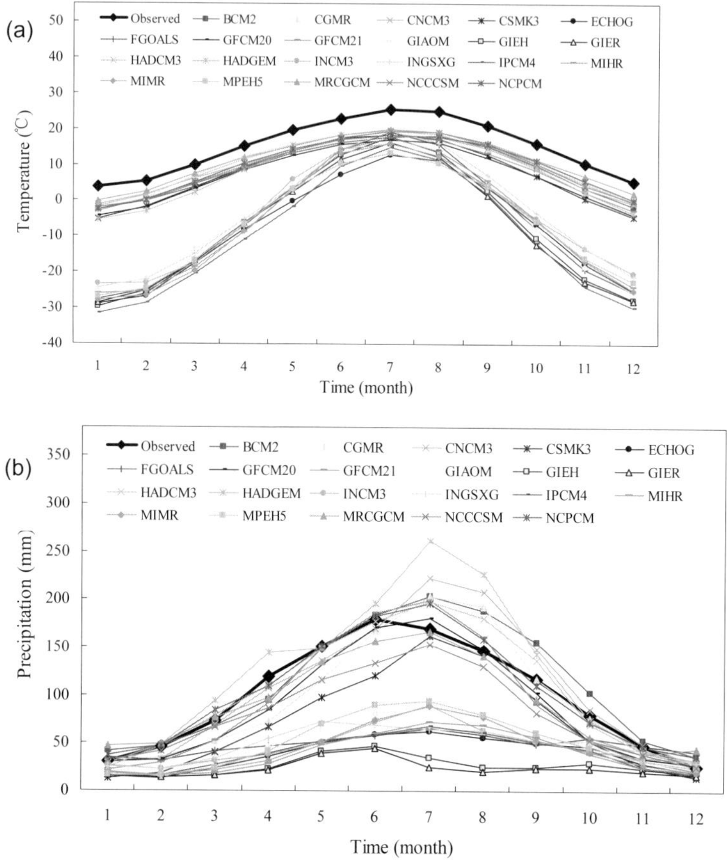

Fig. 3 Comparison between simulation and observation for temperatures (a), and precipitation (b) of the Yangtze River Basin during 1961–1990.

Table 3 Linear trend of temperature and precipitation from GCM (2000–2099).

Scenario	Precipitation (mm/100 year)			Temperature (°C/100 year)		
Model	A2	A1B	B1	A2	A1B	B1
BCCR_BCM2.0	/	181.3	79.9	3.1	2.8	1.4
CCCMA_CGCM3.1 T47(med-res)	/	106.9	/	/	2.6	/
CNRM_CM3	2.5	38.5	/	3.9	3.1	1.5
GFDL_CM2.1	198.9	–18.5	–19.9	3.3	3.7	2.2
UKMO_HadCM3	269.9	214.9	154.1	5.1	4.	3.0
MRI_CGCM2.3.2	154.8	142.6	95.9	3.7	3.9	2.3
NCAR_CCSM3	152.9	150.7	65.9	4.3	3.0	1.3
NCAR_PCM	34.3	73.9	/	2.6	2.4	/
Average	135.6	111.3	75.2	3.7	3.2	2.0

the smallest warming rate (2.0°C/100-years). Except for five scenarios with no information or incomplete information, the future 19 climate changes scenarios of the Yangtze River Basin are listed in Table 3.

4 CONCLUSION

This paper compares precipitation and temperature simulations for the Yangtze River Basin among 22 global climate models used in the Fourth Assessment of the IPCC. GCM simulations have been assessed by three evaluation indexes: error, correlation coefficient and coefficient of determination. Comparions between the simulated and measured values of the 22 climate models during 1961–1999 suggest that eight models, such as BCCR_BCM2.0, have certain advantages in reproducing temperature and precipitation over the Yangtze River Basin. Model prediction of the future climate indicates that both precipitation and temperature will increase. Based on this analysis, GCM simulations suitable for the Yangtze River have been selected. We will use the GCM temperature and precipitation forecasts for 2010–2099 under the three typical emission scenarios to combine with a hydrological model. This will allow us to forecast basin water resources changes in the future.

Acknowledgements This work is supported by the National Basic Research Program of China (2010CB951101); the National Natural Science Foundation of China (40830639, 50879016); the Special Fund of State Key Laboratory of Hydrology-Water Resources and Hydraulic Engineering(1069-50985512). In addition, we would like to thank the reviewer, for valuable comments and suggestions to improve this paper.

REFERENCES

Chiew, F. S. & McMahon, T. A. (2002) Modelling the impacts of climate change on Australian streamflow. *Hydrol. Processes* **16**(6), 1235–1245.

Christensen, N. S. & Lettenmaier, D. P. (2007) A multimodel ensemble approach to assessment of climate change impacts on the hydrology and water resources of the Colorado River Basin. *Hydrol. Earth. Syst. Sci.* **11**(4), 1417–1434.

Edwin, P. M. (2007) Uncertainty in hydrologic impacts of climate change in the Sierra Nevada, California, under two emissions scenarios. *Climatic Change* **82**(3-4), 309–325.

Ellis, A. W., Hawkins, T. W., Balling, R. C. & Gober, P. (2008) Estimating future runoff levels for a semi-arid fluvial system of central Arizona, USA. *Climate Res.* **35**, 227–239.

Feng, M., Ji, C., Wang, L., Wu, Y. & Wang, S. (2006) Climate change and its effect on hydrology and water resources of the Yangtze River of Hubei. *J. Wuhan. University* **39**(1), 1–5.

Hao, Z. & Su, F. (2000) The study of distributed monthly hydrological model and its application in Huaihe River. *The Progress in Water Science* **11**(supp), 36–43.

Hao, Z., Wang, J., Li, L., Wang, Z. & Wang, L. (2006) Climate change influence water resources of The Yellow River. *J. Glaciology and Geocryology* **28**(1), 1–7.

Hao, Z., Ju, Q., Yu, Z., Wang, L. & Jiang, W. (2010) The IPCC AR4 climate model was evaluated the simulating performance of the temperature and precipitation in The Yangtze River Basin and pre-estimated for the future situation. *Quaternary Res.* **30**(1), 127–137.

IPCC Report AR4 (2007) *Climate Change 2007, The Physical Sciences Basis*. Cambridge University Press, UK.

Jia, Y., Gao, H., Niu, C. & Qiu, Y. (2008) Climate change influence runoff processes in Yellow River. *J. Hydraul. Eng.* **39**(1), 52–58.

Ju, Q. (2009) The Yangtze River Basin climate change and study response of water circulation. PhD Thesis. Hohai University, Nanjing, China.

Nash, L. & Gleick, P. H. (1991) Sensitivity of streamflow in the. Colorado Basin to climate change. *J. Hydrol.* **125**, 221–241.

Nemec, J. & Schaake, J. (1982) Sensitivity of water resources systems to climate variation. *Science* **27**(3), 327–343.

Nijssen, B., O'Donnell, G. M., Hamlet, A. F. & Lettenmaier, D. P. (2001) Hydrologic sensitivity of global rivers to climate change. *Climatic Change* **50**, 143–175.

Ozkul, S. (2009) Assessment of climate change effects in Aegean river basins: the case of Gediz and Buyuk Menderes Basins. *Climatic Change* **97**(1-2), 253–283.

Wang, G., Wang, Y. & Kang, L. (2002) The sensitivity analysis runoff on climate change at middle and upper reaches of the Yellow River. *J. Appl. Met.* **13**(1), 117–121.

Wetherald, R. T. & Manabe, S. (2002) Simulation of hydrologic changes associated with global warming. *J. Geophys. Res.* **107**(D19), 4379, doi:10.1029/2001JD001195.

Zhang, J. & Wang, G. (2007) *The Study for Climate Change Impact on Hydrology and Water Resources*. Science Press, Beijing, China.

Zhu, L. & Zhang, W. (2005) Based on the runoff simulated of the water resources at upper reaches of the Hanjing River. *Resource Science* **27**(2), 16–22.

4 Groundwater

Stopping runaway wells in permafrost: the cryogenic freezeback method

D. M. FILLER[1] & R. PETERSON[2]

1 *Department of Civil and Environmental Engineering, University of Alaska Fairbanks, Fairbanks, Alaska, USA*
dmfiller@alaska.edu

2 *Department of Mechanical Engineering, University of Alaska Fairbanks, Fairbanks, Alaska, USA*

Abstract Artesian wells are often encountered in permafrost valleys where aquifer pressures beneath confining sub-permafrost vary between 135 and 1035 kPa (20 to 150 psi). These wells must be heated to prevent freeze-up. However, there are no standards for well heating in North America, and overheating can thaw the permafrost around the casing and lead to loss of control of the well. Further, Arctic warming may be playing a role in the increased frequency of occurrence of uncontrolled wells. With runaway wells, impacts to property and infrastructure can be catastrophic, and the costs to regain control of the well and mitigate damages high. Methods to regain control of artesian wells in permafrost are not well developed and are risky. A new method, cryogenic freezeback with liquid nitrogen, was successfully used to mitigate a runaway artesian well in a permafrost valley north of Fairbanks, Alaska. The well was stopped and infrastructure saved and restored to pre-icing conditions for approximately 63% of the insured property value. Three years of heat exchange and thermal monitoring indicate permafrost restoration and permanent freezeback. The event is documented from massive icing, emergency action to save the residence, well mitigation, to damage assessment and foundation restoration. The cryogenic freezeback method is presented complete with seepage and thermal analyses, well conversion, and thermal monitoring data. Remediation costs and lessons learned are summarized.

Key words permafrost hydrology; artesian wells; cryogenic freezeback; climate change

1 INTRODUCTION

There are few documented instances of uncontrolled artesian flows in permafrost in the literature (Muller, 1945; USACE, 1950; Linell, 1973; Péwé, 1982; Wheaton, 1990). Muller (1945) describes the case of groundwater moving downslope under hydrostatic pressure between a thickening seasonal-frost layer and underlying permafrost, until it reaches the thawed zone beneath a heated house and exits upward and through the structure, ultimately encasing the house in ice. USACE (1950) and Linell (1973) document the COE Well, an artesian well that was problematic from 1946 to 1949 at a research station near Fairbanks, Alaska. The reports chronicle attempts to regain control of the well after installation into a sub-permafrost aquifer with frost tubes and passive refrigeration, and observations of surface icing, frost blisters, ground subsidence, and erosion and thermokarst development around the wellhead. Wheaton (1990) documents the infamous runaway Steese Well that was problematic in Fairbanks during 1976–1977, when a test hole drilled into a sub-permafrost aquifer under high pressure resulted in significant residential damage, several law suits, and a remediation cost of ~US $1.2 million (1977 dollars). Perhaps the most salient points in the Wheaton paper relate to the descriptions of the various grouting and passive and mechanical cooling schemes tried over 18 months to stop the flow. North American records suggest few instances of runaway permafrost wells since 1946, that is, until 2005.

Between 2005 and 2010, four instances of runaway artesian wells occurred in rural developments in interior Alaska (Fig. 1). Well overheating is attributed as the cause of two of the well failures, and the cause of a third is still under investigation, as of this writing. However, the fourth *Propwash Well* failed mysteriously after 23 years of normal operation. Through investigation of these wells, cryogenic freezeback was developed as a new cost-effective method for stopping runaway wells in permafrost. The method entails wellhead access and preparation (i.e. removal of heating elements and scale build-up inside the casing), insertion of a constructed evaporator for injection of liquid nitrogen (LN2), cryogenic freezeback of the well over time, and well retrofit with a thermosyphon for long-term ground heat liberation. The Propwash Well experience and Cryogenic Freezeback Method (CFM) are summarized here.

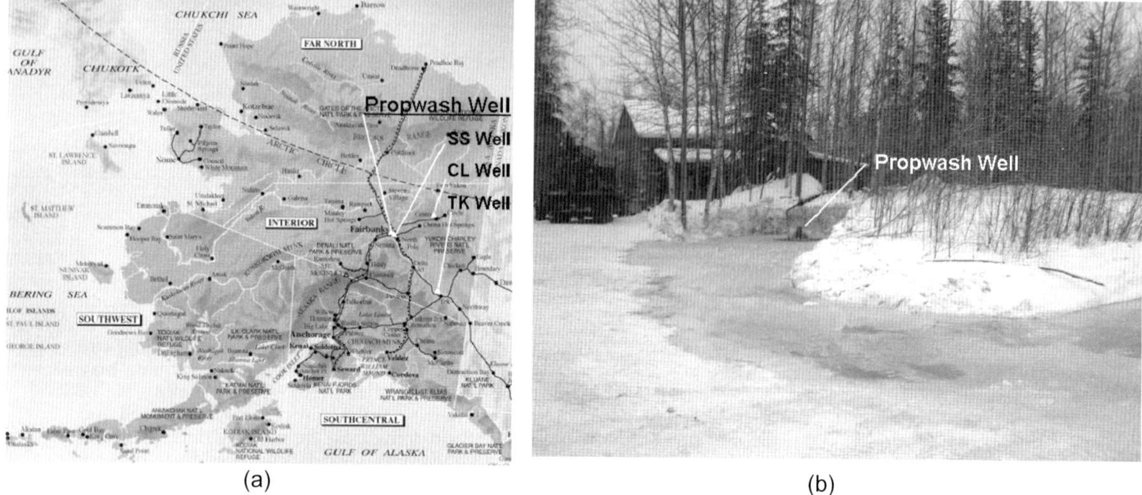

(a) (b)

Fig. 1 (a) Runaway artesian wells in Alaska (2005–2010), and (b) the Propwash Well setting.

2 THE PROPWASH RUNAWAY WELL

The Propwash Well was installed at a Goldstream Valley residence in 1982. Cased through thaw-unstable permafrost into a schistose bedrock artesian aquifer (Table 1), it operated with regulated heat for 23 years before seepage began to emerge from around the wellhead in December 2005. The homeowner called their insurance company, an appraiser inspected and reported 12 days later, and the insurer spent the next 42 days deciding limits of liability and response. All the while, a 9500-m^3 icing developed over two acres of insured property during the winter's deep freeze at –40°C to –45°C air temperatures.

Table 1 Well geology and operating characteristics.

Geology (m below ground surface)		Characteristics	
0–2.1 m	Wet organic silt	Aquifer pressure:	140 kPa (~20 psi)
2.1–5.5 m	Frozen silt	Well flow rate:	11.4 m^3/h (50 gpm)
5.5–8.9 m	Massive ice	Well dimensions:	15.2 cm diameter steel casing
8.9–34.3 m	Permafrost (unsaturated silt with gravel)		– length 46.3 m – screened from 42.6 to 46.3 m
34.3–46.3 m	Schistose bedrock	Well heating: 0–3 m	Regulated heat tape through the seasonal frost layer
		0–46.3 m	Unregulated heat cable

The icing progressed from the wellhead outward and upward in layers over the landscape initially as a function of ground-surface relief, and then according to water movements beneath and over the ice. Once the ice was sufficiently thick to insulate the ground beneath it, seepage water travelled outward under the icing to its perimeters, and upward around tree trunks and through pressure cracks to grow the *naled*. The naled grew to an average height of 1.9 m, reached a maximum thickness of 3.4 m, and created an icescape of ice terraces, frost mounds (or frost blisters), ice domes around trees, and hydraulic pressure ridges and cracks among its many features (Fig. 2). Ultimately, the icing filled the metre-high crawl space (under the house) and encased the home up to within centimetres of the windows, and encroached on the frontage road and two adjacent properties. Although infrastructure impacts were confined to the Propwash property, area well pressures were reduced by as much as 60% in response to a depleted aquifer.

Fig. 2 Icing features.

Emergency corrective action

Work at the Propwash site was conducted in four stages: emergency corrective action, well mitigation, ice abatement, and infrastructure assessment and rehabilitation. Emergency corrective action to save the house was first initiated because the icing had engulfed the heating-oil tank and risen to near window level, water had begun to seep through the floor of the house, and because of concerns for foundation undermining. The water intrusion issue, a consequence of house heating and melting of the crawl space ice, was alleviated by drilling weep-holes through the ice at the corners of the crawl space (Fig. 3(a)). Icing progression at the house was then interrupted with an elaborate system of sloped intercept trenches and heated collection pipes (Fig. 3(b)) that diverted water away from the house to down-slope woodlands. Ice excavation with a frost bucket exposed the wellhead and seepage zone to assess and develop a mitigation strategy.

Fig. 3 Emergency corrective action measures: (a) crawl space drainage, and (b) intercept trench.

3 CRYOGENIC FREEZEBACK METHODOLOGY

Cryogenic freezeback makes use of liquid nitrogen (LN2) at –196°C to create a massive ice plug inside the well, refreeze the seepage zone outside the well casing, and restore the surrounding permafrost. Freezeback design is a function of well, permafrost, and aquifer characteristics (Table 1), seepage and thermal analyses, and determination of LN2 requirements. Site-specific information is critical to the analyses, and accurate determination of LN2 requirements for the desired freezeback. After freezeback, soil erosion zones are grouted and the well retrofit with a thermosyphon to ensure lasting freezeback.

Seepage analysis

Paramount to stopping the runaway well were seepage analysis and a feasibility study of viable well mitigation methods. Seepage analysis required an understanding of the well installation and permafrost characteristics, and development of plausible seepage progression scenarios. Two seepage scenarios were hypothesized for the Propwash Well: (1) narrow thaw along the full length of well casing above bedrock, and (2) conical thaw. Narrow thaw results from over-heating, wherein the well *short-circuited* from prolonged use of the heat cable and resulted in rapid uniform thaw of permafrost around the casing. Conical thaw implies greater permafrost degradation (i.e. thaw and erosion) at the schistose-permafrost interface, having been eroded under the influence of water for the longest time. Conical thaw is not an anthropogenic failure scenario, but rather a consequence of hydrological erosion of basal permafrost, in which intermittent and infrequent well heating may or may not have played a minor contributing role. These seepage erosion scenarios are illustrated in Fig. 4. Conical thaw is the worst case scenario because of its greater extents of permafrost degradation.

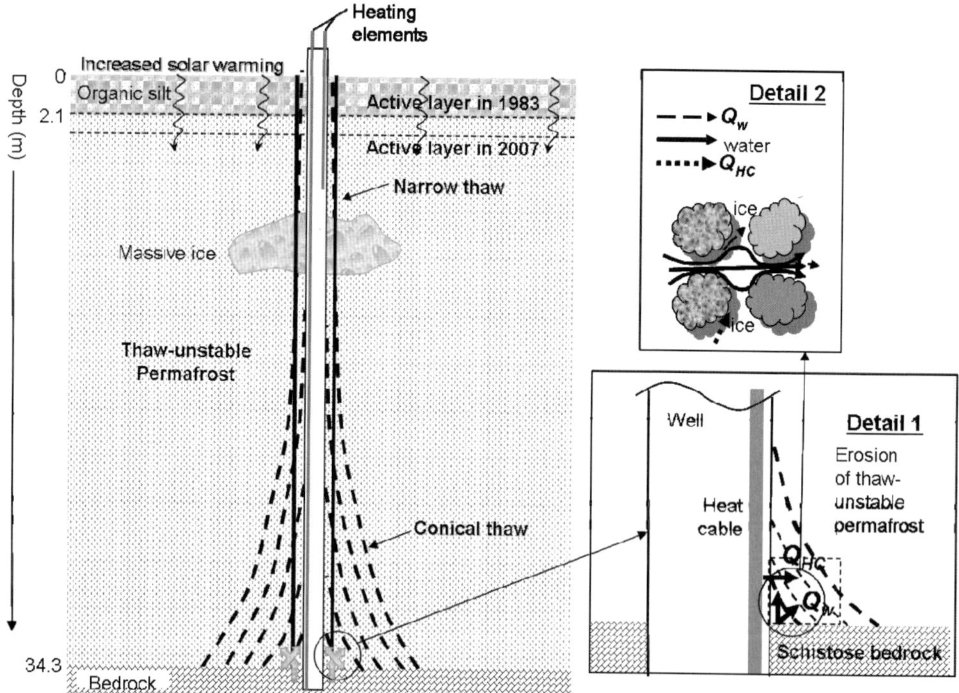

Fig. 4 Well seepage erosion scenarios. Detail 1 depicts the inception of conical thaw, and Detail 2 the influence of water sensible heat and that from the heat cable to thaw erosion of permafrost.

Four methods were evaluated based on potential for success, availability or resources, and cost. Top-down and bottom-up grouting methods were deemed too risky for the conical thaw scenario; chance of seepage re-emergence was too great. Passive refrigeration using freeze pipes, a proven method for well freezeback, was too expensive given that resources were not locally available at the time. Cryogenic freezeback, with its moderate cost and potential high chance of success, was chosen as most resources were locally available. CFM was developed and implemented with a multi-disciplinary team of engineers and an expert well installer. Freezeback success would hinge upon the critical thermal analysis.

Thermal analysis

An energy-balance analysis across the well-permafrost regime was performed to determine the well's heat removal rate, soil thaw radii, and freezeback parameters (i.e. LN2 requirements and freezeback radius). First, analysis of the phase change interface between frozen and thawed soil as a function of heat conduction and rate of thaw from the well was considered. For a well in permafrost, the energy balance can be expressed in cylindrical coordinates as:

$$q = \frac{2\pi k(T_w - T_r)}{\ln(r/r_w)} = 2\pi r L \frac{dr}{dt} \tag{1}$$

where q is the energy transferred (or heat conducted), T_w and T_r are the respective temperatures at the well casing and soil freeze-thaw interface, r_w and r are the outside radius of the well and the soil thaw radius, k is soil thermal conductivity (the inverse of thermal resistance), and L is volumetric latent heat (Freitag & McFadden, 1997). Sensible heat is neglected since the Stefan number for freezing soil is small, and the heat tape temperature is constant at r_w when turned on. By rearranging and integrating both sides of the energy balance equation, we obtain an expression for thaw radii as a function of the freezing index, I_f, (Andersland & Ladanyi, 2004):

$$I_f = \int (T_w - T_r)dt = \frac{L}{k}\left\{\frac{r^2}{2}\ln(r/r_w) - \left(\frac{r^2 - r_w^2}{4}\right)\right\} \tag{2}$$

For the Propwash Well, the confining layer above bedrock was frozen Fairbanks silt with massive ice. An average value of thermal conductivity for Fairbanks silt is 1.1 W/m/°C, and the volumetric latent heat of Fairbanks silt and pure water/ice are 150 000 kJ/m^3 and 333 000 kJ/m^3, respectively. The equation was then used to plot thaw radius as a function of the freezing index (Fig. 5) for freezeback design. From this plot we see that as the freezing index increases so does the thaw radius, which implies that the longer thawing occurs the more extensive is permafrost

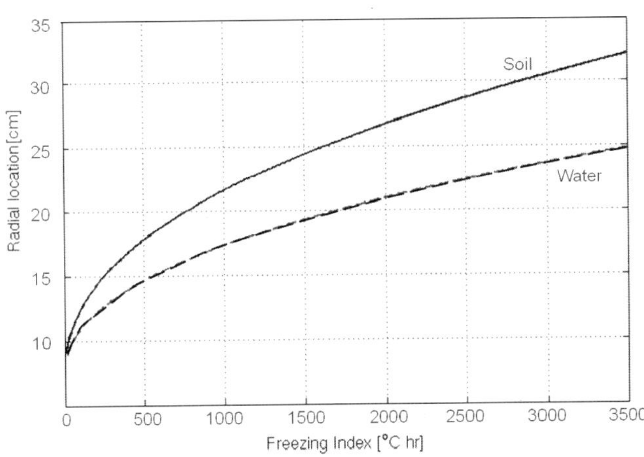

Fig. 5 Permafrost thaw radius as a function of freezing index.

degradation. Further, a larger thaw radius results from warmer seepage water. For example, the difference between 2°C and 4°C water seeping up the well casing for four weeks is an increase in soil thaw radius from about 23 cm to 29 cm.

Next, determination of seepage parameters and observations of the seepage area at the wellhead were required for conservative freezeback design. The seepage rate was estimated at ~1 kg/s (~15 gpm) and the water temperature measured at 2.5°C. The freezing index for 54 days of seepage flow (the time from seepage inception at the ground surface to implementation of mitigation measures) was calculated at 3240°C-h, resulting in a minimum desired freezeback radius of 25 cm.

Finally, LN2 requirements were then computed based on heat removal rates of water and Fairbanks silt. The heat removal rate for flowing water was quantified from $q_w = m_w C_p \Delta T$ as 10.5 kJ/s. The energy required to re-freeze soil to the desired radius of 32 cm, from the ground surface down to 11.6 m (the bottom of the LN2 evaporator), was calculated from $E = \pi(r^2 - r_w^2)HL$ as 530 MJ, or a rate of 6.1 kJ/s for 24 h. Thus, the total heat removal rate required for ground freezing was about 17 kJ/s, and the heat removal rate provided by LN2, as the product of the average LN2 injection rate (0.1 kg/s) and nitrogen's vaporization energy (~200 kJ/kg), was 20 kJ/s. With consideration for subsurface heterogeneity, uncertainties about the soil thaw and freezeback profiles, and potential for seepage re-emergence, a factor of safety of 4 was incorporated for freezeback design. The desired freezeback radius was achieved with 10 400 kg of LN2 injected over 24 h.

Well preparation, freezeback and monitoring

Well preparation required de-icing, removal of obstructions, and water management. With aged wells, scale (i.e. build-up of mineral precipitates) on the inside casing walls is common. Scale was removed from the casing's upper reach with a metal brush; the LN2 evaporator was used to further detach scale at greater depths to accommodate its installation. Water management during the preparation work was accomplished with a sump pump and heated over-the-ice pipeline to drain the wellhead reservoir (Fig. 1(b)) to downslope woodlands. Subsequent to de-icing, removal of the heating elements and scale, and water management implementation, the well was ready for permanent installation of the evaporator.

The evaporator was a discharge cylinder for LN2 injection. Designed to optimize freezeback effectiveness, important design parameters are length and diameter, clearance, and well seating. Constructed as a 10-cm diameter and 13-m long steel cylinder, welded shut at the bottom, the diameter afforded clearance to bypass residual scale and a bend in the well casing, and was sufficient to accommodate a thermosyphon. The evaporator was sufficiently long to span the active layer and massive ice (Fig. 4) zone. Once properly seated on the wellhead with a thick rubber doughnut gasket, the well's discharge flow rate was sufficiently diverted to prevent interference with LN2 discharge (Fig. 6(a)).

A cryogenic tanker regulated LN2 discharge into the evaporator through a 2.5-cm diameter copper *tremmie* tube (Fig. 6(b)), with its end suspended two metres above the bottom of the evaporator to accommodate LN2 boiling, which occurs at –196°C. The LN2 discharge rate, initially set at 1350 kg/h to quickly freeze-off the well flow, was optimized to a constant rate for uniform discharge over 24 h. The thermal gradient across the evaporator opening with the atmosphere was monitored to ensure uniform freezeback with depth.

Once freezeback was complete and the evaporator frozen in place inside the well, the annulus between the evaporator and well casing was filled with a thermal bentonite-grout seal. A thermosyphon 8.9-cm in diameter and 11.6-m long, charged with CO_2, and fitted with thermistors and a data logger was then inserted and sealed in the evaporator with a custom low-temperature thermal grout. The well was abandoned and converted to a heat exchanger to enhance freezeback over the next few years (Fig. 6(c)). The data logger provided a continuous hourly record of ambient air and the temperature at the base of the evaporator inside the abandoned well.

Fig. 6 Cryogenic freezeback and well conversion: (a) evaporator seated in well, (b) LN2 truck and tremmie tube set for LN2 discharge, and (c) thermosyphon installed.

4 DAMAGES, RESTORATION AND MITIGATION COSTS

Damage to infrastructure was largely confined to the home's foundation. The house was supported by 34 independent post-and-pier foundations that formed an open-air crawl space beneath the structure. (It is common practice to raise a house above the frozen ground to preserve underlying permafrost.) The concrete piers of various sizes supported stacked timbers or single square or circular posts. Some posts were anchored to their piers; others were not. Depending on configuration and anchoring, impacted posts and piers experienced shove, tilt and/or rotation, separation and lift, or frost heave and collapse. Because the foundation system was completely encased in ice, extreme care was given to foundation stability during ice abatement to minimize potential for additional distress to the house. Temporary shoring replaced toppled foundations before comprehensive foundation rehabilitation with adjustable foundations. Settling was monitored for two years and the new foundations adjusted accordingly as supersaturated ground drained and resettled.

Several mitigation strategies were considered to stop the runaway well. These included bottom-up grouting, top-down grouting, passive refrigeration using freeze pipes, and cryogenic freezeback with LN2. Evaluation criteria used for decision making were cost, chance of success, and consequences. Although CFM had the highest cost at \$150 000, the method offered the greatest chance of success. Evaluation criteria and implementation costs for the various mitigation strategies considered are summarized in Table 2.

The total cost to stop the Propwash Well and restore the house and grounds to pre-icing conditions was approx. US \$220 000, or about 63% of the insured property value. This includes \$25 000 for emergency corrective action, \$150 000 for well preparation, cryogenic freezeback, and well retrofit, and \$45 000 for ice abatement and foundation assessment and rehabilitation. Emergency corrective action limited structural damage to the home's foundation system. Other infrastructure losses were the well and septic system. The well was replaced with an above-ground insulated holding-tank water supply system. The underground septic was abandoned and replaced with an above-ground, arctic-grade residential wastewater treatment plant.

Table 2 Feasibility study of Propwash Well mitigation strategies.

Strategy/ Method	Chance of success	Cost to implement (US $)*	Consequences
Do nothing	Low	0	Loss of house and other infra-structure; decreased property value; high potential for extraneous liability.
Bottom-up grouting	Low	65 000	Injection rate must be >> aquifer pressure; potential for grout dissociation; entire seepage zone may not be sealed; no permafrost restoration potential; high potential for seepage reoccurrence.
Top-down grouting	Low to moderate	45 000	Injection holes could collapse; possible formation of alternate seepage pathways and blowout (method has a history of blowout failure); no permafrost restoration potential; high potential for seepage re-occurrence.
Passive refrigeration	Moderate	95 000	Requires multiple freeze pipes; limited penetration depth; well abandoned; expertise and equipment not locally available. How long will it last?
Cryogenic freezeback (CFM)	Moderate to high	150 000	Sufficient freezeback radius and depth; 3-yr low-maintenance enhanced freezeback; some permafrost restoration; greatest potential for permanent solution; well abandoned.

*Includes planning and all labour, equipment and material costs through implementation.

Potential changes in permafrost hydrology

Of the three runaway wells investigated in Alaska since 2005, only with the Propwash Well can we attribute loss of control to natural causes with confidence. The Propwash Well homeowner was adamant that the well was operated consistently as designed, and that without exception, the heat cable had not been used during the three years prior to the failure. Further, the shorter heat tape had influence only through the seasonal frost realm. The presumption that well heating was not a significant factor in the Propwash Well failure bears the question: Could climate change be a factor in permafrost well failures?

The authors sought regional and local data to investigate potential influence of climate change on local hydrogeological regimes, permafrost hydrology in particular. Two data sets relevant to the Propwash site are an unbroken 100-year record of meteorological data for Fairbanks (Wendler & Shulski, 2009), and a 27-year uninterrupted record of permafrost temperatures along the International Geosphere–Biosphere Program Alaskan transect, a north–south transect of permafrost monitoring stations from Prudhoe Bay to Gulkana. The Propwash Well site lies along this transect, between Livengood and Gulkana, in interior Alaska.

The Fairbanks meteorological data set includes a time series of mean annual air temperature (MAAT) in Fairbanks from 1906 to 2006. This data shows that the MAAT rose 1.4°C over the century (compared to 0.8°C worldwide), the last three decades combined have on average the highest temperature of the record, and winter monthly changes in MAAT were greatest for December and January (2.4°C and 2.6°C, respectively). Further, the number of days with air temperatures less than –40°C decreased on average from 14 to 8 days annually, and warm days with temperatures above 26.7°C increased from 11 to 12 days. Finally, the length of the growing season, which is the time period when the air temperature in summer never dips below the freezing point, increased from 85 to 123 days (a 45% increase) over the century, with an earlier spring and a later autumn contributing about equally to the overall increase. However, an 11% decrease in annual precipitation and corresponding decrease in winter precipitation since 1916 does not help us to understand the hydrological changes that may have occurred.

Evidence of warming and thawing of discontinuous permafrost has been measured in Alaska since the 1980s. Estimates of the magnitude of warming at the top of permafrost in interior Alaska since the mid-1980s range from 0.5°C to 1.5°C (Osterkamp & Romanovsky, 1999; Osterkamp, 2008; Brown & Romanovsky, 2008). Based on the prevailing warming trend, time scales on the order of a century to thaw the top 10 m of ice-rich permafrost, and an order of magnitude smaller at the permafrost base have been estimated (Osterkamp & Romanovsky, 1999).

The Fairbanks meteorological data corresponds well with the 0.4 m thickening of the active layer in the vicinity of the Propwash Well. While we can conclude that the permafrost table is degrading and there is a corresponding recharge contribution to the overall hydrological system in the Goldstream Valley, little can be said about the impacts from these changes on the permafrost base. Despite predictions made about climate-change impacts in permafrost regions, i.e. more active recharge and discharge to hydrogeological regimes and warmer aquifers (Michel & VanEverdingen, 2006), the science of permafrost hydrology is not yet sufficiently developed to fully understand permafrost degradation processes, time scales, and geophysical relationships, e.g. topographic to sub-permafrost hydrological links (Woo *et al.*, 2008). Nevertheless, we will have to deal with the impacts as they occur, and sometimes at great expense.

5 CONCLUSION AND DISCUSSION

Previous permafrost well failures resulted in catastrophic losses and great expense to regain their control. The Propwash Well experience demonstrated that CFM can be a viable and cost-effective method to mitigate a runaway artesian well in permafrost with comparatively light damage and property devaluation. Further, the experience attests to the importance of problem recognition and resolution early on. By doing so, the risk of catastrophic loss and mitigation costs can be significantly reduced.

Although artesian wells in permafrost can operate with moderate but continual maintenance for many years, it is clear that uncontrolled seepage outside of the well due to thaw can be catastrophic. There is need for design principles that might aid the preservation of a functioning well while avoiding thermal degradation of permafrost. This type of design begins with a first-order model of the thermal regime surrounding the well.

The steady-state thermal regime outside the casing of an operating well is described by:

$$\frac{\partial T}{\partial t} = 0 = \frac{1}{r}\frac{\partial}{dr}\left(rk\frac{\partial T}{\partial r}\right) \tag{3}$$

The non-zero thermal gradient is maintained by a heat flux from the well into the permafrost, which is supplied by both the liquid water brought from depth and the heat cable. This equation can be solved with the boundary conditions of heat flux (q) and well temperature (T_w) at the inside radius, and constant permafrost temperature (T_r) at some distance r away, resulting in:

$$q = \frac{k}{r_w}\frac{T_w - T_r}{\ln(r/r_w)} \tag{4}$$

Figure 7 shows the heat flux as a function of r, the distance at which the outer thermal boundary is at the permafrost temperature (≈ –1°C). It is reasonable to assume that this distance is on the order of 1 to 2 metres beyond the well, which requires a flux of the order of 5 W/m^2 to maintain. For a 15-cm diameter well, this corresponds to a linear heat flux of 2.4 W/m that is supplied from the heat cable, well water, or combination of both.

This dynamic balance of heat supplied from the well to the permafrost poses the interesting question of where the 0°C isotherm resides in a permafrost well? Clearly it should not be outside of the well casing, because that implies the uncontrolled flow problem discussed here. It must therefore reside somewhere inside the well. (The large thermal conductivity of steel relative to ice and water means it is highly unlikely to reside within the casing itself.) Figure 8 shows the radial liquid/ice interface location as a function of heat flux, with an asterisk at the well casing radius (7.5 cm) and a flux of 4.85 W/m^2 (see Fig. 7).

For this particular situation, the well should be operated with a maximum average heat flux of 4.85 W/m^2 so as not to induce thawing outside of the casing. A slightly lower heat flux leads to a layer of ice forming on the inside of the casing, which is the desired operational point. Most

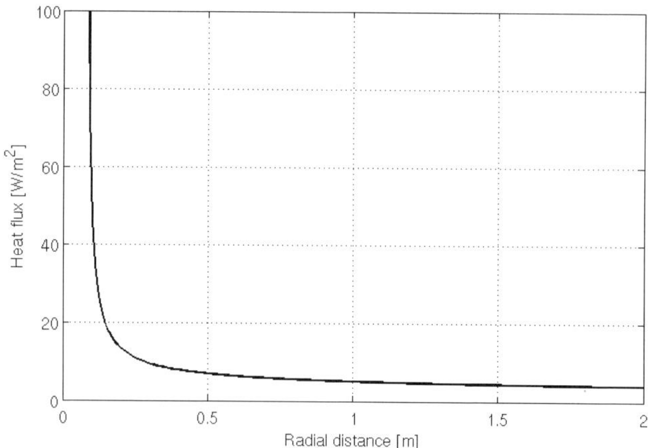

Fig. 7 Heat flux as a function of the radial distance away from a permafrost well.

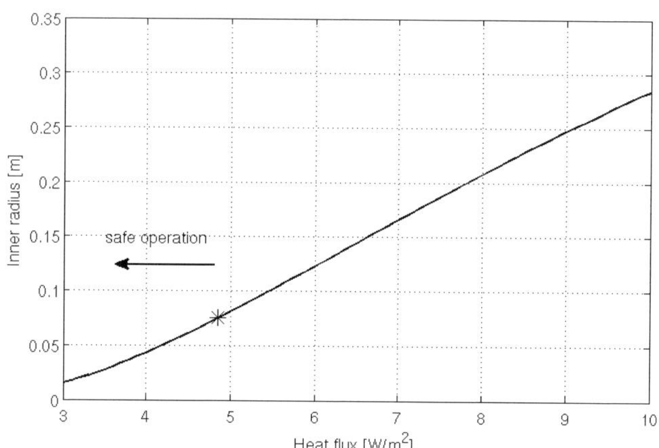

Fig. 8 Safe heating of an artesian well in permafrost.

commercial heat tape provides energy at a higher rate and therefore cycling the heat tape is required. Alternatively, if the energy is provided only by the sensible heat of the well's liquid water, an energy balance indicates a 45-m deep well drawing water at +2°C from depth would need to flow at 91.5 L/h (or 2.2 m^3/d) to prevent well freeze-up. Most residential wells would likely operate somewhere between these two extremes. Fortunately, the large latent heat of water allows for significant fluctuations in the instantaneous heat flux before complete well freezing (or catastrophic thawing) would occur.

Although the quantitative numbers discussed here apply only to this particular situation of permafrost temperature and well diameter, the general design principle should carry over to other artesian wells in permafrost. Maintaining the dynamic energy balance that avoids both complete freeze-up and thaw outside the casing could be assisted by using a double-walled casing with insulation. For this design, it would be easier to maintain the 0°C isotherm either within the well or the double-wall casing. However, the additional material and logistical costs of such an installation are likely a significant disincentive when the probability of well failure is difficult to assess.

REFERENCES

Andersland, O. B. & Ladanyi, B. (2004) *Frozen Ground Engineering*. John Wiley & Sons Inc., Hoboken New Jersey, USA. 363 pp.

Brown, J. & Romanovsky, V. E. (2008) Report from the International Permafrost Association: State of Permafrost in the First Decade of the 21st Century. *Permafrost and Periglacial Processes* **19**, 255–260.

Freitag, D. R. & McFadden, T. (1997) *Introduction to Cold Regions Engineering*. ASCE Press, New York, USA. 738 pp.

Linell, K. A. (1973) Risk of uncontrolled flow from wells through permafrost. *Proc. Second Int. Conf. on Permafrost* (Yakutsk, USSR), North American Contribution, Washington DC, National Academy of Sciences, 462–468.

Michel, F.A. and VanEverdingen, R.O. (2006) Changes in hydrogeologic regimes in permafrost regions due to climate change. *Permafrost and Periglacial Processes* **5**(3), 191–195.

Muller, S. W. (1945) *Permafrost or Permanently Frozen Ground and Related Engineering Problems*. Washington, DC: Office of Chief Engineers, US Army, 231 p. (Lithoprinted 1947, Edwards Bros., Ann Arbor, MI).

Osterkamp, T. E. (2008) Thermal state of permafrost in Alaska. In *Proceedings, Ninth Int. Conf. on Permafrost*, 29 June–3 July, 2008, Fairbanks, Alaska, University of Alaska Fairbanks, Alaska, USA.

Osterkamp, T. E. & Romanovsky, V. E. (1999) Evidence for warming and thawing of discontinuous permafrost in Alaska. *Permafrost and Periglacial Processes* **10**(1), 17–37.

Permafrost Division, United States Army Corps of Engineers (1950) Observations of an uncontrolled artesian well, Field Research Area, Fairbanks, Alaska. US Army Cold Regions Research and Engineering Laboratory Internal Report, prepared by Soils Laboratory, Field Operations Office, Fairbanks, Alaska, 22 p.

Péwé, T. L. (1982) Geologic hazards of the Fairbanks area, Alaska. Alaska Geological & Geophysical Surveys, Special Report 15, 109 p.

Wendler, G. & Shulski, M. (2009) A century of climate change in Fairbanks, Alaska. *Arctic* **32**(3), 295–300.

Wheaton, S. R. (1990) Flowing artesian wells in permafrost regions. In: *Cold Regions Hydrology and Hydraulics* (ed. by W. L. Ryan & R. D. Crissman, 721–737. American Society of Civil Engineers, New York, USA.

Woo, M.-K, Kane, D. L., Carey, S. K. & Yang, D. (2008) Progress in permafrost hydrology in the new millennium. *Permafrost and Periglac. Processes* **19**(2), 237–254.

Numerical simulation of seepage processes in permafrost near a hydro unit

SVET MILANOVSKIY[1], ALEXEY PETRUNIN[1,2], SERGEY VELIKIN[3] & VIATCHESLAV ISTRATOV[4]

1 *Institute Physics of the Earth RAS, 123995, Moscow, Bolshaya Gruzinskaya 10, Russia*
svetmil@mail.ru

2 *GeoForschungsZentrum, Potsdam, Germany*

3 *Vilyui Permafrost Station of the Permafrost Institute RAS Siberian Branch, Chernishevskii, Russia*

4 *Radionda Ltd Company, Moscow, Russia*

Abstract In the territory of Western Yakutia, during the last 20 years, complex geophysical monitoring of hydraulic engineering units has been applied. Alongside field studies, numerical evaluation of permeable talik zone (thawing) origination and development in a broad zone around a dam was made. The non-steady problem of heat-mass transfer in fractured-porous saturated frozen media, interbedded in frozen impermeable strata is discussed. The model takes into consideration the main conditions causing initiation and development of talik near a reservoir: annual temperature and snow cover variation, seasonal water temperature distribution with depth in the storage basin adjacent to the dam, and evolution of permeability in rock due to thaw-freeze processes. The proposed model can be used to analyse more complex situations.

Key words permafrost; talik; hydro unit; geophysical monitoring; numerical modelling; Western Yakutia

INTRODUCTION

An artificial water reservoir in permafrost creates conditions for talik zone (thawing) formation and development along the shores flanking the reservoir. The stability of the dam and reservoir shores is the key to the safety of reservoirs (such as power station pools, water supply, tailing pits, etc.). Similar problems due to thaw may also occur in natural basins in permafrost regions – climate change may activate lake drainage causing hazardous situations. To avoid water loss and maintain the hydro unit's stability in a permafrost zone we need to use geophysical tools, including long-term monitoring for detecting a talik formation. Along with the required temperature control, different geophysical methods give information about variation of rock physical properties caused by thaw–freeze processes. Integrated geophysical monitoring allows the observation of time–space variations of physical fields, reflecting the evolution of thaw–freeze processes in the dam and the reservoir's flank shores. To better understand the situation of talik formation we have developed a numerical model describing a system that includes a water storage basin (with annual temperature variation), a frozen mass (with vein ice in fractures), snow cover (insulating the ground), and annual temperature variation.

GOALS AND OBJECTIVES OF THE STUDY

The study was conducted in the Aichal-Mirni region of Western Yakutia, a region of a potential development of different types of anthropogenic loads. There are many hydraulic engineering facilities of different function: hydropower installations, reservoirs for water supply, surface and underground storages for technogenic brines, etc. The permafrost rocks condition is very sensitive to natural and human-caused impacts. Consequently numerous thaw–filtration processes have been observed during the last two decades at a number of hydro-technical objects in Western Yakutia at Marha, Irelyah, Vilyui, Anabar, Sitikan, Kieng, Iyraaas-Yuryah (Velikin & Snegirev, 2004). The studies were focused on the detection and location of talik zones and seepage in a dam body and adjacent to reservoirs and dam areas as well as estimating the dynamics of seepage processes. At present, special attention should be focused on climate impacts on permafrost stability and particularly on engineering objects in permafrost zones as the objects are very sensitive "trigger points" for hazardous processes. Similar problems may also occur with natural basins in cold regions – climate change may activate lake drainage and irreversible changes in permafrost (Frauenfeld *et al.*, 2003). Our geophysical surveys (Milanovskiy *et al.*, 2008) were focused on:

I detection and location of inflow seepage near dams and areas adjacent to dams, detection and location of places of the most intensive permafrost thaw and seepage from the reservoir;

II investigation of talik geometry in the frozen mass;

III monitoring the dynamics of the progressive seepage in space and time.

TALIK THERMAL MODELLING

Talik can be formed in permeable soils, and spread over impermeable soils. Its intensity is mostly controlled by the freezing conditions; e.g. if the talik localized under riverbeds, the talik may change to part-permeable type. A specific feature of artesian (head) aquifers is the predominantly upward movement of pressure head in permeable zones (disjunctives, karsts zones in carbonates, fractured permeable zones in rocks). In the case of surface water basins, the pressure headwaters are constrained by the frozen dam with a set of engineering buildings, and including a by-pass channel (spillway) with waterway. This channel often has the role of a permanently-acting heat source in relation to the permafrost adjoining it. It is also possible to guess the presence of fractured water-saturated frozen rock strata (e.g. limestone) the rocks flanking the basin. Under frozen conditions the rocks are impermeable. Under thawing conditions, these strata may potentially become permeable, allowing filtration.

Problem definition

The non-steady problem of heat-mass transfer in a fractured-porous saturated frozen Stratum (II), interbedded in frozen impermeable adjacent strata (I, III) is discussed assuming a pressure head in an aquifer, explaining thawing in the saturated Stratum (II) and considering seasonal variations of temperatures of air and aquifer (Fig. 1)

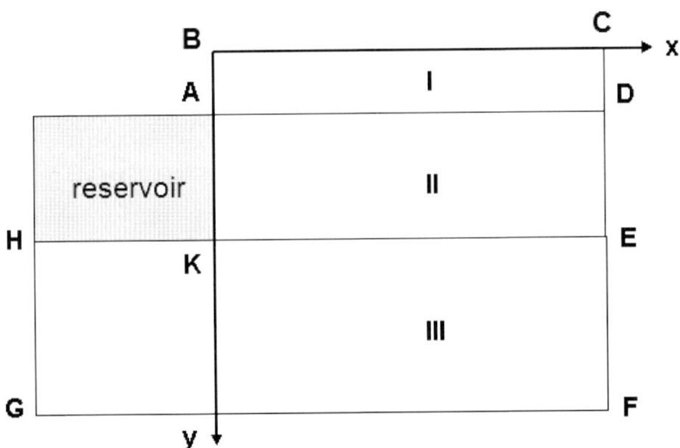

Fig. 1 Schematic geometry of the model.

The following assumptions for simplification of the pattern were made: strata I and III are dry, impermeable, rigid, homogeneous and isotropic, with conductive heat transfer only. It is supposed, that in Stratum II the matrix and fluid (ice) are in a temperature balance (local temperatures of the matrix and fluid are equal at any time), the fluid and matrix are incompressible, the only cause of filtration is water-head pressure, and vertical mass-transfer is neglected whereas the horizontal fluid flow is considered as stationary and incompressible.

Taking into consideration the above-mentioned limitations, the problem can be formulated by a general two-dimensional equation of energy:

$$\left(\rho_m c_{pm} + L_w \Theta \frac{\partial \Theta_u}{\partial T}\right)\frac{\partial T}{\partial t} + \rho_w c_{pw} V_x \frac{\partial T}{\partial x} = \lambda_m \left(\frac{\partial^2 T}{\partial x^2} + \frac{\partial^2 T}{\partial y^2}\right) \quad (1)$$

where ρ, c_p and λ are density, heat capacity and thermal conductivity, respectively (the subscripts m and w refer solid matrix and fluid, respectively), V_x is the Darcy velocity component:

$$V_x = -K \frac{\mathrm{d}H}{\mathrm{d}x} \quad (2)$$

where K is the coefficient of water permeability and H is water head in a saturated layer. Following the limitations described above, the equation of continuity will be written in the form:

$$\frac{\partial V_x}{\partial x} = 0 \quad (3)$$

The term $L_w \Theta(\partial \Theta_u / \partial T)$ on the left of equation (1) describes a heat quantity generated or absorbed by the media at variation of its temperature on ∂T as a result of phase change. Here L_w is the latent heat of melting of ice, Θ is the volumetric content of fluid in soil, and $(\partial \Theta_u / \partial T)$ is variation of the non-frozen water content with temperature variation of the medium (Fig. 2) Newman, (1996).

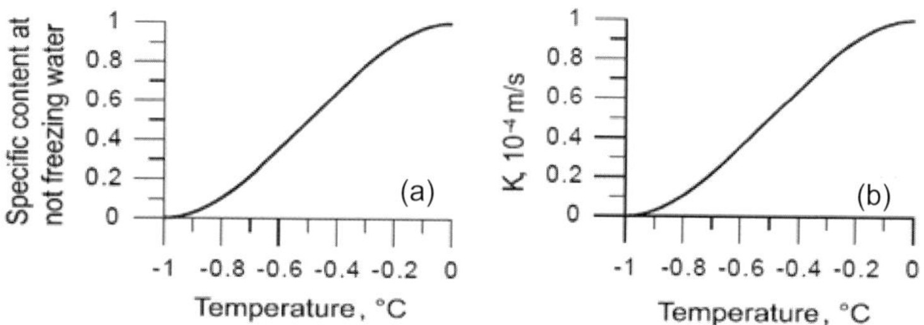

Fig. 2 Variation of the specific contents of non-frozen water (a) and coefficient of permeability K, 10^{-4} m/s (b) as a function of temperature in a zone of phase transfer (–1°C up to 0°C).

Thus, it is possible to prevent numerical instabilities and take into account the fact, that the phase change takes place not instantaneously, but in certain temperature range (in our case from –1°C to 0°C). In strata I and III, the equation (1) can be reduced to a simpler form:

$$\rho_m c_{pm} \frac{\partial T}{\partial t} = \lambda_m \left(\frac{\partial^2 T}{\partial x^2} + \frac{\partial^2 T}{\partial y^2}\right) \quad (4)$$

where values of density, heat capacity and thermal conductivity correspond to the physical properties of strata I and III.

As seen in Table 1, for Stratum II the heightened thermal conductivity relative to the adjacent strata is accepted. Such a stratum can be for example, a layer of fractured limestone or dolomite. Its substrate is the impermeable layer III (clay materials), with thermal properties identical to layer I (loams and other impermeable rocks are typical in the upper part of cross-sections). The

Table 1 Values of some physical parameters used in calculations.

	Thickness (m)	$\rho_m c_{pm}$ (MJ/m^3/K)	λ_m (W/m/K)	K (m/s)
Stratum I	5	1.49	1	0
Stratum II, frozen	10	1.96	3.5	0
Stratum II, thawed	10	2.52	3	10^{-4}
Stratum III	60	1.49	1	0

impermeable layer III underlies permeable Stratum II, bounding the lower level of filtration. Note that the permeability coefficient K of a Stratum II is equal to zero at temperatures below –1°C and gradually increases to magnitude 10^{-4} within the temperature range –1°C to 0°C. Therefore, if at the boundary D–E of Stratum II in a cold season the temperature drops below –1°C, which results, in theory, in impermeability there.

Initial and boundary conditions

At the surfaces (A–B, B–C, C–D, D–E and E–F in Fig. 1) we used the following expression for temperature, T_a (°C); equation (5) describes observed data well and is consistent with theoretical estimations by Shipitcina (1983):

$$T_a = -2° + 17\sin\left(\frac{2\pi}{8760}t_h\right) \tag{5}$$

where t_h is time in hours.

The temperature of an aquifer T_w (°C) at the boundary A–K (Fig. 1) characterizes seasonal fluctuations, related to summer warming, winter cooling, the influencing of cold flood water and periods of rainfall, during which there is active mixing of water in the aquifer resulting in levelling of temperature with depth. A similar seasonal distribution of temperature with depth in an aquifer of a permafrost zone in Canada is given by Lai Yuanming *et al.* (2002). Here the seasonal temperature distribution (T_w) in water storage was described by the expression:

$$T_w = 5° + \left(2.5 + 7.5\sin\left(\frac{2\pi}{8760}t_h\right)\right)\cdot\exp\left(-\frac{h}{2}\right) \tag{6}$$

where h is depth from the surface of a basin in metres and t_h is time in hours. As apparent from expression (6), temperature at a surface of an aquifer has seasonal fluctuations from 0°C to +15°C, which decreases with depth. The boundaries F–G, G–H, H–K are at constant temperature –3°C. It is necessary to note that both the initial value of temperature of the system, and the value of temperature on boundary F–G–H–K do not essentially influence the results. In a numerical experiment for initial and boundary conditions, at the lower boundaries various values down to –10°C were used. Despite that the talik evolution was similar, although this process evolved over the time. It demonstrates that for talik origination, the thermal properties and the geometry of the upper layer are key. At boundaries A–D and K–E the condition of continuity of heat flow (amount of heat input equals heat output) is assumed. A constant water layer in the water storage basin was assumed. Actually it has considerable oscillations, related to natural (flood, high water, rains) and technological (discharge water, off take, water exchange) reasons. An example of seasonal level variations for water storage in northern Quebec is presented by Lai Yuanming *et al.* (2002) that basically can be taken into account in the given formulation by input of a function of head H (t). In our case it was accepted that the drop of levels over Stratum II was fixed and equal to 10 m (base of Stratum II coincides with the bottom of water storage). Initially the system (including layer II) was at –3°C; later on there is an instantaneous infill of the aquifer with temperature T_w, and then the system moves to thermodynamic equilibrium.

To solve the problem, the finite difference approach was applied. The explicit scheme with accuracy $O(h^2 + \tau)$, where h and τ are steps in space and time. The stability was defined by the Courant condition.

RESULTS

We delivered a problem of determination of the conditions of talik origination, and also estimation of the dynamics of progress of a thawed zone in a permeable pressure head layer adjoining an aquifer. The results visually illustrating the process of origination and progress of the talik front were obtained and also the basic factors causing its initiation, were determined. The results show that the process of thermal evolution of frozen strata can conditionally be divided into two basic and two transient stages (Fig. 3). The first stage starts from the moment of infill of the water storage and represents installation of thermodynamic equilibrium in the absence of convection.

Rather quickly, in 2–3 years, a settled quasi-state conductive heat regime develops (Fig. 3), in which seasonal variations of surface temperature quickly decay with depth according to Fourier law, and the summer fire-setting boundary (thawing) migrates a little deeper each summer season.

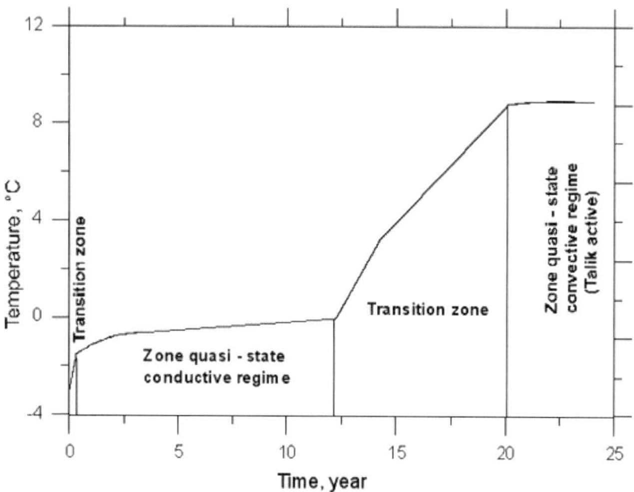

Fig. 3 Variation of seasonal T-maximum at the reference-point situated in the middle of the top of the permeable zone (talik zone) with time. (A–D length on Fig. 1 was 50 m).

Figure 3 shows that for 2–12 years the maximum seasonal temperature of a reference point at the top of the layer (middle of A–D) of Stratum II stays in the limits –1°C to 0°C, i.e. within the temperature limits when the ice filling in pores starts to thaw and the permeability of the environment becomes higher than zero. In this case, water pressure head can produce an advective (convective) heat-mass transfer, which is much more effective than conductive transfer. This is a key factor in our model of talik origination. In this case, the main parameters of a Stratum I – its thickness and thermal diffusivity – and also the average annual temperature of a near-surface stratum are extremely important. According to our numerical experiment, a lowering of average annual temperature of only 2°C results in no talik development at the given thickness of Stratum I. This leads us to the conclusion that for applied problems the basic critical (extreme) thermal and geometric parameters of the environment can be determined by solving a 1-D non-steady problem of heat-mass transfer in half-space with periodic perturbations of temperature on a surface (Turcotte & Schubert, 2002). A lot of complicated models exist for Stratum I, including phase transition for different degrees of water saturation of the ground. We used the simplest model which permitted us to obtain the principal estimations.

The duration of the transient period between the conductive and convective regimes of a system (Fig. 3) depends mainly on the temperature of water in the water storage, permeability coefficient of the thawed rocks and hydro-head. As expected, the frontal movement of talik takes place primarily during summer (warm) months and at first, primarily in the upper zone of Stratum II. Heated water very effectively transfers heat and results in fast talik progress. Over 3–4 summer seasons, the talik starts to occupy all of saturated Stratum II. Then, in quite a short time the system reaches a quasi-state regime with convective heat-transfer dominating. As mentioned above, in a cold period filtration is "locked" in the model; this follows from the K (permeability) value of Stratum II, which equals zero if temperature is lower than –1°C. Actually, filtration may occur in local zones of fractures, which remain permeable even in a cold winter period. This circumstance was taken into account hereinafter to construct a more advanced model that involves the special conditions of permeability at the boundary D–E. This model includes: (a) open crack(s) at the D–E boundary of limited length (5 m) penetrating into the frozen/thawed mass (Stratum II), or (b) the presence of a through-mass narrow permeable channel (crack) in frozen Stratum II. We can observe both special cases. For case (a), after forming a thawed zone we shall have year-round filtration without the winter freezing "lock" in the layer. For case (b), talik development for winter periods with a pre-existing transparent crack in the basement of Stratum II; the filtration process runs much faster. An interesting case is the "domino" or "puff cake" model, including more than one layer with Stratum II properties. If we have an existing water head in the "domino" model the thawing process runs deep into the frozen mass due to a combination of conductive and convective heating. We have also made long-term field observations of the thermal regime and radiowave monitoring of effective electric resistance and relative dielectric transmissivity inside forming talik zone on the shores of Sitikan Reservoir. (Milanovskiy *et al.*, 2008) These data are consistent with these modelling results and illustrate the evolution of the thaw–freeze process in time.

DISCUSSIONS AND CONCLUSIONS

We have carried out numerical modelling of the origination and development of talik in stratified media. According the modelling results, the maximum depth of a zero isotherm in relation to the top of the Stratum II is a key point for talik origination. It is noticeable that the dynamics of heating of the thawed zone is essentially controlled by the permeability coefficient (which depends on properties of the rock matrix) and its temperature (phase state of water). The most effective means to control the progress of filtration (seepage) could be artificial decrease of the permeability coefficient of the saturated layer. This can be obtained by a forced temperature drop in the environment (freezing), or by mechanical reduction of the permeability of the matrix at the boundary A–K using fine-grained clay material, piling with impermeable shields on the infiltration zone of Stratum II, or using special cryogenic liquids – cryogels based on polymer solutions with electrolytes addition. It seems that the second approach is most effective for several reasons: first, cooling a high-permeability pressure head layer needs much of energy, and secondly, the maintenance of strata in the frozen state will require incessant energy consumption. The second approach solves the problem at the diverse quality level and requires much less energy expense compared with freezing. Geophysical data show the non-uniform types (morphology) of talik zones (Milanovskiy *et al.*, 2008); however, it has a common characteristic feature – the presence of a triggering mechanism for talik initialization. For realization of this mechanism we need to know the pre-talik temperature history (for example – put global or regional temperature warming trends into the model), leading to preheating of the strata, existing fracture or pore channels (partly-open or ice cemented) and water head from reservoir. An important factor for the stability of long-existing constructions like hydro units or waste burial in permafrost zones is global climate change. The proposed simple model can be used for more complex situations. We can imagine a situation of similar geometry with alternation of thinner layers with physical properties of Stratum I and Stratum II instead of layer II with water head existing at the layer's boundary. In this case, development of seepage in Stratum II.1 will heat Stratum I.2 and then the domino effect will provide seepage in lower Stratum II.2 until the water head and thermal conditions permit. The real problem is essentially three-dimensional.

Acknowledgements The authors are grateful to an anonymous reviewer for acquaintance with the work and helpful comments.

REFERENCES

Frauenfeld, O. W., Zhang, T., Barry, R. G., Gilichinsky, D. & Etringer, A. J. (2003) Effects of freeze/thaw index, air temperature, and snow cover on seasonal freeze and thaw depths in Russia. AGU, Fall Meeting 2003, abstract #C11A-03 Bibliographic Code: 2003AGUFM.C11A.03F.

Lai Yuanming, Liu Songyu, U Ziwang, Wu Yaping, G & Konrad, J. M. (2002) Numerical simulation for the coupled problem of temperature and seepage fields in cold region dams. *J. Hydraul. Res.* **40**(5), 631–635.

Milanovskiy S., Velikin S. & Istratov V. (2008) Geophysical study of talik zones (Western Yakutia) In: *Proceedings of the Ninth International Conference on Permafrost* (ed. by D. L. Kane & K. M. Hinkel) (University of Alaska, Fairbanks), vol. 2, 1221–1226.

Newman, G. P. (1996) Heat and mass transfer in unsaturated soils during freezing. MSc Thesis, University of Saskatchewan, Saskatoon, Canada.

Shipitcina, L. I. (1983) Calculation of dynamics of a temperature regime of soils under snow taking into account dependence of its response from temperature In: *Thermal Studies of Cryosphere of Siberia.* Novosibirsk, "Science", 187–203.

Turcotte, D. L. & Shubert, G. (2002) *Geodynamics*. Cambridge University Press.

Velikin, S. A. & Snegirev, A. M. (2004) Study of the right-bank contiguity of the dam at the Viluy hydroelectric power plant: complex of techniques. *J. Glaciology and Geocriology* **26**, 142–150.

Modelling the impact of climatic variability on groundwater and surface flows from a mountainous catchment in the Chilean Andes

D. RUELLAND[1], N. BRISSET[2], H. JOURDE[2,3] & R. OYARZUN[3,4]

1 *CNRS*, 2 *UM2 – UMR HydroSciences Montpellier, Place E. Bataillon, 34395 Montpellier Cedex 5, France*
denis.ruelland@um2.fr

3 *CEAZA*, 4 *Departamento Ingeniería de Minas, Colina El Pino, Universidad La Serena, La Serena, Chile*

Abstract This study aims to simulate the relationship between climate forcing and the dynamics of both water table levels and runoff from the upper Elqui catchment (5660 km^2, Chile). Simulations are performed with a daily conceptual model that takes into account: (i) a shallow reservoir supplied by precipitation and feeding evapotranspiration, surface/sub-surface runoff and infiltration, and (ii) a deep reservoir fed by infiltration and generating the baseflow. A third reservoir, in which fluxes are controlled by temperature, has been introduced to account for the snowmelt regime of the catchment. A nearly 30-year period (1977–2008) was chosen to capture long-term hydro-climatic variability due to alternating ENSO and LNSO events. Calibration and validation were performed on the basis of a multi-objective function that aggregates a variety of goodness-of-fit criteria. The model correctly reproduces the observed discharge at the basin outlet, for either lumped or semi-distributed applications. Nash coefficients are about 0.9 over the calibration period (1979–1990) and 0.75 over the validation period (1991–2008). The volume error between observation and simulation is lower than 11% over the whole period studied. The dynamics of both the water level in the deep conceptual reservoir and the water table in a piezometer at the basin outlet are also in good agreement. The model thus provides encouraging simulations of groundwater and surface flows when applied to various climatic conditions. However, improvements are still needed before forecasting water availability using medium-term climatic scenarios.

Key words hydro-climatic variability; hydrological modelling; snowmelt regime; groundwater/surface exchanges; River Elqui, Chile

INTRODUCTION

The Intergovernmental Panel on Climate Change recently published a report confirming global climate change (IPCC, 2007). In South America, this change has resulted in a decrease of precipitation in the Western Andes Cordillera, while an increase is observed in the east. Half of the semi-arid Norte Chico region of Chile (26–33°S) is thus threatened by desertification, according to the United Nations Convention to Combat Desertification (UNCCD, 2006). Mean annual precipitation observed at La Serena has indeed decreased by a factor of two over the 20th century (Squeo *et al.*, 1999). Apart from this general trend, the climate of the region is controlled by alternating ENSO (*El Niño* Southern Oscillation) and LNSO (*La Niña* Southern Oscillation) events that induce significant interannual hydro-climatic variability. While the former generally lead to extreme precipitation across regions on the South Pacific coast, the latter cause significant precipitation deficits (Kane, 1999; McPhaden, 2004). In those areas where agriculture, constantly expanding, holds a predominant place, water resources management has thus become essential to meet current and future supply needs.

Facing the global changes that particularly affect the Andean regions with agricultural activities requires further insight into flow phenomena and improved capacity for water resources forecasting. Hydrological modelling can provide a better understanding of the impact of these changes on water resources by representing the relationship between climatic variability and changes in groundwater and surface flow regimes. However, Andean basins are characterized by a very high altitudinal gradient and their hydrological regime is essentially driven by snowmelt. Water table levels and river flows thus do not result directly from heavy rain, but from the melting of snow and ice. Moreover, since precipitation is extremely variable over space and time in these mountainous areas (Barry, 2008), water balance is extremely difficult to respect at the catchment scale (Favier *et al.*, 2009). Thus, the SHETRAN physically-based model, implemented over 1980–

2000 in the upper Elqui catchment, failed to obtain satisfactory simulations because of poor reproduction of the snowmelt regime (Araya & Hunt, 2003). A four-reservoir model also has been applied over the 1985–1999 period in this catchment (Trigos & Munizaga, 2006), but the simulations remain difficult to assess since they are available only in graphic form. The HEC-HMS model has also been implemented (Souvignet, 2007) with relatively satisfactory results, but on limited portions of the basin (1510 km^2) and shorter periods (1990–2000).

Before a model can be used for prediction, it must be adapted to various climatic conditions, which implies use of data over a wide spatiotemporal scale. The scarcity of descriptive data on such a scale militates in favour of a conceptual approach for studying large, poorly gauged catchments. However, the question remains whether such an approach is adequate for simulating the long-term rainfall–runoff relationship in catchments that are: (i) subject to significant hydro-climatic variability, and (ii) characterized by a snowmelt regime. This also calls for investigation of whether better simulations can be achieved at the basin outlet by accounting for the spatial variability of physical characteristics within the basin. Answering these questions constitutes the essential condition in a possible use of the models for evaluating the availability of water resources in the future.

This paper thus aims to adapt, calibrate and validate a daily conceptual model in order to simulate the impact of climatic variability on groundwater and surface flows from a mountainous catchment in the Chilean Andes. It assesses the ability of the model in reproducing interannual and annual flow volumes, seasonal regimes, peaks and groundwater storage in the specific context of the Elqui catchment.

STUDY AREA

The Elqui basin is one of the main river systems of the Norte Chico region in Chile (Fig. 1). With an area of 5660 km^2 at the Algarrobal gauging station, this catchment presents a large altitudinal gradient with elevations ranging from 750 to 6200 m a.s.l. in the Andes Cordillera. Annual precipitation varies from 80 mm in the littoral plain to 220 mm in the Cordillera, resulting in a semi-arid climate (Kalthoff *et al.*, 2002). The catchment comprises four principal formations: granite rock, volcanic rock, volcano-sedimentary deposits and Quaternary deposits. The latter are composed of alluvium and colluvium and constitute the main aquifers of the study area.

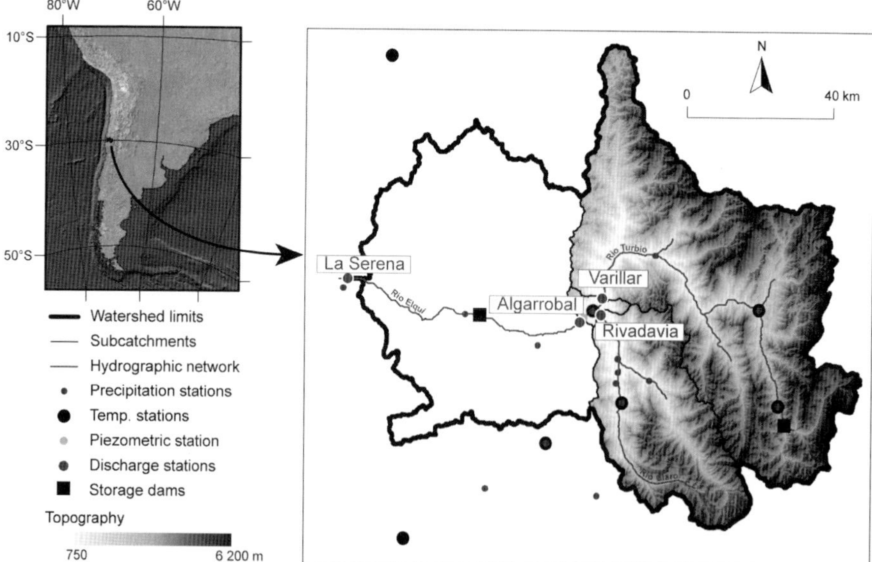

Fig. 1 The Elqui catchment at Algarrobal (5600 km^2) in Chile.

The main activity in the basin is production of grapes and other fruits. Cultivated land, which has spread continuously in the catchment, is located mainly on alluvial terraces and over closed hillsides. Agriculture accounts for more than 90% of the total water volume used in the basin. Water supply is provided by two storage dams, one of which is located in the upstream part of the basin and has been in operation since 1937 (Fig. 1). This dam (capacity 40×10^9 m^3) drains an area of about 500 km^2, which contributes 23% on average of the outlet flow, hence delaying the flow process along the river valley. The other dam is located downstream of the basin outlet and has been in operation since 1999, with a capacity of 200×10^9 m^3; it is used to store water for agricultural needs.

The hydrological functioning of the catchment is based on a snowmelt regime. Precipitation occurs principally in winter and is mainly stored as snow and ice in the high mountain areas. The melting of snow and ice causes high flows and high water tables during summer. The basin is also subject to alternating ENSO and LNSO events, which can result in extremely wet or dry years. There has been a decreasing trend in precipitation over the last century (Squeo *et al.*, 1999). At the same time, the population has increased (+28% in the last decade), intensifying the pressure on water resources.

MATERIALS AND METHODS

Adaptation of a hydrological model to a snowmelt regime

The HydroStrahler model (Ruelland *et al.*, 2008, 2009, 2010) was chosen to represent the seasonal and interannual variations in runoff and baseflow from the catchment. This daily conceptual model accounts originally for two reservoirs: (i) a shallow reservoir supplied by precipitation and feeding evapotranspiration, surface/sub-surface runoff and infiltration, and (ii) a deep reservoir fed by infiltration and generating the baseflow (Ruelland *et al.*, 2009).

A third reservoir, in which fluxes are controlled by temperature, has been introduced to account for the snowmelt regime of the catchment. The model structure is presented in Fig. 2. Below a temperature threshold *tmpt*, a fraction *snof* of precipitation is considered as snow; this fraction feeds a snow reservoir that generates no flows as long as the daily temperature is below *tmpt*. Above the threshold *tmpt*, a fraction *snor*, weighted by the difference between the daily

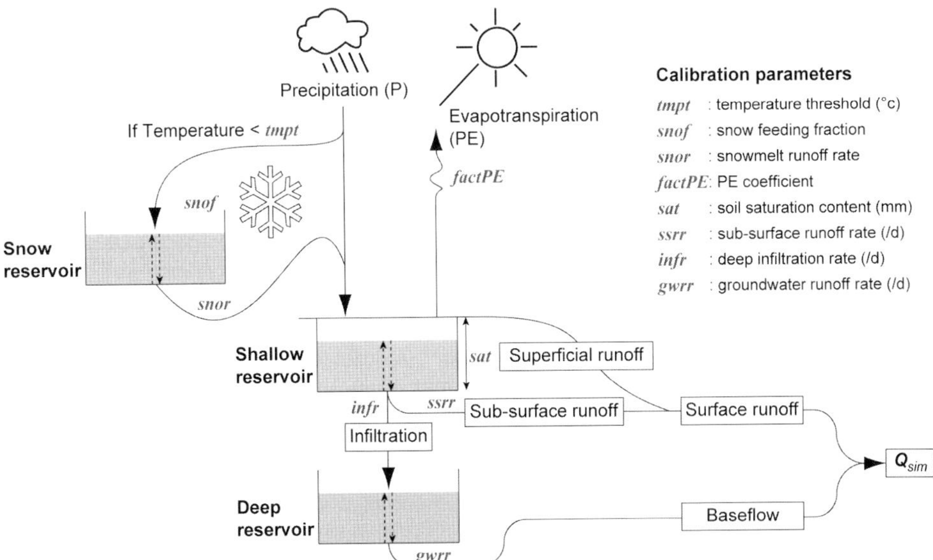

Fig. 2 Adaptation of the HydroStrahler model to a snowmelt regime (based on Ruelland *et al.*, 2008).

temperature and the threshold *tmpt*, is taken in the snow reservoir to constitute a snowmelt runoff. Each day, this runoff is added to the precipitation part, which is considered as liquid by the parameter *snof*, in order to feed a shallow reservoir. Potential evapotranspiration (PE) is controlled by a coefficient *factPE* and is active as long as the shallow reservoir is not exhausted. When the level of the shallow reservoir exceeds a saturation threshold *sat*, the surplus of water over *sat* immediately goes to surface runoff. The shallow reservoir has two other outlets: one of them producing sub-surface flow *via* a parameter *ssrr*; the other producing deep infiltration *via* a parameter *infr*. Each day, both parameters draw off a fraction from the water level in the shallow reservoir. While the sub-surface flow adds to runoff to form surface runoff, infiltration feeds a deep reservoir, which produces baseflow, drawing off a fraction *gwrr* of its fill level. Surface runoff and baseflow are summed to compose discharge simulated at the basin outlet. The eight parameters mentioned (*tmpt*, *snof*, *snor*, *factPE*, *sat*, *ssrr*, *infr* and *gwrr*) must be calibrated in order to account for four types of runoff: snowmelt, immediate, rapid and delayed runoff.

Hydro-climatic data

In order to represent the hydro-climatic variability over the catchment, a nearly 30-year period (1977–2008) was chosen according to data availability. Precipitation and temperature data were interpolated based on 15 and 10 stations, respectively (Fig. 1) using the inverse distance weighted method, which proved to be sufficiently accurate in the region among other methods for data reconstruction (Morales *et al.*, 2007). Note, however, that no measurements were available outside the river valleys (Fig. 1) and elevation effects on precipitation and temperature distribution was not considered (see Conclusions and Recommendations). Since the only data available for calculating PE were temperature data, a formula relying on solar radiation and mean temperature was selected (Oudin *et al.*, 2005). This method was shown to provide estimates similar to those obtained with more complex methods over many catchments. Lastly, three discharge gauging stations (Algarrobal, Varillar and Rivadavia) corresponding to the sub-catchments of the Elqui River were selected on the basis of the number and quality of the time series available (Fig. 1).

Model calibration and validation

Model calibration was based on a multi-objective function that aggregates a variety of goodness-of-fit indices (for more details on this function, see Ruelland *et al.*, 2009, 2010):

$$F_{agg} = (1 - NSE) + |VE| + VE_{avg} + PE_{avg} \quad (1)$$

where NSE is the Nash-Sutcliffe efficiency (Nash & Sutcliffe, 1970) criterion, VE the cumulative volume error, VE_{avg} the annual average relative volume error, and PE_{avg} the annual average peak error.

Calibration was then performed in an 8-D parameter space by searching for the minimum value of F_{agg}. The simulation period was divided into two parts: calibration was performed over 1979–1990 and validation over 1991–2008. These periods are distinguished by contrasted climatic behaviours, the calibration period being rather wet and the validation one being rather dry, which allowed the model to be tested for its suitability in differing climatic conditions. The 1977–1978 period was used as a spin-up to limit the influence of initial conditions in the model reservoirs. Simulations were performed according to two spatialization modes. In lumped mode (LP), a single set of parameters was used to calibrate the model. Distributed climate forcings were aggregated over the entire basin at the outlet gauging station of Algarrobal (Fig. 1). In semi-distributed mode (SD), two sub-catchments were considered: the Claro catchment at Rivadavia and the Turbio catchment at Varillar (Fig. 1). A parameter set was calibrated for each sub-catchment. Note that a small portion of the catchment (70 km^2) exists between the gauging stations. However, given its small area, it has been neglected, and discharge simulated at the two sub-catchments outlet was summed and compared directly to the discharge observed at Algarrobal. Calibration and validation were performed according to a 10-day time step in order to limit problems related to time transfer.

RESULTS

Model efficiency

Analysis of the fit of the hydrological model (Fig. 3) shows that discharge is simulated with a fair degree of realism at the basin outlet. With either lumped or semi-distributed simulations, NSE coefficients are over 0.89 and the other goodness-of-fit criteria present low values over the calibration period. Over the validation period, simulations tend to deteriorate due to the early occurrence of peak flows and depletion phases (Fig. 3(a)). In particular, some peak flows in the 2000s are not well simulated (Fig. 3(b)). Moreover, low flows between April and October are slightly underestimated (Fig. 3(a)). This results in a general underestimation of total runoff volume over the hydrological year. Hence, if the model can simulate streamflow volume over a two decade period (VE values of 0.10–0.18), it can barely simulate annual streamflow volume (VE_{avg} values around 0.26) and does not correctly reproduce peak flows (PE_{avg} values (0.37–0.40).

Nevertheless, the snowmelt regime for the catchment, which is characterized by a nearly four-month delay between precipitation and discharge, is reproduced accurately by the model (Fig. 3(a)). Moreover, the volume error values are about 11% and 8% for the lumped and semi-distributed simulations, respectively, over the whole 1979–2008 period (Fig. 3(b)). Finally, it can be noted that moving from a lumped to a semi-distributed application of the model does not greatly improve the simulated discharge at the basin outlet.

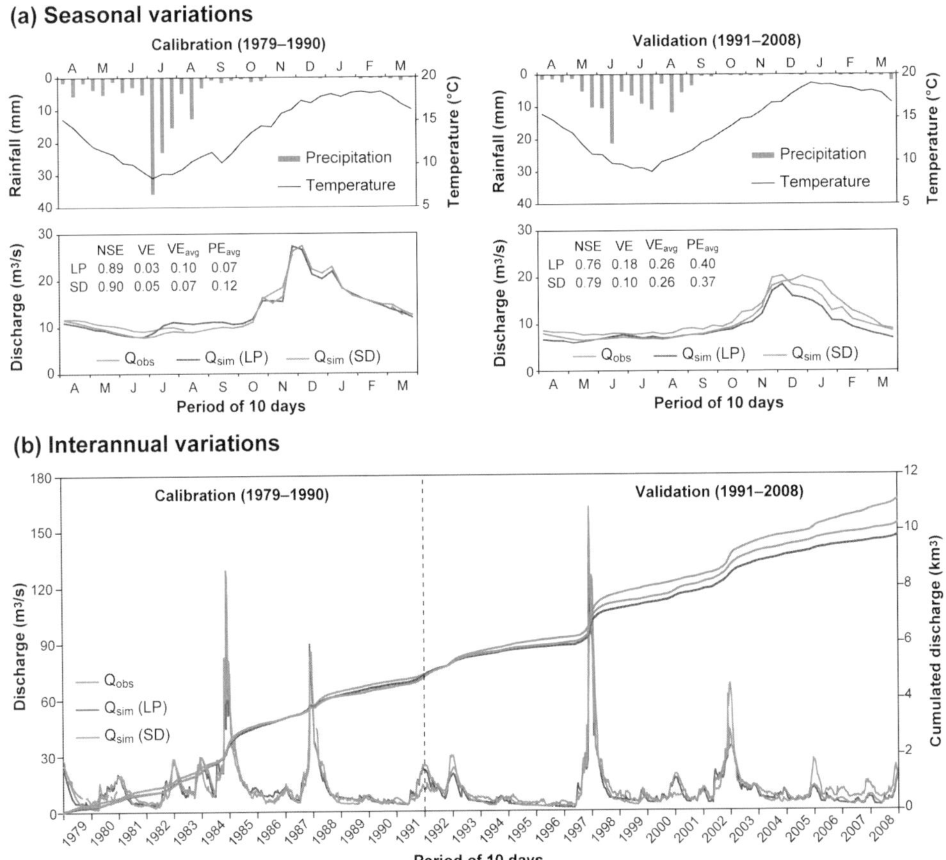

Fig. 3 Model efficiency at the Algarrobal gauging station according to lumped (LP) and semi-distributed (SD) simulations: (a) seasonal variations in climate forcings (precipitation, temperature) and in observed and simulated discharge over the calibration and validation periods; (b) comparison of the observed and simulated interannual hydrographs.

Analysis of the calibrated parameters

The upper Elqui catchment, constituted from the Turbio and Claro sub-catchments (Fig. 1), is a heterogeneous basin in terms of climate, geology and topography. The optimized calibration parameters from the semi-distributed modelling reflect indeed different hydrological properties (Table 1). Interpretation of these parameters must, however, be conducted very cautiously due to the imperfect hydrological simulations. Moreover, the calibration parameters are essentially theoretical; their values have a limited physical meaning and can result from numeric solutions that have few links with reality. For instance, the temperature threshold *tmpt*, which controls the portion of precipitation attributed to the liquid and snow states, as well as the runoff from the snow reservoir, presents values (12.2–13.8°C) that are much higher than the freezing mark. This can be explained by the fact the temperature forcing is mainly based on observations at valley bottoms and that it results from a daily average over the entire catchment or sub-catchments with significant elevation gradient. This threshold is higher for the Claro catchment. In fact, the Turbio catchment presents higher elevations and consequently lower mean temperatures than those of the Claro catchment (Table 2). The differences in elevation also affect the parameter *snof* (higher in the Turbio catchment), which determines the part of precipitation attributed to snow below the temperature threshold *tmpt*. On the other hand, the fraction *snor*, which generates snowmelt runoff, presents similar values for both catchments. However, note that this parameter is eventually weighted by the difference between the daily temperature and the temperature threshold *tmpt*. The *factPE* coefficient, which serves to limit the role of PE, has very low values for both catchments. This can be explained by the fact that evapotranspiration can be ignored above 3000 m a.s.l. in the study area (see Favier *et al.*, 2009). The value of the soil saturation level *sat* is higher in the Claro catchment, which could be linked to lower-permeability formations at surface level due to rock alteration of the granite. In contrast, the parameter *ssrr*, which generates sub-surface flow, is higher in the Turbio basin. The values of *infr* and *gwrr* are also much larger in the Turbio catchment, which suggests that baseflow makes a much larger contribution to discharge in the Turbio catchment. Indeed, this catchment has a much higher proportion of sedimentary formations (24% as against 13% in the Claro basin) and little of the low-permeability granitic rock that constitutes 80% of the Claro basin. Slopes, less pronounced in the Turbio basin (Table 2), also play a major role in the differences in infiltration.

The lumped calibration parameters (Table 1) generally represent intermediate values of these various properties. According to the parameter *snof*, 89% of the precipitation is attributed to snow below the temperature threshold. The model thus simulates a snow accumulation of 99.4 mm/year on average, which represents 78.5% of mean annual precipitation over 1979–2008. This highlights the importance of snowmelt runoff for the basin. For other parameters (*factPE*, *ssrr*), values are closer to those calibrated in the Turbio catchment, which contributes much more to runoff because

Table 1 Lumped (LP) and semi-distributed (SD) calibrated parameters over 1979–1990.

Catchment	*Tmpt* (°C)	*Snof* (%)	*Snor* (d^{-1})	*factPE* (%)	*Sat* (mm)	*Ssrr* (d^{-1})	*Infr* (d^{-1})	*Gwrr* (d^{-1})
Elqui (LP)	13.0	89	0.006	10	76	0.013	0.019	0.0030
Turbio (SD)	12.2	93	0.004	12	70	0.013	0.002	0.0007
Claro (SD)	13.8	86	0.004	3	83	0.008	0.001	0.0004

Table 2 Physical characteristics of the three catchments modelled.

Catchment	Outlet gauging station	Catchment area (km^2)	Mean precipitation (mm/year)	Mean temp. (°C)	Mean annual discharge (m^3/s)	Elevation range (m)	Mean slopes (%)
Elqui	Algarrobal	5660	130.8	13.9	11.7	750–6200	48.8
Turbio	Varillar	4080	130.6	13.5	6.6	875–6200	47.4
Claro	Rivadavia	1510	132.5	15.0	4.4	825–5550	52.1

of its larger area (Table 2). Only *infr* and *gwrr* present values much higher than those obtained in the sub-catchments. The role of baseflow (and therefore of the aquifer) thus appears to be strengthened by the lumped model.

Dynamics in the deep reservoir of the model

Since the lumped simulation relies on a single deep reservoir, the dynamics of the water levels in this reservoir were compared to the water table levels observed in the piezometer at the outlet of the basin (Fig. 4). In this comparison of one well against a modelled groundwater store meant to represent an entire basin, there are scale issues that were ignored. However, it is interesting to observe a significant correlation ($R^2 = 0.79$) between the water table levels and dynamics in filling/emptying of the deep reservoir over the 1979–2008 period. Despite its conceptual approach, the hydrological model thus seems to accurately represent the water quantity in the aquifer in the Algarrobal area.

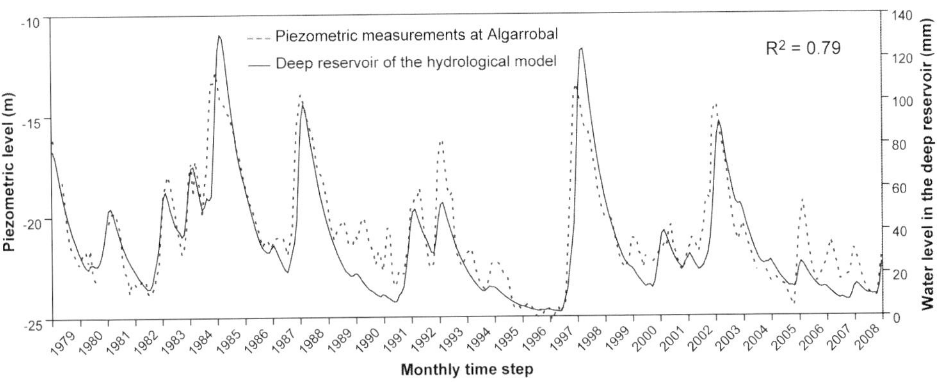

Fig. 4 Comparison of the monthly dynamics of water levels in the conceptual deep reservoir of the lumped hydrological model, and the piezometric measurement at Algarrobal over 1979–2008.

CONCLUSIONS AND RECOMMENDATIONS

Located in Chile between the Pacific Ocean and the Andean Cordillera, the Norte Chico region is characterized by a strong altitudinal gradient. This results in a snowmelt regime of the region's catchments, where precipitation occurs mainly as snow during winter, and discharge and water table levels increase during summer due to the increase in temperatures. These mountainous catchments are also subject to significant hydro-climatic variability due to alternating *El Niño* and *La Niña* events. In the Elqui catchment, these extreme events can vary river discharge by a factor of 10. Since agricultural activities and population have increased continuously over the last decades, the availability of water resources to meet current and future needs in this context of climatic variability has become a crucial issue.

A hydrological model was thus constructed and run to simulate the rainfall–runoff relationship and to improve our understanding of the hydrological processes of this catchment. A snow reservoir was introduced to account for the snowmelt regime of the catchment. Over a nearly 30-year period subject to significant hydro-climatic variability, the simulation results agree rather well with the measured discharge at the basin outlet, with either the lumped or the semi-distributed application of the model. The simulated cumulated volume over 1979–2008 is only slightly underestimated (8–11% depending on the simulations), which highlights the model's consistency for assessing water resources evolution in the long term. The results also show clearly that the deep reservoir of the model reproduces the water table dynamics in a piezometer at the basin outlet with a fair degree of realism. This reveals the importance of interactions between groundwater and surface flows in the basin, but also indicates an interesting result from the model, as the piezometric measurements were not used as data forcings for calibration. Incidentally, these measurements could be used during the parameterization in order to limit some well-known equifinality problems that may exist because of the significant number of model parameters and that could be highlighted by a sensitivity analysis.

The semi-distributed application of the model highlights the physical heterogeneity of the catchment, as illustrated by the different parameter sets. Nevertheless, even though the semi-distributed modelling accounts better for the hydrological processes inside the basin, moving from a lumped to a semi-distributed approach does not greatly improve the simulated hydrograph at the basin outlet. These findings contradict earlier research (e.g. Baudez *et al.*, 1999; Bourqui *et al.*, 2006; Das *et al.*, 2008) and reinforce the idea that semi-distributed approaches demonstrate no clear practical superiority over lumped ones in simulating streamflow, as already shown by Ruelland *et al.* (2008, 2010). A lumped modelling should thus be preferred in this context due to the fair degree of realism of its simulations and its easier implementation.

Nevertheless, it should be noted that the quality of the simulations (notably peak flows) is limited by a number of factors. First, another interpolation technique accounting for the altitudinal gradient (e.g. bivariate kriging) could improve the distribution of precipitation and temperature over the catchment (see e.g. Valéry, 2009). Second, as shown by Favier *et al.* (2009), while PE can be ignored beyond 3000 m a.s.l., snow and ice sublimation can reach a value of 80 mm/year to 4000 m a.s.l. This indicates the complexity of estimating PE in this mountainous context and also argues in favour of accounting for elevations in this estimate. Finally, even in the upstream part of the catchment, natural flows can be disturbed by anthropic activity. In particular, further modelling should take into account the hydraulic works of the sector: the dam of La Laguna in the upper part of the catchment and water withdrawals *via* irrigation canals in the downstream part. These improvements are essential so that the model can be used to forecast the impact of global climate changes on the regional hydrology and overcome the major obstacles that water scarcity poses to its agriculture-based development.

Acknowledgements This work was supported by the HydroSciences Montpellier laboratory, the CEAZA and Agropolis, within the framework of a partnership between the *Conseil général de l'Hérault* and the Coquimbo region. The authors would like to thank the *Dirección General de Aguas* (DGA, Chile) for providing the necessary data for this study.

REFERENCES

Araya, A. & Hunt, J. (2003) Aplicación del Sistema de Modelación Hidrologica SHERTAN a la cuenca del rio Elqui. Memoria de Titulo para optar al titulo de Ingenior Civil, Universidad de La Serena.

Barry, R. G. (ed.) (2008) *Mountain Weather and Climate* (3rd edn). Cambridge University Press, Cambridge, UK. 506 p.

Baudez, J.-C., Loumagne, C., Michel, C., Palagos, B., Gomendy, V. & Bartolli, F. (1999) Modélisation hydrologique et hétérogénéité spatiale des bassins : Vers une comparaison de l'approche globale et de l'approche distribuée. *Etude et Gestion des Sols* **6**, 165–184.

Bourqui, M., Loumagne, C., Chahinian, N. & Plantier, M. (2006) Accounting for spatial variability: a way to improve lumped modelling approaches? An assessment on 3300 chimera catchments. In: *Large Sample Basin Experiments for Hydrological Model Parameterization*, 300–310. IAHS Publ. 307. IAHS Press, Wallingford, UK.

Das, T., Bardossy, A., Zehe, E. & He, Y. (2008) Comparison of conceptual model performance using different representations of spatial variability. *J. Hydrol.* **356**, 106–118.

Favier, V., Falvey, M., Rabatel, A., Praderio, E. & Lopez, D. (2009) Interpreting discrepancies between discharge and precipitation in high-altitude area of Chile's Norte Chico region (26-32 degrees S). *Water Resour. Res.* **45**, W02424.

Kalthoff, N., Bischoff-Gauss, I., Fiebig-Wittmaack, M., Fiedler, F., Thurau, J., Novoa, E., Pizarro, C., Gallardo, L. & Rondanelli, R. (2002) Mesoscale wind regimes in Chile at 30° S. *J. Appl. Met.* **41**, 953–970.

Kane, R.P. (1999) Some characteristics and precipitation effects of the El Nino of 1997–1998. *J. Atmos. and Solar-Terrestrial Physics*. **61**(18), 1325–1346.

IPCC (2007) *Climate change 2007: The Physical Science Basis. Contribution of Working Group I to the Fourth Assessment Report of the Intergovernmental Panel on Climate Change* (ed. by S. Solomon, D. Qin, M. Manning, Z. Chen, M. Marquis, K. B. Averyt, M. Tignor & H. L. Miller). Cambridge University Press, Cambridge, UK, 996 p.

McPhaden, M. J. (2004) Evolution of the 2002/03 El Niño. *Bull. Am. Met. Soc.* **85**, 677–695.

Morales, L., Canessa, F., Mattar, C. & Orrego, R. (2007). Comparison of interpolation methods for detection of microclimatic areas. *5th Int. Symp. Spatial Data Quality* (ITC, Enschede, The Netherlands, 12–15 June 2007).

Nash, J. E. & Sutcliffe, J. V. (1970) River flow forecasting through conceptual models, a discussion of principles. *J. Hydrol.* **10**, 282–290.

Oudin, L., Hervieu, F., Michel, C., Perrin, C., Andréassian, V., Anctil, F. & Loumagne, C. (2005) Which potential evapotranspiration input for a lumped rainfall-runoff model? Part 2: towards a simple and efficient potential evapotranspiration model for rainfall–runoff modelling. *J. Hydrol.* **303**, 290–306.

Ruelland, D., Ardoin-Bardin, S., Billen, G. & Servat, E. (2008) Sensitivity of a lumped and semi-distributed hydrological model to several methods of rain interpolation on a large basin in West Africa. *J. Hydrol.* **361**, 96–117.

Ruelland, D., Guinot, V., Levavasseur, F. & Cappelaere, B. (2009) Modelling the long-term impact of climate change on rainfall–runoff processes over a large Sudano-Sahelian catchment. In: *New Approaches to Hydrological Prediction in Data Sparse Regions* (Proc. Symp. HS.2 at the Joint IAHS & IAH Convention, Hyderabad, India, 6–12 September 2009) 59–68. IAHS Publ. 333. IAHS Press, Wallingford, UK.

Ruelland, D., Larrat, V. & Guinot, V. (2010) A comparison of two conceptual models for the simulation of hydro-climatic variability over 50 years in a large Sudano-Sahelian catchment. In: *Global Change: Facing Risks and Threats to Water Resources* (Proc. 6th FRIEND Int. Conf., Fez, Morocco, 25–29 Oct. 2010), 668–678. IAHS Publ. 340. IAHS Press, Wallingford, UK.

Souvignet, M. (2007) Climate Change Impacts on Water Availability in the Semiarid Elqui Valley, Chile. PhD Thesis, Cologne University of Applied Sciences, Institute for Technology in the Tropics, 110 p.

Squeo, F. A., Olivares, N., Olivares, S., Pollastri, A., Aguirre E., Aravena, R., Jorquera, C. & Ehleringer, J.R. (1999) Grupos funcionales en arbustos desérticos definidos en base a las fuentes de agua utilizadas. *Gayana Botánica* **56**, 1–15.

Trigos, H. & Munizaga, I. (2006) Análisis de modelos hidrologicos de bases fisicas para cuencas demi-aridas y estructuracion sistema de simulación cuenca rio Elqui. Memoria para optar al titulo de ingenior civil.

UNCCD-Chile (2006) Implementación en Chile de la Convención de Naciones Unidas de Lucha Contra la Desertificación en los países afectados por sequía grave o desertificación. Santiago de Chile, 47 p.

Valéry, A. (2009) Modélisation précipitations – débit sous influence nivale. Élaboration d'un module neige et évaluation sur 380 bassins versants. PhD Thesis, Cemagref (Antony), AgroParisTech (Paris). 405 p.

Relative contribution of groundwater and surface water fluxes in response to climate variability over a mountainous catchment in the Chilean Andes

H. JOURDE[1,3], R. ROCHETTE[1,3], M. BLANC[1,3], N. BRISSET[1,3], D. RUELLAND[2], G. FREIXAS[4] & R. OYARZUN[3,5]

1 *UM2 – HydroSciences Montpellier, Place E. Bataillon, 34395 Montpellier Cedex 5, France*
herve.jourde@um2.fr

2 *CNRS -UMR HydroSciences Montpellier, Place E. Bataillon, 34395 Montpellier Cedex 5, France*

3 *CEAZA, Colina El Pino, Universidad La Serena, La Serena, Chile*

4 *DGA, Direccion General de Aguas,Plaza de Armas;*

5 *Departamiento Ingeniería de Minas , Universidad La Serena, La Serena, Chile*

Abstract In the semi-arid region of Norte Chico (Chile), climate variability, mainly controlled by ENSO and LNSO events, generates a high variability of both surface water and groundwater fluxes. Taking the upper Elqui catchment as an example, this study found that, during LNSO events, the abnormally high values (>200%) of the runoff coefficient may be the consequence of a groundwater contribution to surface water flow. During ENSO events, however, the lower values (<100%) of the runoff coefficient and the dynamics of the water table level highlight the recharge of the subsurface compartment. For the hydrological years characterized by a high Pluviometric Index during the 1977–2008 period, three dynamics of interaction between groundwater and surface water are identified: (i) the water table increases before the river discharge, and its logarithmic increase highlights a rapid recharge related to the concomitance of snowmelt and rainfall events; (ii) the water table increases after the river discharge and its exponential increase shows a progressive intensification of the recharge over time; and (iii) the water table and the river discharge increase are concomitant. Dynamics (i) and (ii) are observed during the ENSO events, when precipitation occurs over a long period; dynamic (iii) is observed during the neutral years, when high intensity precipitations occur over short periods. Accordingly, if the present climate trend marked by an increased frequency of El Niño events in recent decades (IPCC, 2007) persists, this should favour dynamics (i) and (ii), and thus enhances the relative importance of the groundwater resource with respect to surface water resource. However, both the present positive trend in temperature and the difference of trends at the scale of the catchment may favour the less efficient of these two dynamics in terms of groundwater recharge.

Key words hydro-climatic variability; water resource; surface/subsurface interactions; snowmelt; Río Elqui, North-Central Chile

INTRODUCTION

In North-Central Chile, climate variability controlled mainly by the El Niño and La Niña events causes substantial disparity in precipitation intensity and distribution. The water balance is rarely respected at an annual scale in the catchments of the region (Favier *et al.*, 2009), mainly due to lack of understanding of orographic enhancement of precipitation, and of glacier contributions. However, the imbalance of the water budget could also arise from an underestimation of the subsurface compartment's contribution to surface runoff. A better characterization of the surface and subsurface compartments' functions and interactions to each other is therefore of prime importance to better understand Andean hydrological systems. Using the upper Elqui catchment as an example, surface and groundwater fluxes are analysed in this study, according to the spatio-temporal climate variability over 32 years (1977–2008).

BASIN DESCRIPTION AND METHODS

The Elqui catchment (9700 km^2, Fig. 1(a)), is one of the three main hydrographic networks of the Coquimbo region (Chile). It lies between latitudes 29°27′S and 30°34′S and longitudes 71°22′W and 69°52′W, and runs between the Pacific coast and the summit ridge of the Andes, which exceeds 6000 metres in some places. Like most rivers of this region, the Río Elqui exhibits a strong interannual variability in river discharge (mean annual discharge between 2 and 33 m^3/s). This river starts at the confluence of the Río Turbio and Río Claro (altitude 815 m) – its two main tributaries having catchment areas of 4200 and 1550 km^2, respectively. They both have a nival hydrological regime, typical of arid mountainous areas, with the highest river discharges recorded from November to February. From a geological point of view, the catchment consists of four major types of formation: granite, volcanic rock, volcano-sedimentary deposits, and Quaternary sedimentary cover. The latter, mainly alluvial and colluvial, can be as thick as 200 m and forms the main aquifers in the study zone (USDI, 1955). This aquifer can store and release large volumes of water (Miranda & Llanca, 2003). In the Río Turbio and Río Claro catchments as far as 3 km downstream of their confluence (Algarrobal), deposits can be more than 100 m thick, and their permeability have been assessed at 10m/d, which allow rapid groundwater transfers. Regarding the substratum, the plutonic rocks display different hydraulic conductivities as a result of fracturing at various scales (USDI, 1955).

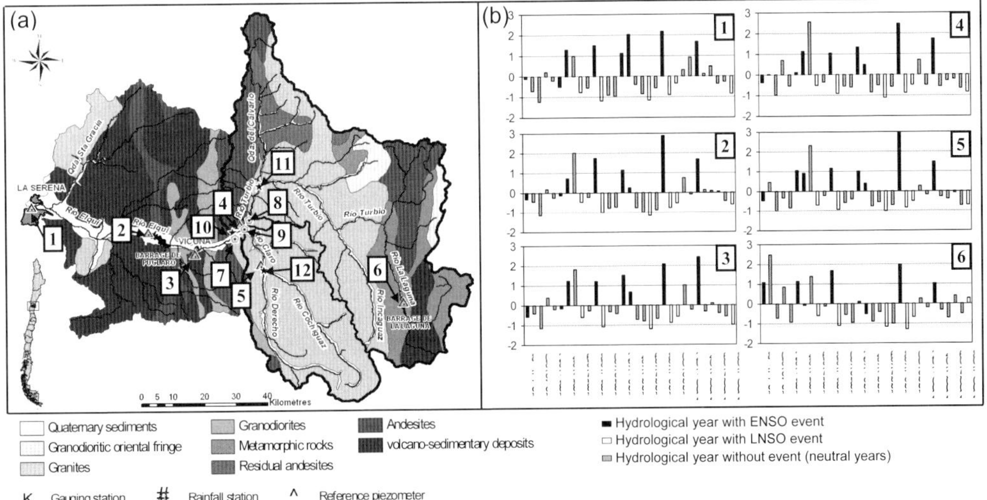

Fig. 1 (a) General context of the Río Elqui catchment (Cabezas *et al.*, 2007) and monitoring network: meteorological stations (La Serena (1), Almendral (2), Vicuña (3), Rivadavia (4), Monte Grande (5) and La Laguna (6)), gauging stations (Río Elqui at Algarrobal (7), Río Turbio at Varillar (8) and Río Claro at Rivadavia (9)), and reference piezometers (Algarrobal (10), Bocalume (11) and Monte Grande (12)); (b) Pluviometric indices (PI) calculated for the period 1977–2008 at the meteorological stations.

Climate is semi-arid in this region. Precipitation is less than 100 mm along the coast and more than 400 mm in mountain areas, where snow depths can reach several metres in exceptional years. The uneven precipitation over the catchment is analysed with pluviometric indices (PI), defined as centred reduced values for each hydrological year (April to March):

$$PI = \frac{xi - \bar{x}}{\sigma} \tag{1}$$

where xi is precipitation of year i, $\bar{x}$ is mean annual precipitation over the reference period, and σ is standard deviation of annual precipitation over the reference period.

PI, identifying excess or shortage of precipitation relative to the reference period, is compared over the last 30 years with ONI (Oceanic Niño Index) data based on Pacific Ocean surface temperature anomalies in the NIÑO 3.4 region.

For each catchment, we considered a reference piezometer representative of the natural dynamics of the subsurface compartment (no pumping and strong inter annual variability) and a hydrometric station (Table 1). Hydrological years with a high PI were selected for analysis of the existing relations between surface and subsurface compartments. Four of these years are marked by El Niño events: 1987–88, 1992–93, 1997–98 and 2002–03. Two others are neutral years (no events): 1978–79 and 1984–85. As for precipitation characterization, hydrological indexes *HI* (centred reduced discharge values) and piezometric indices *PZI* (centred reduced piezometric values) are calculated to assess variation of discharge and groundwater flow in relation to the reference period, respectively.

Table 1 Altitude of the reference piezometers and hydrometric stations for the three catchments.

	Reference piezometers	Hydrometric station
Río Turbio catchment	Bocalume (altitude 1025 m)	Varillar (altitude 860 m)
Río Claro catchment	Monte Grande (altitude 1111 m)	Rivadavia (altitude 820 m)
Upper Elqui catchment	Algarrobal (altitude 767 m)	Algarrobal (altitude 760 m)

To complete the analysis, runoff coefficient (*RC*) is also calculated :

$$RC = \frac{R}{P} \qquad (2)$$

where *R* is the cumulated runoff flow volume at the outlet of the catchment, and *P* the volume of precipitation over the catchment during the hydrological year. For each catchment (Table 2), the cumulated runoff flow accounts for irrigation and canal intake (available data until 1998–99), and *P* is calculated with precipitation data interpolated at 200 × 200 m resolution using the inverse distance weighting (IDW) method. Runoff coefficient *RC* is influenced by geology, topography, soil type, land use and plant cover. It gives an idea of soil infiltration capacity and evapotranspiration, but also allows identifying the contribution of subsurface flow to surface flow when abnormal *RC* values (> 60–70%), are obtained (Jourde *et al.*, 2007). Then, comparing these values with the expected *RC* values according to the characteristics of the watershed, allows quantification of the subsurface flow contribution.

Table 2 Runoff coefficients (*RC*) in the Río Turbio, Río Claro, and upper Elqui catchments for each hydrological year. Grey and light grey shading indicate years corresponding to El Niño and La Niña events, respectively. Greyed runoff coefficients were calculated with data series slightly incomplete.

		1977~78	1978~79	1979~80	1980~81	1981~82	1982~83	1983~84	1984~85	1985~86	1986~87	1987~88
Río Turbio catchment	P (mm)	160	253	37	201	54	210	206	350	69	108	296
	RC	26%	46%	119%	29%	69%	29%	30%	41%	79%	45%	47%
Río Claro catchment	P (mm)	129	193	21	191	47	200	222	359	60	99	299
	RC	88%	111%	544%	89%	236%	77%	91%	96%	258%	137%	131%
Upper Elqui catchment	P (mm)	150	235	33	198	52	206	210	352	66	105	296
	RC	15%	29%	132%	30%	80%	34%	41%	55%	104%	44%	39%

		1988~89	1989~90	1990~91	1991~92	1992~93	1993~94	1994~95	1995~96	1996~97	1997~98	1998~99
Río Turbio catchment	P (mm)	22	77	54	222	122	40	75	14	46	387	35
	RC	293%	48%	56%	16%	58%	119%	47%	204%	48%	38%	200%
Río Claro catchment	P (mm)	14	74	66	232	157	42	65	8	59	423	34
	RC	1027%	137%	122%	57%	127%	277%	167%	1041%	70%	93%	434%
Upper Elqui catchment	P (mm)	20	76	57	225	132	40	72	12	49	396	35
	RC	332%	49%	47%	21%	70%	119%	50%	208%	36%	56%	205%

RESULTS AND DISCUSSION

In the Río Turbio catchment, analysis performed at annual time scale shows that *HI* is always greater than *PZI* (Fig. 2) in the wet years (high *PI*). This results from a faster hydrological response of surface water compartment than subsurface groundwater compartment. However, *PZI* remains high in the year following a wet year, and is in most cases higher than that of the preceding year while *HI* decreases strongly. This is related to the greater inertia of the subsurface compartment, highlighted by a long recession period (1 to 2 years). Finally, shortage of rainfall results in both small *HI* and *PZI* values in years following the dry years. This yearly dynamics is less obvious in the Río Claro catchment where irrigation water feeds canals and reservoirs; this can

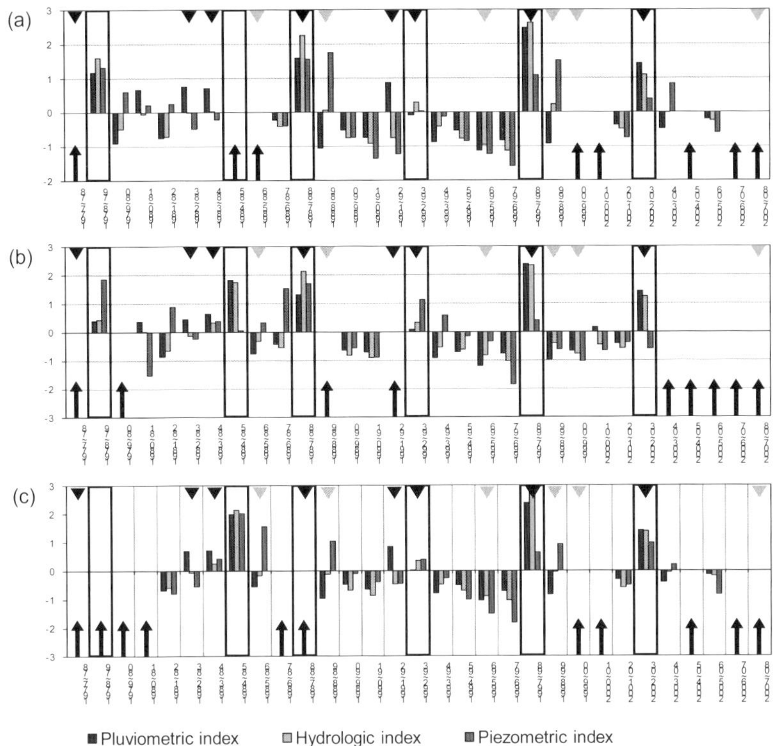

Fig. 2 Pluviometric Index, Hydrological Index, and Piezometric Index for: (a) the Río Turbio catchment; (b) the Río Claro catchment, and (c) the upper Elqui catchment. Top black and grey symbols indicate hydrological years associated to *El Niño* and *La Niña* events, respectively. Bottom black arrows indicate years with missing data.

contribute to recharge and generate both a relative *PZI* increase and an *HI* decrease (as the irrigation water comes from the Río Claro). *PZI* values higher than *HI* values are determined (Fig. 2) in dry years (low *PI*) like 1986–87, 1994–95 and 1995–96, but also in particularly wet years (high *PI*), such as 1978–79 and 1992–93. In both cases, this indicates a preponderant recharge of the subsurface compartment with respect to surface runoff; in dry years (low *PI*), this behaviour could be explained by the geology, with recharge of the subsurface compartment coming from the underlying fractured granite bedrock that, for this catchment, constitutes more than 80% of its area. Another possibility might be the contribution of water from permafrost or snow melt coming from former precipitations. However, this behaviour should occur in a more noticeable manner in the Rio Turbio catchment than in the Rio Claro catchment where elevation is lower, and thus water storage as snowpack or permafrost are less evident; this is not the case. Besides, these dry years (1986–87, 1994–95 and 1995–96) also immediately follow other dry years without noticeable precipitation, which tends to exclude this hypothesis.

In the upper Elqui catchment, indices display the same trends as those of the Río Turbio, which highlights the dominance of this major tributary on the overall dynamics of interactions between surface and subsurface compartments.

During the wet years (high *PI*), the small runoff coefficients (<100%) observed in the Río Turbio and upper Elqui catchments show that the relative contribution of the subsurface compartments to surface flows is lower than in dry years (low *PI*). This also indicates a large recharge of the subsurface compartment, as illustrated by considerable piezometric variations of nearly 50 m at the Bocalume piezometer (see Fig 4(a)). Concerning this recharge, it can take place in various places: (i) recharge in the major bed of the river when a discharge threshold is reached and river overflows; (ii) direct inflow from the subsurface compartments (aquifers) of the river tributaries; (iii) delayed recharge from the fractured bedrock, as seen above.

During the dry years (low *PI*), runoff coefficients are greater than 100% (>200% during *La Niña* events), which can be explained by the strong contribution of groundwater to surface flow. Another possibility may be the contribution of water from permafrost melt or from snow accumulation of former precipitations, as stated previously. However, as explained above, this behaviour should occur in a more noticeable manner in the Rio Turbio catchment than in the Rio Claro catchment; this is not the case. Accordingly, groundwater contribution to surface flow is the most likely hypothesis to explain these high runoff coefficients values during dry years.

Analysis of runoff coefficients in a succession of several dry years shows that the contribution of the subsurface compartment to surface flow considerably increases during the year immediately following an event at the origin of high *PI*, and then decreases gradually. Accordingly, the overall relation between surface and subsurface compartments will be a fingerprint of climate variability.

The relationship between runoff coefficients *RC* and mean annual precipitation (Fig. 3) shows that when precipitation exceeds approximately 60–70 mm, *RC* values remain between 45 and 50%; this indicates a relative equilibrium between runoff and recharge processes. Below this threshold, *RC* values increase exponentially when precipitation decreases. Figure 3 also allows us to identify two categories of hydrological years: (i) hydrological years 1980–81, 1982–83, 1983–84 and 1991–92 (light grey circles) with *PZI* > *HI*, which indicates an aquifer recharge preponderant with respect to surface runoff; (ii) hydrological years 1990–91, 1995–96 and 1996–97 (arrows) with a noticeable contribution of the subsurface compartment to surface flow (high *RC*) as a result of several dry years. Hence, the above mentioned threshold may be used as a warning for drought; indeed, when precipitation remains below this threshold, recharge to the subsurface compartments is zero and the subsurface compartment feeds the river; the consequences are high hydric deficits, highlighted by a severe piezometric decrease in the aquifers.

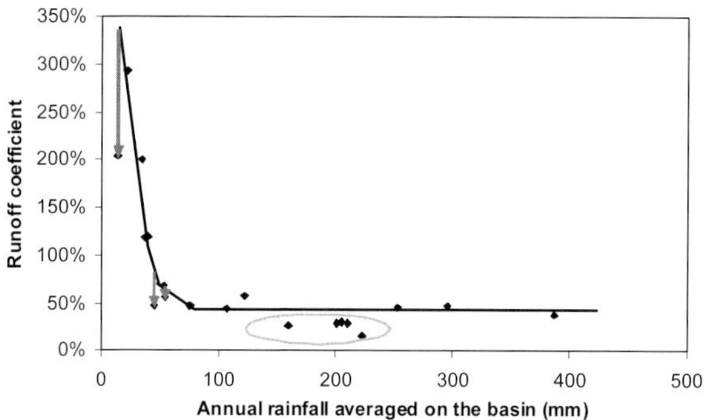

Fig. 3 Variation of the Río Turbio runoff coefficients *versus* mean annual precipitations over the Río Turbio catchment.

Regarding the spatial variability of runoff coefficients, *RC* values are much higher in the Río Claro catchment than in the Río Turbio and upper Elqui catchment; this highlights a greater contribution of the subsurface compartment in the Río Claro catchment, as confirmed by hydrological modelling (Ruelland *et al.*, 2011). This could be explained by the smaller proportion of alluvium in the Río Claro than in the Río Turbio catchment (12.6% cf. 23.5% of catchment area). As a result, the volume of water stored in the subsurface compartment (alluvium) would be potentially smaller in the Río Claro catchment and generate higher cumulated runoff flow volumes (and thus *RC* values). However, as seen above, the most likely hypothesis may be related to the geology, with the recharge of the subsurface compartment coming from the underlying fractured granite bedrock that covers 80% of the catchment area. For hydrological years characterized by

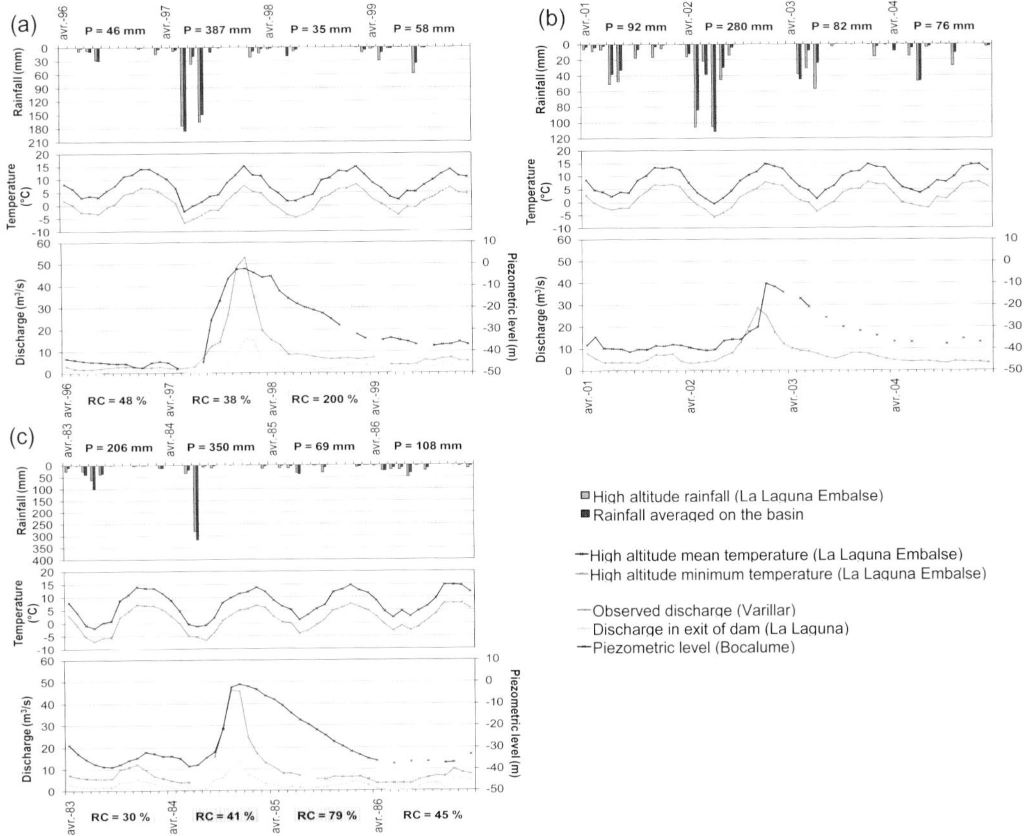

Fig. 4 Hydroclimatic monthly variations for hydrological years: (a) 1996–1999 associated with dynamics 1, (b) 2001–2004 associated to dynamics 2, and (c) 1983–1986 associated to dynamics 3.

high *PI*, it is seen from monthly data (Fig. 4) that three dynamics of interactions between the surface and subsurface compartments can be identified: (i) *dynamic 1* corresponds to a rise in piezometric level that precedes the river discharge increase. It is observed for two *El Niño* events (1987–88 and 1997–98) with high precipitation for two to three months; (ii) *dynamic 2* corresponds to a rise in piezometric level that takes place after river discharge increase. This dynamic is observed for two *El Niño* events (1992–93 and 2002–03), with high precipitation for a longer period (4–6 months); (iii) *dynamic 3* corresponds to a simultaneous increase of piezometric level and river discharge, showing the same transfer dynamics between surface and subsurface compartments. Dynamic 3 is observed in neutral years (1978–79 and 1984–85), with large precipitation during a very short time (one week). If the climate trend marked by an increased frequency of *El Niño* events in recent decades (IPCC, 2007) persists, this should favour dynamics 1 and 2, and enhance the relative importance of groundwater resources with respect to surface water resources. Indeed, these dynamics are associated with large piezometric level rises: + 45 m and + 35 m, for dynamics 1 and 3, respectively, although it is "only" + 30 m for dynamic 2.

The intervals of minimum and mean temperature corresponding to the onset of discharge increase and piezometric levels rise are shown in Fig. 5. These intervals are defined by the monthly temperatures before and after the onset of discharge increase and piezometric level rise, respectively. The intervals of temperature corresponding to the onset of discharge are relatively stable (Fig. 5(a)). In contrast, the temperature intervals corresponding to the onset of piezometric level rise exhibit a high variability (Fig. 5(b)), with dynamic 2 associated with noticeably

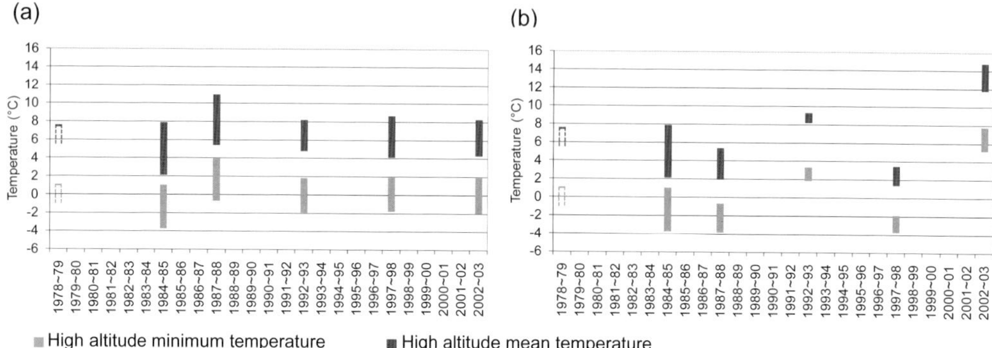

Fig. 5 Temperature intervals corresponding to the onset of (a) discharge increase, and (b) piezometric level rise. Intervals are given for minimum and mean temperatures at La Laguna station (3160 m).

higher temperature. Aquifer recharge therefore involves processes depending on factors other than temperature.

The logarithmic increase of piezometric levels *versus* time (dynamics 1, Fig. 4(a)) indicates a rapid recharge of the subsurface compartment, which subsequently decreases. This dynamics could be the consequence of a gradual temperature rise (Fig. 4(a)) that causes a slower melt of the snow cover and thus enhances infiltration with respect to runoff in a first stage. In contrast, the exponential increase of piezometric levels *versus* time (dynamics 2 and 3, Fig 4) indicates a delayed recharge of the subsurface compartment. For dynamic 2, this delayed rise in piezometric level than in discharge could be related to aquifer recharge resulting from the overflowing of the Río into its major bed. Besides, precipitation over 4–6 months (1992–93 and 2002–03) affects the structure of the snow cover (hardening and freezing), which can stabilize the cover and delay the snowmelt processes. For dynamic 2, the intervals of temperature corresponding to the onset of discharge, but especially to the onset of recharge, are higher than that for other dynamics (Fig. 5). Figure 6 shows that this dynamic is dominated by surface runoff processes related to a rapid snowmelt, which prevents a large infiltration towards the subsurface compartment.

An analysis of temperature changes in this region (Souvignet *et al.*, 2010) shows that both maximum and minimum monthly temperatures have positive trends. Besides, different trends in temperature between the higher and middle elevations indicate a possible variation of the lapse rate—an important feature in the snow accumulation and melt processes. If the present trend in temperature persists, the frequency of dynamic 2, less efficient than other dynamics with respect to groundwater recharge, may thus increase in the future.

At the daily timescale, a strong correlation between temperature and river discharge can be observed for the three dynamics (Fig. 6).

As for the monthly time scale, river discharge increases in two stages: the first stage can be attributed to snowmelt, and the second stage is the main increase of flow, resulting from substantial snowmelt when the *high altitude minimum temperature* exceeds the threshold of 0°C. In dynamic 1 (Fig 6(a)), the main increase in the piezometric level preceding that of river discharge takes place when *high altitude minimum temperatures* remain above –3°C and below 0°C, previously mentioned threshold for a sufficiently long period (estimated to be about a month). Within this temperature range, snow melts and tends to infiltrate the soil layers rather than forming surface runoff. River discharge therefore remains stable while the piezometric level rises. In dynamics 2 (Fig 6(b)), the main river discharge increase, which precedes the rise in piezometric level, takes place when *high altitude minimum temperatures* exceed the –3°C threshold, and do not remain lower than the 0°C threshold for long enough. The snow thus melts more quickly, with melt rates greater than filtration rate, resulting in a strong river discharge increase.

The much later rise in piezometric level might be related to aquifer recharge resulting from the overflow of the Río Turbio from its minor to its major bed, when a 30 m^3/s river discharge

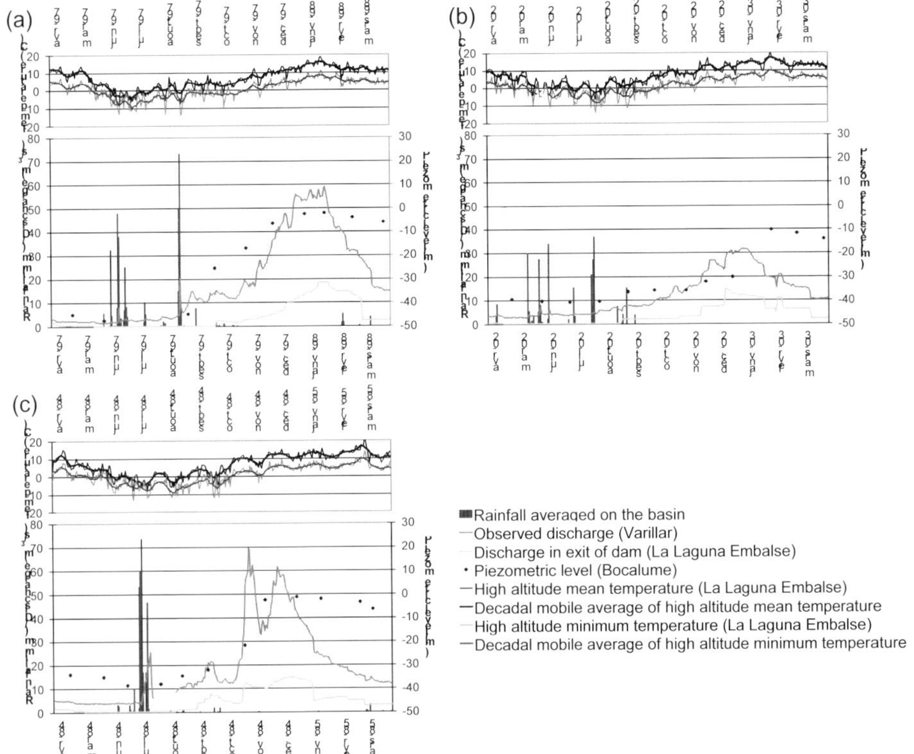

Fig. 6 Daily records of climatic and hydrologic variables for selected hydrological years: (a) 1997–1998 characteristics of dynamics 1, (b) 2001–2004 characteristics of dynamics 2, and (c) 1983–1986 characteristics of dynamics 3.

threshold is reached (Fig 6(b)). Another reason, as proposed in the monthly analysis, would be the change in the structure of the snow cover, more resistant to melt and thus delaying the contribution of melt water to surface flows.

In dynamic 3 (Fig 6(c)), the increase in both the piezometric level and river discharge occurs when high altitude minimum temperatures exceed –3°C and remain within the range previously identified (–3°C and 0°C) but for a shorter period than in dynamic 1. After exceeding the –3°C threshold for several days, a fall in high altitude minimum temperatures to below –3°C causes the refreezing of melt water. When high altitude minimum temperatures overpass the 0°C threshold again, this triggers an increase in river discharge and then a rise in piezometric level, when the 30 m^3/s discharge threshold corresponding to flooding into the major bed is attained.

These temperature thresholds and ranges may thus be used as warning pointers for flood risks downstream in the watershed, especially when dynamic 3, associated with large flood peaks, is likely to occur.

CONCLUSION

Analysis of runoff coefficients over the last decades shows that when annual precipitation is lower than approximately 60–70 mm, the relative equilibrium between runoff and recharge processes is broken, which may generate a large groundwater deficit that will result in a high hydric deficit at the watershed scale. Accordingly, this threshold may be used as a warning for drought. The spatial variability of the runoff coefficients shows higher contributions of the subsurface compartment in zones where granites are the most represented, thus highlighting the recharge of the aquifer from the underlying fractured granite bedrock. Accordingly, such zones would be suitable for sustainable groundwater resource exploitation.

Analyses at the monthly time scale performed for the Río Turbio catchment over the 1977–2008 period allowed us to identify three distinct dynamics of interaction between surface and subsurface compartments. Each of these dynamics is associated with a large groundwater recharge followed by a relatively long recession of the piezometric levels (1–2 years) and a shorter recession of the river (4–6 months). As climate trend is marked by an increased frequency of *El Niño* events in recent decades (IPCC, 2007), this should favour dynamics 1 and 2, associated with such events, and thus enhance the relative importance of the groundwater resource with respect to surface water resource. However, the trend in temperature at the catchment scale may favour the less efficient dynamics (dynamic 2) in terms of groundwater recharge.

Finally, a daily analysis allowed identification of temperature thresholds and ranges that may explain both the onset of surface runoff and groundwater recharge; they could be used as warning pointers for flood risks downstream the watershed, especially when dynamic 3 is likely to occur.

Acknowledgements This work was realized within the framework of the action ECOS no. C10U01. It was supported by the HydroSciences Laboratory, but also by the CEAZA and Agropolis international within the framework a partnership between the Conseil general de l'Hérault and the Coquimbo region. The authors would like to thank the *Dirección General de Aguas* (DGA, Chile) for providing the necessary data for this study.

REFERENCES

Cabezas, R., Cepeda, J. & Bodini, A. (2007) Descripción cartográfica de la hoya hidrográfica del río Elqui (Región de Coquimbo Chile). Universidad de La Serena.

Favier, V., Falvey, M., Rabatel, A., PradeRío, E. & Lopez, D. (2009) Interpreting discrepancies between discharge and precipitation in high-altitude area of Chile's Norte Chico region (26–32°S). *Water Res.* **45**, W02424.

IPCC (2007) *Climate Change 2007: The Physical Science Basis*. Contribution of Working Group I to the Fourth Assessment Report of the Intergovernmental Panel on Climate Change (ed. by S. Solomon, D. Qin, M. Manning, Z. Chen, M. Marquis, K. B. Averyt, M. Tignor & H. L. Miller). Cambridge University Press, Cambridge, United Kingdom and New York, NY, USA, 996 p.

Jourde, H., Roesch, A., Guinot, V. & Bailly-Comte, V. (2007). Dynamics and contribution of karst groundwater to surface flow during Mediterranean flood. *Environmental Geology J.* **51**, 725–730.

Luengo, P., Oyarzún, R., Oyarzún, J., Alvarez, P. & Canut de Bon, C. (2006) Aguas subterráneas en macizos rocosos fracturados : su utilización en zonas rurales montañosas del Centro Norte de Chile. VIII Congreso Latinoamericano de Hidrología Subterránea, Asunción, Paraguay.

Miranda, E. & Llanca, J. C. (2003) Estudio de las aguas subterráneas en la cuenca del Río Elqui. Memoria de título para optar al titulo de Ingenior Civil. Universidad de La Serena, 203 p.

Ruelland, D., Brisset, N, Jourde, H. & Oyarzun, R. (2011) Modelling the long-term impact of climatic variability on the groundwater and surface flows from a mountainous catchment in the Chilean Andes. In: *Cold Regions Hydrology in a Changing Climate* (Proc. of symp. H02 held during IUGG2011 in Melbourne, Australia, July 2011). IAHS Publ. 346, this volume.

Souvignet, M., Heinrich, J. & Gaese, H. (2010) Recent hydro-climatic trends in the arid Northern-Central Chile: assessing climate variability for policy makers. In: *Global Change: Facing Risks and Threats to Water Resources* (Proc. of the 6th World FRIEND Conf., Fez, Morocco, October 2010). IAHS Publ. 340, 688–694.

USDI (1955) Elqui Valley Chile, Ground Water Investigation, 92 p. United States Department of The Interior – Bureau of Reclamation.

5 Ecosystems and Methods

Man-made oasis change and its effects on the hydrological regime of the Aksu River basin

SUXIA LIU, CHUN ZHANG, SHOUHONG ZHANG & XINGGUO MO

Key Laboratory of Water Cycle & Related Land Surface Processes, Institute of Geographic Sciences and Natural Resources Research, Chinese Academy of Sciences, Beijing 100101, China

liusx@igsnrr.ac.cn

Abstract By carefully classifying the NDVI spatial information retrieved from MODIS 13 over the Aksu Basin (China) into seven categories based on fractional vegetation cover, with a careful division of the whole study region (WS) into man-made sub-region (MMS) and natural sub-region (NS), and with special consideration of the seasonal difference between summer and winter, a new index, called the man-made oasis index (MMOI), to describe the extent of man-made oasis (EMMO), is proposed. It is expressed as the linear weighted combination of the area ratio of each class from III to VI to the total area, with the higher the class number the higher the weight. The reason to choose classes from III to VI is that in winter they can be only found in MMS. MMOI in winter in MMS shows an increasing trend over the last 10 years, which matches well with the increase of EMMO found from the documented study. A transfer function between MMOI in winter in MMS and EMMO is then proposed to calculate EMMO based on MMOI. As paddy field was only found located in MMS, evapotranspiration over the paddy field (ET_P) simulated by the VIP distributed eco-hydrological dynamic model was chosen as the rate representative of water consumption by man-made oasis (WCMMO) per unit of EMMO. WCMMO is then calculated yearly based on the ET_p information multiplied with EMMO based on the index MMOI. The simulated results of yearly WCMMO are useful in exploring the effects of the oasis on the hydrological regime of the Aksu River.

Key words index; man-made oasis; VIP distributed eco-hydrological model; water consumption; the Aksu River, China

INTRODUCTION

The Aksu River Basin (Fig. 1(a)), with the headwater rivers Toxkan and Kumarik flowing into the Aksu River, is located in the arid area of northwest China, to the south side of west Tianshan Mountain and northwestern part of Tarim Basin. The basin's area is 5.9×10^4 km^2, extending from 75°35′ to 80°59′E and from 40°17′ to 42°27′N. It is a sensitive geographical zone with glaciers in the mountain headwaters, the oasis in the midstream, and the desert in the downstream with very low precipitation, high evapotranspiration and scarce water resources. With the development of the economy, the area of natural oasis, including natural grassland, natural forest, wetland, river and lake, is changing, generating some new land units. These are called man-made oasis and include cultivated land, man-made forest, man-made grassland, urban and reservoirs (Liu *et al.*, 2008). With the increase of the area of man-made oasis, it exerts influence on the hydrology of the Aksu River. All man-made oasis units need water. Each year large amounts of water are withdrawn and diverted from the river to irrigate the crops, trees and grass. Generally, the higher the level of economic development, the larger the extent of man-made oasis (EMMO), and the more the water consumed by the man-made oasis (WCMMO).

As water flowing by the Aksu region is not just for the use of the Aksu's economy, but also for the downstream environment, it is necessary to determine the rational EMMO of the Aksu region. The higher the EMMO, the more serious stress the downstream environment will face. A rational point of EMMO for the Aksu region is important for the sustainable development for the whole region.

In order to determine this rational point of EMMO, it is important to explore the relationship between EMMO and WCMMO. However, this is not an easy task as the method to exactly estimate both EMMO and WCMMO is not fully developed. The EMMO can be estimated by remote sensing image (e.g. Liu *et al.*, 2008). However, it is time consuming to determine the EMMO on a yearly basis, which is essential building the relationship between EMMO and WCMMO. Usually the WCMMO is roughly estimated by the irrigation quota. However, this single value is not enough to establish the relationship. Furthermore, it may be possible to obtain the WCMMO information from river runoff changes. However, as the basin is incurring climate change (Shi *et al.*, 2002; Zhang *et al.*, 2010), it is difficult at present to separate the runoff changes caused by climate change and man-made oasis.

This paper proposes a new index, called the man-made oasis index (MMOI) to calculate the EMMO, based on the NDVI data derived from the Modis13 data (MODIS/Terra Vegetation Indices 16-Day L3 Global 250m SIN Grid V005). Daily runoff data collected at the Sharikilank station in the Toxkan River, Xehera station in the Kumarik River, Xidaqiao and Aral stations in the Aksu River from 1964 to 2008 are used to analyse the hydrological trend. There were missing data of runoff at Xehera (1973, 1974, 1990–2007), Xidaqiao (1967–1971, 1990–2000, 2006) and Aral (1968–69, 1972–73, 1975–76, 1990–2000). A process-based eco-hydrological model (VIP) is used to estimate the WCMMO with the information of EMMO. Meteorological data from 1960 to 2009 at 10 weather stations, including pan evaporation, sunshine duration, precipitation, temperature, and wind speed are used to drive the VIP model. The basin-wide average precipitation and temperature data are obtained by using the reverse square interpolation by the VIP model. Finally, the yearly WCMMO is calculated.

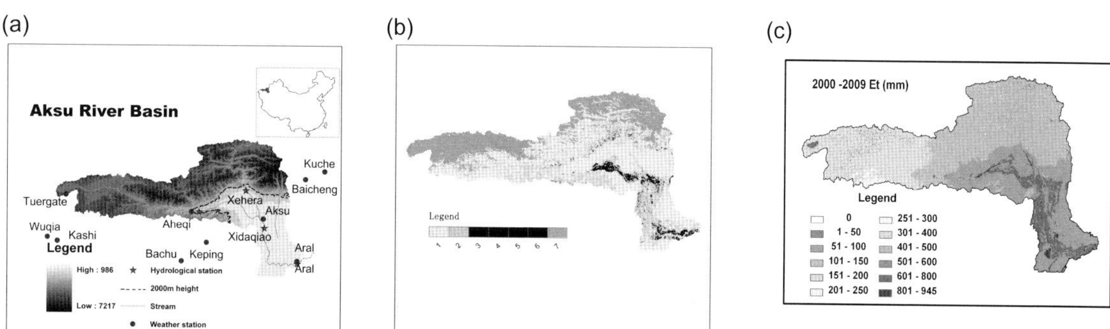

Fig. 1 (a) The Aksu River Basin, (b) distribution of classes III–VI (highlighted in black) in winter in 2009, and (c) the distribution of average evapotranspiration from 2000–2009 simulated by the VIP model.

METHOD

The Vegetation Interface Processes (VIP) model is used to simulate WCMMO. The VIP model (Mo *et al.*, 2009) was developed with the support of geographical information data. Its ecological and hydrological processes are coupled by a one-dimensional soil–vegetation–atmosphere transfer (SVAT) scheme, a distributed runoff routing scheme, and a vegetation dynamic scheme. The water cycle and energy transfer are coupled through evapotranspiration (ET, latent heat flux) process. Overland runoff is estimated with a variable infiltration curve, modified from Zhao (1992). Transfers of both overland and channel runoff are calculated with the kinematical wave equation. The estimation of canopy photosynthesis is based on Farquhar's biochemical model (Farquhar *et al.*, 1980). The canopy leaves are simplified into sunlit and shaded groups to account for their different photosynthesis responses to radiation, and a multilayer scheme is designed to describe the radiation penetration in canopy. Leaf area index (LAI) is the key state variable linking the hydrological processes and vegetation dynamics, which is derived from the foliage biomass.

Usually, after the ET over the grids is calculated based on the VIP model, annual WCMMO can be estimated for each year by summing up the values over those grids with man-made oasis recognized from the land-use map. However, this is not easy because the land-use map is only available for one limited period or at one single time point. Often we cannot obtain the yearly data of land-use cover. This encourages us to explore a new way to calculate yearly WCMMO.

Julian days 49 and 225 have been chosen to represent the typical conditions in the winter and summer seasons. NDVI distribution is obtained from the MODIS 13 for these two days for every year from 2000 to 2009. NDVI is then classified into seven categories shown in Table 1 according to the fractional vegetation cover (Chen, 2002):

$$F = (N - N_{min}) / (N_{max} - N_{min}) \tag{1}$$

where N is the value of NDVI multiplied by 10 000. Based on the vegetation information over the basin, we take $N_{min} = 10$ for winter, $N_{min} = 56$ for summer, and $N_{max} = 9000$ for both winter and summer.

Table 1 The definition of seven categories for the NDVI classification.

Fractional vegetation cover	0–10%	10–20%	20–40%	40–60%	60–80%	80–100%	Ice, snow and water
Class	I	II	III	IV	V	VI	VII

From the land use map obtained from the Environment Centre of Chinese Academy of Sciences (http://www.resdc.cn/first.asp) and the MODIS product, it is seen that most man-made oasis areas are located in areas of elevation below 2000 m (Fig. 1(a)). We thus divide the whole study region (WS) into two sub-regions, i.e. natural sub-region (NS) where elevation is above 2000 m, and man-made sub-region (MMS) where elevation is below 2000 m. Figure 2 shows the ratio of the area for each class to the total area for summer and winter averaged over the 10 years for WS, MMS, and NS. It is clear that in NS, classes from III to VI disappeared in winter, while in MMS they can be found. This is a strong indication that classes III to VI can be used to index human activity.

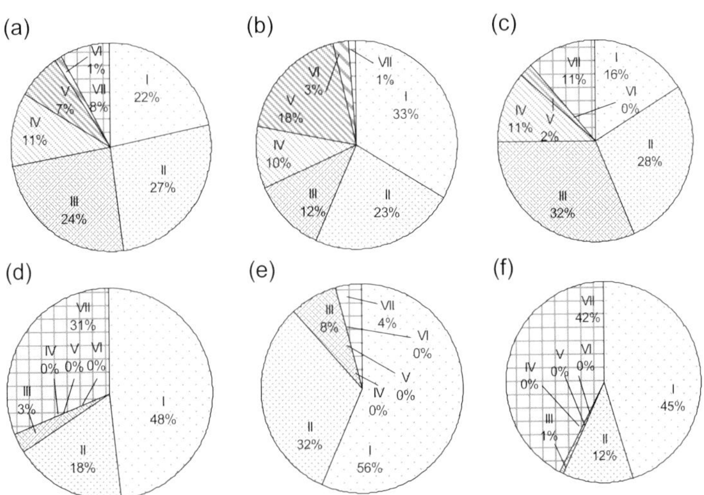

Fig. 2 The ratio of the area for each class to the total area for summer (upper panel) and winter (lower panel) averaged over the 10 years for WS (a), (d), MMS (b), (e) and NS (c), (f), respectively.

By comparing the distribution of classes III–VI in winter in the basin (Fig. 1(b)), it is clear that classes III–VI basically are distributed in the MMS area. It behaves like the footprint of the man-made oasis and can be used to index human activity. We thus propose the following index, called man-made oasis index (MMOI) to calculate EMMO:

$$O_q = 0.1 \times PLAND_{c3} + 0.2 \times PLAND_{c4} + 0.3 \times PLAND_{c5} + 0.4 \times PLAND_{c6} \tag{2}$$

where:

$$PLAND = Class\ area / Total\ area \times 100\% \tag{3}$$

c_3 to c_6 are classes III to VI, respectively. O_q is MMOI. The higher the class number, the higher the fractional vegetation cover. The larger the weight, the larger the EMMO. Based on MMOI plus the ET information from the VIP model, WCMMO can be obtained.

RESULTS

Figure 3(a) shows the annual precipitation record and its change over the Aksu River basin for the past 50 years. Precipitation slightly increased during 1955–2009, but significantly decreased in the last 10 years. Figure 3(b) shows the annual average temperature data and its change. There is a clear warming trend over the study period. In the last 10 years, temperature is rising with a relatively higher trend.

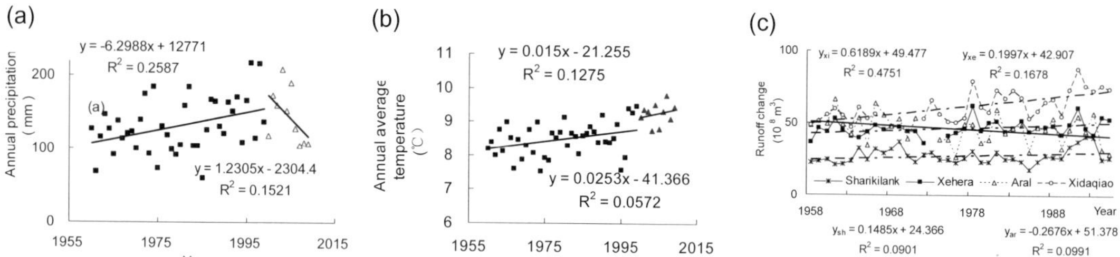

Fig. 3 (a) Precipitation data and change, (b) temperature data and change, and (c) annual runoff and changes, respectively, at the four hydrological stations.

Figure 3(c) shows the runoff data and changes during the past 51 years. Three hydrological stations, i.e. Sharikilank (Sh), Xidaqiao (Xi) and Xehera (Xe), have increasing trends, while the Aral station shows a weak decrease. The basin above the Xehera station receives a large supply of glacier melt water, as much as 52.4% (Wang, 2008). Runoff increase in this part of the basin may indicate snow and glacier melt water increases associated with temperature change (warming). On the other hand, the Aral (Ar) station, located at the downstream of the basin, shows a decreasing trend. This suggests that the effect of man-made oasis is larger than the effect of climate change, which counters its essential increasing trend with the increase of temperature. Studies (Shi *et al.*, 2002; Jiang *et al.*, 2005) show that with the increase of temperature, the glacial area is decreasing. More snow melt runoff caused discharge increases at the three stations in the upper basin, where human activity is less.

Figure 4(a) and (b) shows the change of monthly runoff averaged over a decade in the upstream and downstream of the basin. The grey bar represents an increase range for the past 50 years, and the black bar shows a decrease range. At the upstream station (Xehera), the warmer temperature (Fig. 3(b)) has resulted in a larger increase of runoff in summer, due to more water melted from snow, which is another signal of the influence of climate change on the regional hydrology. It is difficult to see this increasing signal at the downstream Aral station in every month due to the large amount of water withdrawn from the rivers (Zhang *et al.*, 2010). In other words, the hydrological regimes in the headwaters of the Aksu Basin are greatly influenced by climate conditions. However, human activity reduces the amplitude of the trend or causes the change in the opposite direction after the river passes through the city of Akesu, which is the main part of man-made oasis in the basin. This explains how, due to the combined effect of climate and human activity, it is not appropriate to use runoff data to determine WCMMO directly.

Figure 5 displays the MMOI data and their change over time. The MMOI in MMS increased in winter and decreased in summer. Over the NS or WS, the R^2 of the increasing trend is lower or the trend is decreasing. Liu *et al.* (2008) shows that the EMMO of Aksu city increased from 1987 km^2 in 1975 to 2865 km^2 in 2000 and 3608 km^2 in 2005. The winter MMOI in MMS matched this trend well. This indicates again that winter MMOI in MMS is a good index for quantifying the extent of man-made oasis. Assuming the change of EMMO, expressed as Y_a, is linear, with the data provided in Liu *et al.* (2008), we can obtain a linear equation for Y_a as follows:

$$Y_a = 148.6X + 2716.4 \tag{4}$$

where X is the order number of the year. Denoting MMOI in winter in MMS as Y_m, the linear

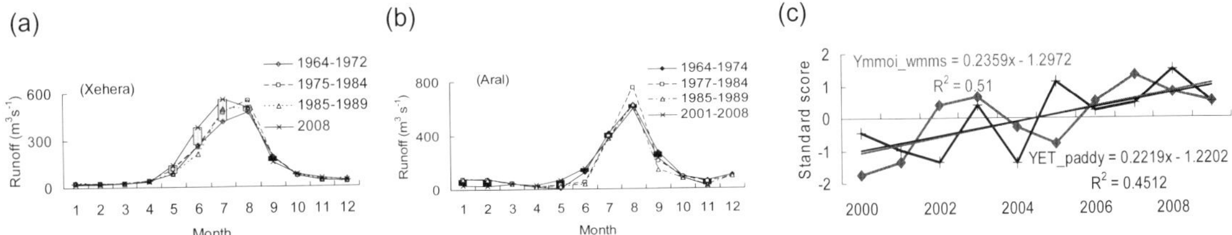

Fig. 4 Comparison of mean monthly runoff at: (a) Xehera, (b) Aral station, and (c) the standard score of ET for paddy field and winter MMO in MMS.

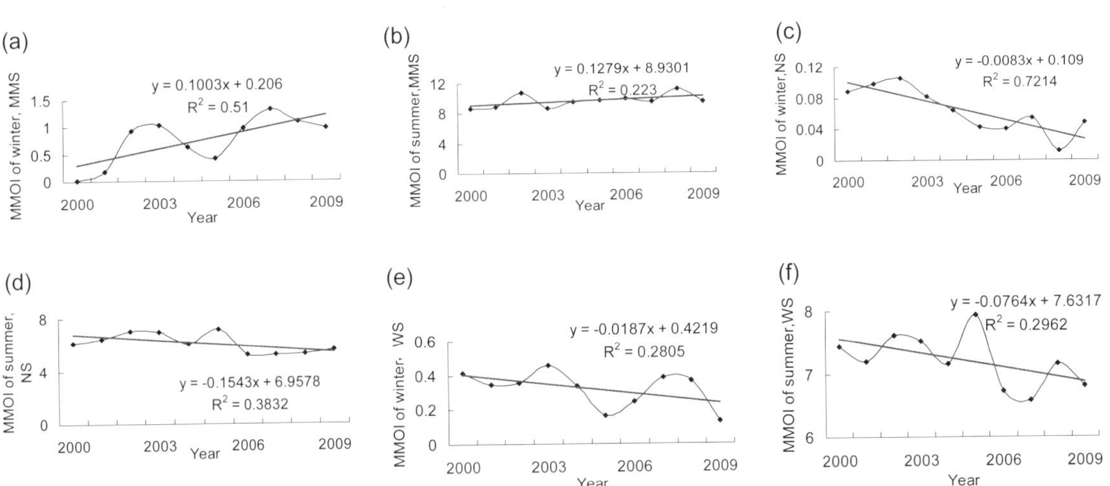

Fig. 5 MMOI in (a) winter, MMS (b) summer, MMS (c) winter, NS (d) summer, NS (e) winter, WS, and (f) summer, WS.

equation from Fig. 5(a) can be written as:

$$Y_m = 0.1003X + 0.206 \tag{5}$$

From the above two linear equations, we obtain the transfer function between Y_m and Y_a as:

$$Y_a = 1482Y_m + 2411 \tag{6}$$

As mentioned in the introduction, it is difficult to calculate the WCMMO from river runoff. We therefore use the VIP model to simulate ET. By combing ET with the information about EMMO as shown in equation (6), we can determine WCMMO.

ET has some advantages over runoff in detecting WCMMO as its spatial pattern is closely related with the land use regimes. However, it is a challenge to validate the modelled ET data. Usually the water balance method is a good tool to validate simulated ET (Mo *et al.*, 2004). In our study basin, this is difficult due to the unknown amount of water withdrawal for various activities. Therefore, we cannot use observed runoff to estimate WCMMO. We thus have to compare our results with other studies. Lei *et al.* (2006) estimated ET over the irrigated land as 686 mm/year. Average annual ET simulated by VIP model is 889 mm for the whole basin and 698 mm for the paddy land (Fig. 1(c)). These results matched well with the documented study.

By overlaying the land-use map and 2000 m dividing-line from the DEM, we obtain the land use distribution for NS and MMS. Then the grid numbers of each land use type in NS, WS and MMS are obtained using ArcGIS. We found that paddy field only exists in the MMS. From the distributed results of ET simulated by VIP, we get the ET from paddy field and an increasing trend (Fig. 4(c)). The increases of winter MMO in MMS matched well with the ET trend for paddy field.

By choosing ET for paddy field as the representative rate of WCMMO per unit of EMMO, we then can obtain a clear picture of WCMMO, by multiplying the annual ET rate with Y_m (equation (6)), as shown in Table 2.

Table 2 WCMMO results based on the ET simulated by VIP model and EMMO based on the new index MMOI.

Year	2000	2001	2002	2003	2004	2005	2006	2007
WCMMO (10^8 m^3)	16.58	17.64	24.25	28.12	21.41	22.60	27.34	31.31

CONCLUSIONS

The NDVI spatial data from the MODIS 13 over the Aksu Basin are classified into seven categories based on fractional vegetation cover. The whole study area (WS) is divided into man-made sub-region (MMS) and natural sub-region (NS), with a special consideration of the seasonal, summer–winter difference. A new index, called the man-made oasis index (MMOI) to describe the extent of man-made oasis (EMMO) is developed. It is expressed as the linear weighted combination of the ratio of the area of each class (from III to VI) to the whole area. The special choice of classes III to VI is based on careful observation of the differences among the seven classes in winter and summer. In NS, classes III to VI all disappear in winter, while all classes exist in MMS during 2000 to 2009. MMOI shows an increasing trend. During this period, the man-made oasis was extending greatly (Liu *et al.*, 2008). This match encourages us to choose the contribution of classes III to VI as the footprint of man-made oasis. A transfer function between MMOI and EMMO was then derived to calculate the EMMO based on MMOI. As paddy field was found only in MMS, evapotranspiration over the paddy field (ET_p) simulated by the VIP eco-hydrological model was chosen as the representative rate of water consumption by man-made oasis (WCMMO) per unit of EMMO. WCMMO is then calculated annually based on the information of Etp, EMMO and the index MMOI. The results of yearly WCMMO are useful for decision makers to monitor the extent of man-made oasis and make a sustainable plan beneficial to local economical development, the downstream environment, and ecosystem protection.

Acknowledgements This study was jointly supported by the Chinese Ministry of Science and Technology for "973" project (2009CB421300, 2010CB428404), International Cooperation project (0911) and water special key project (2009ZX07210-006), Natural Science Foundation of China grant (41071024, 40830636), the Knowledge Innovation Project of Chinese Academy of Sciences (KZCX2-YW-Q06-1-3). Thanks to the great efforts of the reviewers to help improve this paper.

REFERENCES

Chen, J. *et al.* (2001) Sub- pixel model for vegetation fraction estimation based on land cover classification. *J. Remote Sensing* **5**(6), 416–423 (in Chinese).

Farquhar, G. D., von Caemmerer, S. & Berry, J. A. (1980) A biochemical model of photosynthetic CO_2 assimilation in leaves of C3 species. *Planta* **149**, 78–90.

Lei, Z., Hu, H. & Yang, S. (2006) The analysis of oasis water consumption in Tarim basin. *Shuilixuebao* **37**(2), 1470–1475 (in Chinese).

Liu, J., Gao, M. & Su, J. (2008) Analysis of oasis temporal and spatial variation at northern Tarim Basin based on RS and GIS. *Resources Environment and Development* **2**, 13–16 (in Chinese).

Jiang,Y., Zhou, C. H. & Cheng, W. M (2005) Analysis on runoff supply and variation characteristics of Aksu drainage basin. *J. Natural Resources* **20**(1), 27–34 (in Chinese).

Mo, X., Liu, S., Lin, Z. & Zhao, W. (2004) Simulating temporal and spatial variation of evapotranspiration over the Lushi basin. *J. Hydrol.* **285**, 125–142.

Mo, Xingguo, Liu,Suxia, Lin Zhonghui & Guo Ruiping (2009) Regional crop yield, water consumption and water use efficiency and their responses to climate change in the North China Plain. *Agric. Ecosystems & Environ.* **134**, 67–78.

Shi, Y., F. Sheng, Y. P. & Hu, R. J. (2002) Preliminary study on signal, impact and foreground of climatic shift from warm-dry to warm-humid in Northwest China. *J. Glaciology and Geocryology* **24**(3), 219–226 (in Chinese).

Wang, G. Y. (2008) Runoff changes in Aksu River basin during 1956–2006 and there impacts on water availability for Tarim River. *J. Glaciology and Geocryology* **30**(4), 562–569 (in Chinese).

Zhao, R. J. (1992) The Xinanjiang model applied in China. *J. Hydrol.* **135**, 371–381.

Zhang, S. (2010) Assessing the impact of climate change and human activities on the hydrological processes in the Aksu River Basin. MSc Thesis. Institute of Geographic Sciences and Natural Resources Research. Beijing, China.

Zhang, S. H., Liu, S. X., Mo, X. G., Shu, C., Sun, Y. & Zhang, C. (2011) Assessing the impact of climate change on potential evapotranspiration in the Aksu River Basin. *J. Geogr. Sci.* **21**(3), 1–12.

Stream guiding algorithm for deriving flow direction from DEM and location of main streams

JIAHU WANG[1], LI LI[1], ZHENCHUN HAO[1] & JONATHAN J. GOURLEY[2]

1 *State Key Laboratory of Hydrology-Water Resources and Hydraulic Engineering, Hohai University, Nanjing 210098, China*
tigerlly@126.com

2 *NOAA/National Severe Storm Laboratory, Norman, Oklahoma 73072, USA*

Abstract The drainage paths and directions within the drainage basin are important for analyses of the interactions between human and nature. The *stream burning* algorithm is a popular D8-based method and can be effective in the digital reproduction of a known and generally accepted stream network. The *stream guiding* algorithm has been developed in this paper to overcome the stream burning algorithm's disadvantage of locally altering elevation in order to provide the consistency between existing vector hydrography and the DEM. In the new algorithm, flow direction of LMS (location of main streams) grids will be determined first; then possible outlets in non-LMS area will be found; and finally, the flow direction of undetermined area will be calculated by a "filling up" technique. Evaluations for Taiwan Island show that the new algorithm has a similar performance to that of the stream burning algorithm in river network reproduction. The new algorithm obeys the "steepest decent rule" and DEM data more strictly than the stream burning algorithm, especially around the LMS grids.

Key words flow direction derivation; DEM; location of main stream

1 INTRODUCTION

The drainage paths and directions within the drainage basin are important for analyses of the interactions between human and nature. Many methods of judging flow direction and extracting channel network appeared after the 1980s, including the D8 method (O'Ccallaghan & Mark, 1984), the D∞ method (Tarboton, 1997), the multi-direction method (Tarboton, 1997), and other methods to do with pits – the local elevation minima (Hutchinson, 1989). In general, a drainage direction map (DDM) can be derived from a digital elevation map (DEM) by applying standardized and automated procedures. Many software packages, in particular GIS, provide tools to derive the drainage direction for each raster cell of a DEM, by comparing the elevation of the cell to the elevation of its neighbouring cells.

The well-known D8 algorithm (O'Callaghan & Mark, 1984) is the most commonly used method for approximating flow directions on a topographic surface, and this method tracks "flow" from each pixel to one of its eight neighbour pixels. However, it is based on two simplifying assumptions: 1. the use of eight discrete flow angles; and 2. each pixel has a single flow direction (SFD), that does not capture the geometry of divergent flow over hillslopes. That is, all flow leaving any given pixel (or the area contributing flow to the pixel) is assumed to flow into a single downstream neighbour pixel. The D8 method is well-suited to the identification of individual channels, channel networks, and basin boundaries.

Deriving drainage direction from a DEM by the D8 method is a straightforward approach; however, poor quality or simply the inherent generalization of a DEM may cause derived drainage lines to differ from reality (Döll & Lehner, 2002). Thus, a number of methods for DEM improvements have been suggested, such as the removal of spurious sinks (Jenson & Domingue, 1988; Soille *et al.*, 2003), incorporation of vector stream data for stream burning (Maidment, 1996; Mizgalewicz & Maidment, 1996; Saunders, 1999) or surface reconditioning (Hutchinson, 1989, 2004; Hellweger, 1996).

Several studies have addressed this problem through a DEM post-creation modification technique utilizing the vector hydrology layer, a process commonly referred to as the *stream burning* (Mizgalewicz & Maidment, 1996; Saunders & Maidment, 1996; Hellweger, 1997). Where vector hydrography information exists, it can be integrated into the DEM prior to the actual analysis. This process is referred to as the *stream burning* and can be effective in the digital reproduction of a known and generally accepted stream network. However, it has the disadvantage of locally altering elevation in order to provide the consistency between existing vector hydrography and the DEM. Several methods exist (Hutchinson, 1989; Saunders & Maidment, 1996) but greatly differ in their success of improving watershed delineation (Saunders, 1999). The pre-processing of the vector information required often represents an intensive effort.

In this paper we present our new *stream guiding* algorithm to derive the DDM by an interactive analysis of the DEM and the location of main streams (LMS) without altering any DEM data. It inherits the advantage of the *stream burning* algorithm: bringing additional information into the DEM to position main rivers correctly; and overcomes the disadvantage of locally altering elevation. Evaluations are given by both manual inspection and a statistical comparison after detailed description of the algorithm.

2 STREAM GUIDING ALGORITHM

In this algorithm, flow direction of LMS grids will be determined first; then possible outlets in the non-LMS area will be found; and finally, flow direction of the undetermined area will be calculated by a "filling up" technique. The flow diagram in Fig. 1 shows its mechanism. The whole strategy is toward extending indirect outlets to upstream from estuary by LMS data, and making it easier for inland grids, and finding a way to all the outlets. We thus named it the *stream guiding* algorithm.

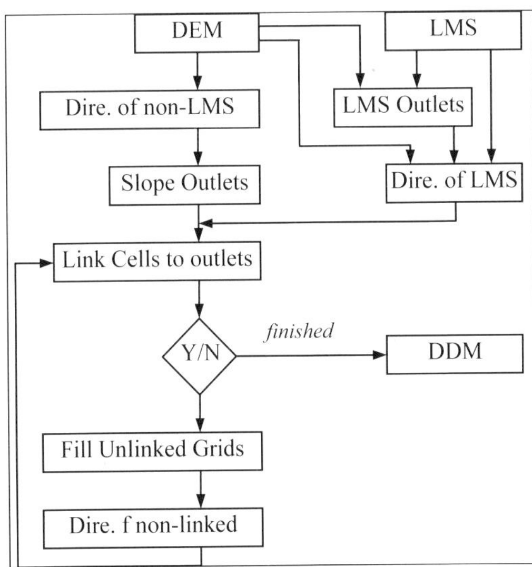

Fig. 1 Flow diagram of stream guiding algorithm (DEM is digital elevation model; LMS is location of main streams; Dire. refers to flow direction; DDM is drainage direction map).

2.1 Flow direction determination

The location of the main stream (LMS) can be derived from maps or from available data sets (e.g. the ArcWorld database) (see Fig. 2(a)). It should be converted into gridded values, as the elevation of LMS grids from DEM data is necessary for interactive analysis (see Fig. 2(b)).

To each LMS grid, the number of its adjoining LMS grids from all eight possible directions (LMS-a8n) needs to be counted first (see table in Fig. 2(c)). Then we can obtain the estuary grid by picking up the one with both minimum elevation and GNT-a8n. This is done based on a

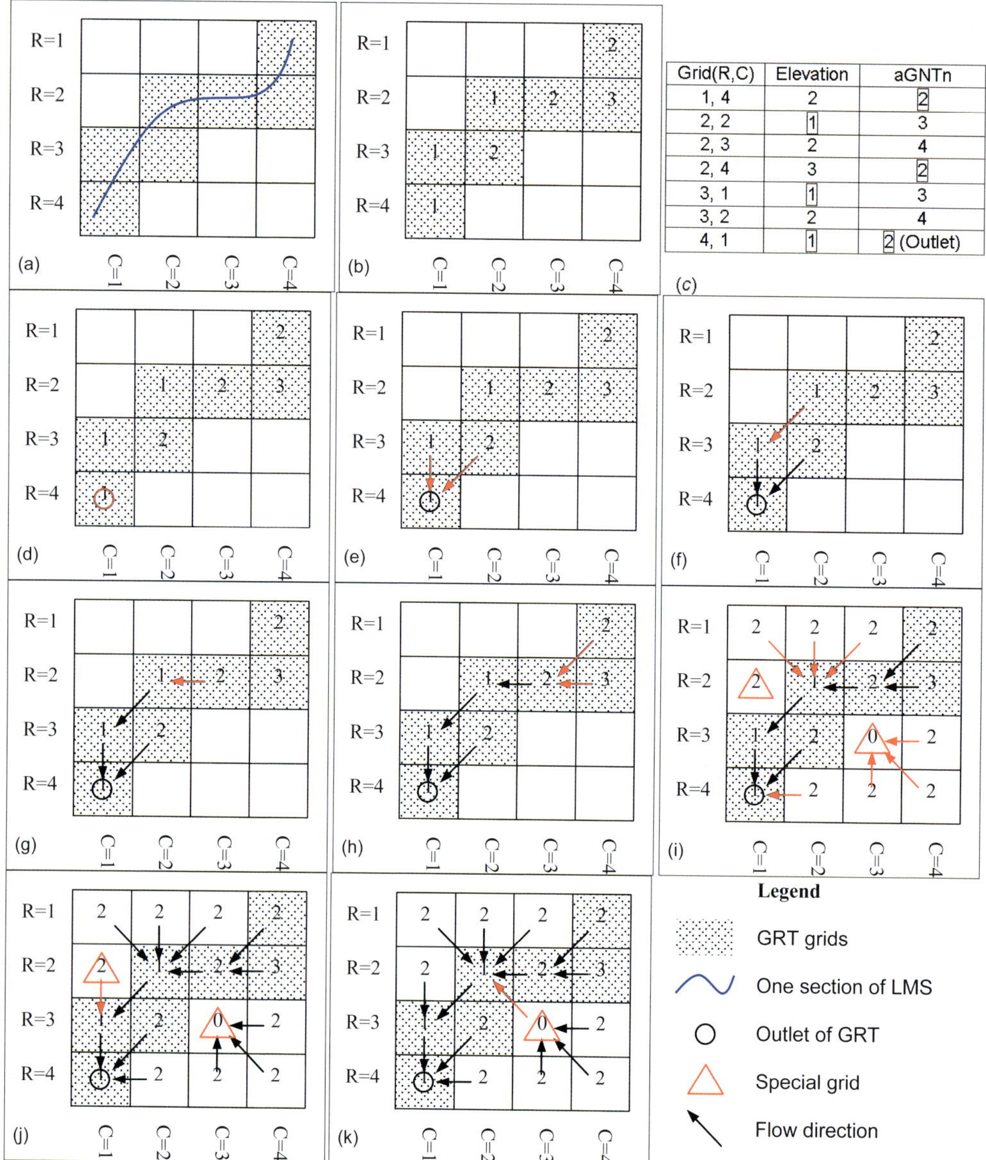

Fig. 2 Theoretical description of deriving flow direction from the DEM and LMS (location of main streams).

comparison of grids in the middle of a LMS to the grid at the end of the LMS (one end of the estuary or several ends of each branch's upstream should have smaller LMS-a8n). In addition, the estuary grid is lower than the other grids. Estuary here means river outlets to both oceans and inland lakes (see Fig. 2(d)). We need to do this repeatedly to find each dendritic river network's estuary in a region.

The flow direction of all adjoining LMS grids of an estuary can be set directly by pointing to it (see Fig. 2(e)). The lower elevation grid has priority to receive water from the upper grid if they have the same upper adjoining LMS grids. For example: $grid_{(3,1)}$ and $grid_{(3,2)}$ have the same upper adjoining LMS grids $grid_{(2,2)}$ in Fig. 2(f), $grid_{(2,2)}$ should flow into $grid_{(3,1)}$ because $grid_{(3,1)}$ is lower than $grid_{(3,2)}$ (elevation 1 m *vs* 2 m). This "steepest descents" processing is repeated so that the flow direction of all the LMS grids can be determined (see Fig. 2(f)–(h)).

2.2 Flow direction determination

Our method scans all non-LMS grids with a 3 × 3 moving window, as used in the other regular D8 method. If their flow direction is pointing to any determined grids, either directly or indirectly, we upgrade them to determined grids. These grids can be found in Fig. 1(i) as $grid_{(1,1)}$, $grid_{(1,2)}$, $grid_{(1,3)}$ and $grid_{(4,2)}$.

Ambiguous flow directions (the same minimum downslope gradient is found in two cells) are usually resolved by an arbitrary assignment. $Grid_{(2,1)}$ in Fig. 1(h) is a typical example; it can flow into $grid_{(2,1)}$ and$_{(2,2)}$ according to elevation. In our method, we set it to flow into the downstream one because the downstream grid should be a bit lower than upstream one, although we can not establish that in the DEM. This means flow direction of $grid_{(2,1)}$ should point to $grid_{(3,1)}$ in Fig. 2(j).

The flow direction of local elevation minima (pits) grids can not be determined directly either. If a pit grid adjoins the ocean grid in any of the eight possible directions, it is considered as a small estuary; otherwise it needs to be filled-up and scanned again. An example of such a pit, $grid_{(3,3)}$, can be found in Fig. 2(j), its flow direction can be determined by just one cycle of filling-up and scan (see Fig. 2(k)). Actually, the process of filling-up and scan often need to be repeated hundreds or thousands times in order to determine all the grids' flow directions.

2.3 Overview of method

The stream guiding algorithm can be summarized by 10 steps as in Fig. 2: (a) determine the LMS and grid with same resolution and span as the DEM; (b) get elevation of LMS grids from the DEM data; (c) compare elevation and LMS-a8n (number of its adjoining LMS grids from eight possible directions); (d) pick up the grid with minimum elevation and LMS-a8n as outlet; (e) set flow direction of all adjoining grids of outlet pointing to outlet; (f) compare elevation of newly determined grids, pick up one with minimum elevation, set flow direction of other adjoining grids pointing to the picked one in step f; (g) repeat step f; (h) repeat step f until the flow directions of all LMS grids have been determined; (i) scan flow direction of the non-LMS grids with a 3 × 3 moving window, and determine the grids either directly or indirectly pointing to determined grids; (j) determine the flow direction of ambiguous grids (the same minimum downslope gradient is found in two cells) by pointing to the downstream grid; and (k) deal with pit(s) by the filling-up method or set it as possible small outlet.

2.4 Example

A real case illustrated in Fig. 3 shows the working of the stream guiding algorithm from a zoom out view. The region is a stochastic rectangular area from Hydro1k (Team, 2003) and the LMS used here was digitized from DEM maps. The processing of this real area is the same as described above and the grid number is the only difference. We obtain reasonable results, such as: (1) LMS map (Fig. 3(a)); (2) determined LMS map (Fig. 3(b)); (3) scan of slope outlets (Fig. 3(c)), and (4) final drainage direction map (Fig. 3(d)).

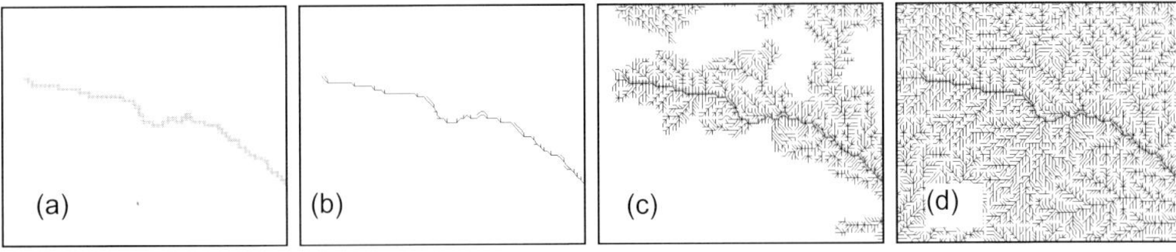

Fig. 3 An example of deriving flow direction from DEM and LMS in a stochastic rectangle area from Hydro1k; and (d) final DDM (drainage direction map).

3 EVALUATION AT AN ISLAND SCALE

A continent is a perfect region to use for an evaluation, but an island is more convenient for testing the results as it can be considered as a shrunken continent because (1) it is surrounded by ocean, and (2) has both mountain and plain areas. Taiwan Island was chosen for this study (Fig. 4(a)). Its land area is approximately 36 000 km^2 with mountains reaching 3952 m in the middle. It is an interesting island for hydrologists because in this relatively small area, the annual mean rainfall is about 2510 mm, about 2.6 times the world annual mean rainfall, but nearly 78% of the rain occurs during the wet months from May to October.

The DEM data for the island are from Hydro1k (Team 2003), and the river network is from the ArcWorld database (ESRI, 1992); 6.1% of total DEM grids were marked as LMS. Deriving flow direction according to the *stream burning* algorithm was executed by ARCGIS (ESRI, 2006); deriving flow direction according to the *stream guiding* algorithm was completed using the Channel Network Tool (Wang *et al*., 2005), a small software developed by us.

3.1 Qualitative evaluation

The performance of our algorithm is difficult to assess directly from drainage direction maps. We therefore checked it by generating a flow accumulation map (FAM). A FAM represents the number of upstream grids, i.e. the number of those grids that drain through a given grid, and is comparable to LMS.

From the FAM (Fig. 5) we can find that both stream burning (see Fig. 4(b)) and stream guiding (see Fig. 4(c)) algorithms generated perfect river networks, and especially the location of main streams. The difference between them is hard to identify visually and the two figures in Fig. 5 are so alike that one could be a duplicate of the other. This shows that the stream guiding algorithm has the same performance in deriving the river network as the stream burning algorithm.

But, their detail is really different because of the different mechanisms used in their determination. The stream burning algorithm gave a regular flow direction around the LMS grids (Fig. 5(a)), but the flow direction in such areas has no relationship with the elevation. However, flow direction generated by the stream guiding algorithm is closely related to the elevation values even that of the LMS grid (see Fig. 5(b)).

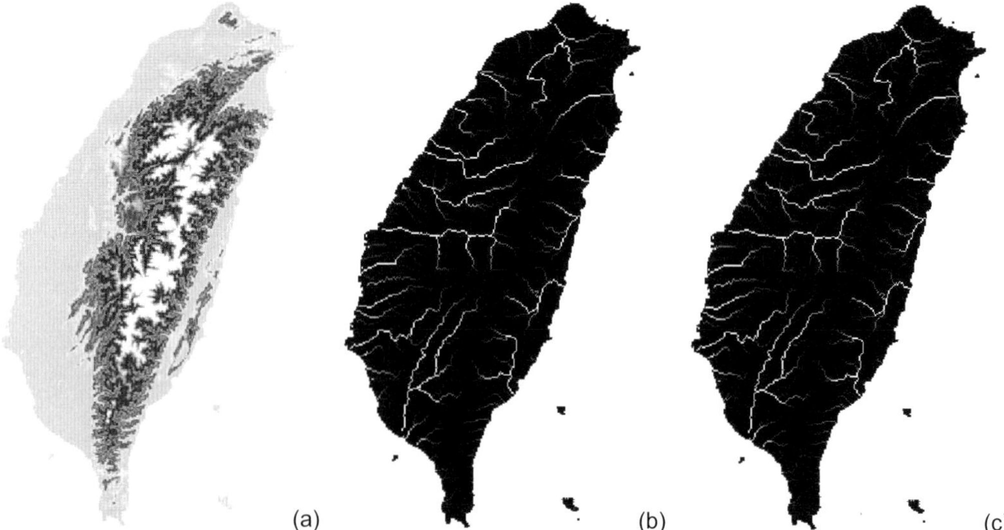

Fig. 4 (a) Study area. Overview of two FAMs (flow accumulation maps): (b) according to stream burning algorithm, and (c) according to stream guiding algorithm.

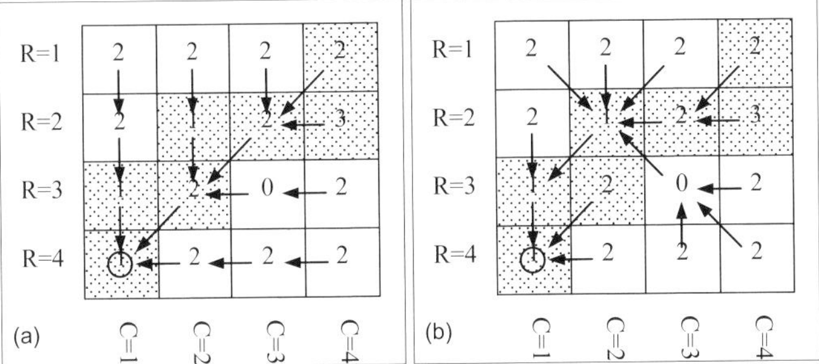

Fig. 5 Detailed view of two DDMs (drainage direction map): (a) flow direction near LMS according to stream burning algorithm has nothing to do with the elevation, and (b) flow direction generated by stream guiding algorithm sticking to the elevation values and "steepest decent rule".

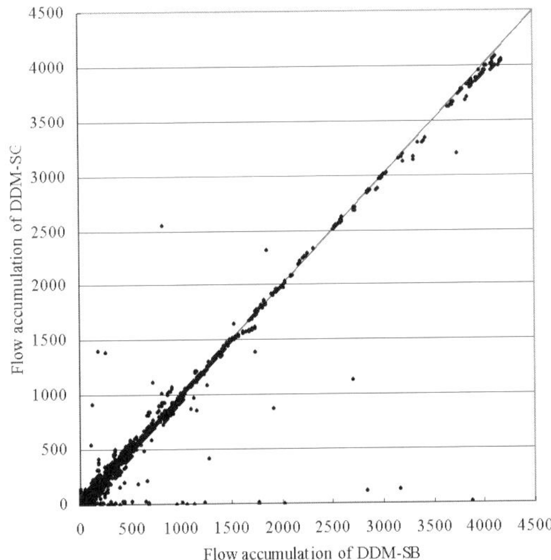

Fig. 6 Grid based comparison of FAMs between stream guiding and stream burning algorithms.

3.2 Quantitative evaluation

Grid based comparison of flow accumulation by the stream burning and stream guiding algorithms in Fig. 6 shows a strong cluster around the 1:1 line. The degree of correspondence is quantified by the modelling efficiency *ME* (Jansen & Heuberger, 1995), which is equivalent to the Nash-Sutcliffe coefficient and measures the goodness-of-fit to the line-of-perfect-fit (the 1:1 line):

$$ME = \frac{\sum_{i=1}^{J}\left(N_i^B - \overline{N}_i^B\right)^2 - \sum_{i=1}^{J}\left(N_i^G - \overline{N}_i^B\right)^2}{\sum_{i=1}^{J}\left(N_i^B - \overline{N}_i^B\right)^2} \tag{1}$$

where N_i^B is the flow accumulation of cell *i* according to the stream burning algorithm; $\overline{N}_i^B$ is average of all N_i^B; N_i^G is the flow accumulation of cell *i* according to the stream guiding algorithm; *J* is the total number of grids covering Taiwan Island, i.e. 46 937.

The ME here is 0.971. If we take flow accumulation according to the stream burning algorithm as the benchmark, about 75.7% grids of flow accumulation determined by the stream guiding algorithm have a bias ratio of less than ±1%, and 86.2% grids have a bias ratio of less than ±5%. This means that the river networks derived by these two algorithms are similar.

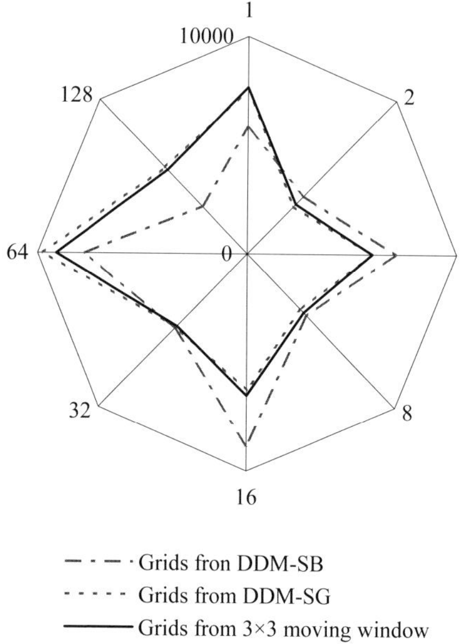

Fig. 7 Grid based comparison of DDM showing the DEM alternation in the stream burning (SB) algorithm and lost information from DEM around LMS grids.

The difference in detail between the two DDMs is also remarkable. The numbers of grids inclining to each direction are shown in Fig. 7. The flow direction of 93.2% grids according to the stream guiding algorithm follow the "steepest decent rule", when compared to the results of a 3 × 3 moving window. Only 75.1% of grids determined by the stream burning algorithm follow this rule, because the elevation of many grids has been lowered through a DEM post-creation modification technique. This result shows that the stream guiding algorithm respects the "steepest decent rule" more strictly than the stream burning algorithm.

4 CONCLUSION

The D8 method is widely used and implemented in many GIS software packages, despite its limitations. It is useful for a number of applications, such as extracting river network maps, longitudinal profiles, and basin boundaries. The s*tream burning* algorithm is a good method to obtain accuracy for dendritic river networks by importing the location of main streams into a DEM. But the alteration of the DEM leads to the loss of some information from the DEM around LMS grids; hence, we develop the *stream guiding* algorithm. Evaluations show that the new algorithm has a similar and good performance to the stream burning algorithm in reproducing river networks. The new algorithm follows the "steepest decent rule" and DEM data more strictly than the stream burning algorithm, and so the detail, especially around the LMS grids, is better.

Acknowledgements The financial support from the National Natural Science Foundation of China (50879016, 40830639, 40801012) is gratefully acknowledged. The authors would like to extend their appreciation to the Fundamental Research Funds for the Central Universities and the Special Fund of the State Key Laboratory of Hydrology-Water Resources and Hydraulic Engineering (1069-50985512).

REFERENCES

Döll, P. & Lehner, B. (2002) Validation of a new global 30-min drainage direction map. *J. Hydrol.* **258**(1-4), 214–231.

ESRI (1992) Global topographic database ArcWorld 1:3M on CDROM, Environmental Systems Research Institute, Inc. (ESRI).

ESRI, I. (2006) ESRI ArcMap 9.2.Redlands, USA.

Hao, Z., Wang, J. *et al.* (2005) Principium and function of Channel Network Tool-I (in Chinese). *Hydrology (China)* **25**(2), 15–19.

Hellweger, F. L. (1997) AGREE – DEM Surface Reconditioning System. http://www.ce.utexas.edu/prof/maidment/gishydro/ferdi/research/agree/agree.html as of July 1999.

Hellweger, R. (1996) Agree.aml. Center for Research in Water Resources, The University of Texas at Austin, Austin, Texas, USA.

Hutchinson, M. (1989) A new procedure for gridding elevation and stream line data with automatic removal of spurious pits. *J. Hydrol.* **106**(3-4), 211–232.

Hutchinson, M. F. (2004) ANUDEM Version 5.1 User Guide. http://cres.anu.edu.au/outputs/software.php. The Australian National University, Centre for Resource and Environmental Studies, Canberra, Australia.

Jenson, S. & Domingue, J. (1988) Extracting topographic structure from digital elevation data for geographic information system analysis. *Photogram. Engng Remote Sens.* **54**(11), 1593–1600.

Maidment, D. R. (1996) GIS and hydrological modeling: an assessment of progress. Presented at the Third International conference on GIS and Enviromental Modeling, Santa Fe, New Mexico, January 1996.

Mizgalewicz, P. J. & Maidment, D. R. (1996). Modeling agrichemical transport in midwest rivers using geographic information systems. Center for Research in Water Resources Online Report 96-6. University of Texas, Austin, Texas, USA.

O'Callaghan, J. &. Mark, D. (1984) The extraction of drainage networks from digital elevation data. *Computer Vision, Graphics, and Image Processing* **28**(3), 323–344.

Saunders, W. (1999) Preparation of DEMs for use in environmental modeling analysis. ESRI User Conference, 24–30 July 1999, San Diego, California (http://www.esri.com/library/userconf/proc99/proceed/papers/pap802/p802.htm).

Saunders, W. & Maidment, D. (1996) A GIS assessment of nonpoint source pollution in the San Antonio-Nueces Coastal Basin. Center for Research in Water Resources Online Report: 96–91.

Soille, P., Vogt, J. V., *et al.* (2003) Carving and adapting drainage enforcement of grid digital elevation models. *Water Resour. Res.* **39**(12), 1366–1378.

Tarboton, D. (1997) A new method for the determination of flow directions and upslope areas in grid digital elevation models. *Water Resour. Res.* **33**(2), 309–319.

Team, H. (2003) Web page of the HYDRO1k Elevation Derivative Database. Technical CLAVR-1 global clear/cloud classification algorithm for the Advanced Very High Resolution.

Wang, J., Hao, Z., *et al.* (2005) Extraction of drainage structure from digital elevation model by combination with raster river network (in Chinese). *J. Hohai University (Natural Sciences)* **33**(2), 119–122.

Key Word Index

HYDROCOMPLEXITY: New Tools for Solving Wicked Water Problems

Edited by *S. Khan, H. H. G. Savenije, S. Demuth & P. Hubert*
IAHS Publ. 338 (*2010*)
ISBN 978-1-907161-11-7,
272 + x pp. Price £55.00

Human activities have become major drivers of change in the Earth's biosphere, resulting in deterioration of water quality, overexploitation of freshwater resources, adverse effects of hydrological hazards and landscape degradation, which make water problems complex and wicked. The same activities also affect the functioning of ecosystems and their ability to provide goods and services on which human well-being depends. There is a need for community-based transdisciplinary management tools to provide a better understanding of water as both an abiotic resource and as a service delivered by ecosystems.

Abstracts of the papers in this volume can be seen at:

www.iahs.info

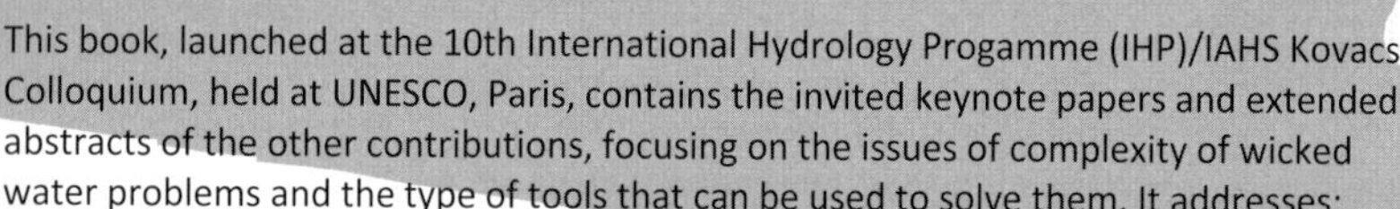

This book, launched at the 10th International Hydrology Progamme (IHP)/IAHS Kovacs Colloquium, held at UNESCO, Paris, contains the invited keynote papers and extended abstracts of the other contributions, focusing on the issues of complexity of wicked water problems and the type of tools that can be used to solve them. It addresses:

1 **Monitoring and Evaluating the Water Cycle**
2 **Linking Climate Change with Water Cycle Management**
3 **Parsimonious *vs* Complicated Approaches**
4 **Whole-of-System and Adaptative Approaches**
5 **Need for Transdisciplinary Issues Approaches to Deal with Water-related Ecosystems**
6 **Integrated Approaches**
7 **Role of Knowledge Platforms for Community Engagement**
8 **From Artificial to Embodied Intelligence**
9 **Water Allocation Dilemma**
10 **Water Quality – a Critical Issue**
11 **Managing Hydrohazards**

Please send book orders and enquiries to:

Mrs Jill Gash at jilly@iahs.demon.co.uk

IAHS Press, Centre for Ecology and Hydrology,
Wallingford OX10 8BB, UK
tel.: + 44 1491 692442
fax: + 44 1491 692448/692424

BENCHMARK PAPERS IN HYDROLOGY

The IAHS Series that collects together, by theme, the outstanding papers that provide the scientific foundations for hydrology today. Each volume reprints 30+ papers selected by a leading expert, who comments on their significance and context.

RAINFALL–RUNOFF MODELLING

Selection and Commentary by Keith Loague

Loague notes that hundreds, if not thousands, of hydrologic-response models have been developed, but that not all were created equal; this volume reprints 30 papers that exemplify the best and charts developments from Mulvany's (1851) rational method for estimating peak flow, probably the first rainfall–runoff model, up to 1989. Papers on other empirical approaches, such as Sherman (1932) and Mockus (1949), are reprinted. As are Richards (1931) and Smith & Parlange (1978), the innovative contributions of Alan Freeze, and later Keith Beven, and the seminal papers of Moore & Clarke (1981) and Abbott *et al.* (1986).

IAHS BM4   ISBN 978-1-907161-06-3 (2010)   A4 format, hardback, 506 + vi pp, £65.00

RIPARIAN ZONE HYDROLOGY AND BIOGEOCHEMISTRY

Selection and Commentary by T. P. Burt, G. Pinay & S. Sabater

Study specifically of riparian zones is relatively new in hydrology, and while the oldest of the 36 benchmark papers selected for this volume dates to 1936, several of the others were published in the 1970s and 1980s. They are grouped under the topics Landscape ecology, Hydrology of the riparian zone, Linking riparian zone hydrology to solute transport, Biogeochemical processes and methods, Riparian buffering of surface and subsurface flows, and In-stream Processes. Together, the reprinted papers and the editors' commentaries map the breakthroughs in the development of this important subdiscipline.

IAHS BM5   ISBN 978-1-907161-09-4 (2010)   A4 format, hardback, 490 + x pp, £65.00, incl. postage

---

HYDRO-GEOMORPHOLOGY, EROSION AND SEDIMENTATION

Selection and Commentary by Michael J. Kirkby

Kirkby presents a systematic analysis of the relationships between hydrology and geomorphology with commentaries on the papers that have been most influential in the development of research at the hydrology/geomorphology interface. Thirty-seven papers are reprinted in full or in part, the majority published pre-1970, including early contributions by Fisher (1866), Davison (1889) and Gilbert (1909), and seminal papers by Hack, Strahler, Wolman & Miller, and Melton, among others.

IAHS BM6   ISBN 978-1-907161-14-8 (2011)   A4 format, hardback, 640 + x pp, £70.00, incl. postage

---

The IAHS Benchmark Papers in Hydrology Series is sponsored by SAHRA

Please send orders and enquiries to:

Mrs Jill Gash
IAHS Press, Centre for Ecology and Hydrology
Wallingford, Oxfordshire OX10 8BB, UK

jilly@iahs.demon.co.uk
tel.: + 44 1491 692442
fax: + 44 1491 692448/692424

IAHS Members may obtain a discount on books bought for their personal use and should ask for it when placing an order.

Climate and the Hydrological Cycle

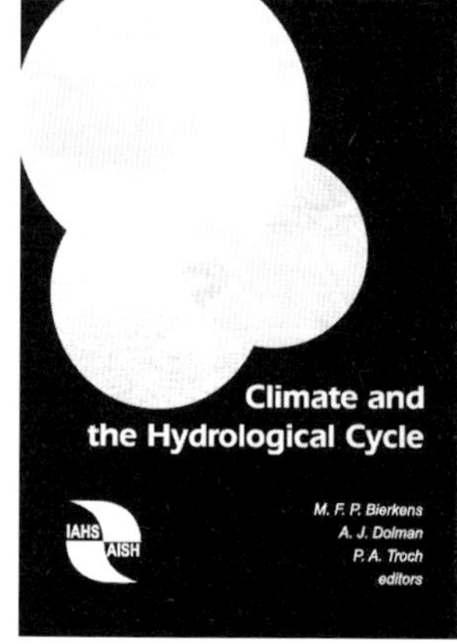

Edited by

Marc Bierkens *Utrecht University, The Netherlands*
Han Dolman *Vrije Universiteit Amsterdam, The Netherlands*
Peter Troch *University of Arizona, USA*

An in-depth overview of the role of the hydrological cycle within the climate system, including climate change impacts on hydrological reserves and fluxes, and the controls of terrestrial hydrology on regional and global climatology. This book, composed of self-contained chapters by specialists in hydrology and climate science, is intended to serve as a text for graduate and postgraduate courses in climate hydrology and hydroclimatology. It will also be of interest to scientists and engineers/practioners interested in the water cycle, weather prediction and climate change.

IAHS Special Publication 8

(*October 2008*) ISBN 978-1-901502-54-1 (Paperback); 344 + xvi pages
Price £50.00

Sponsored by:

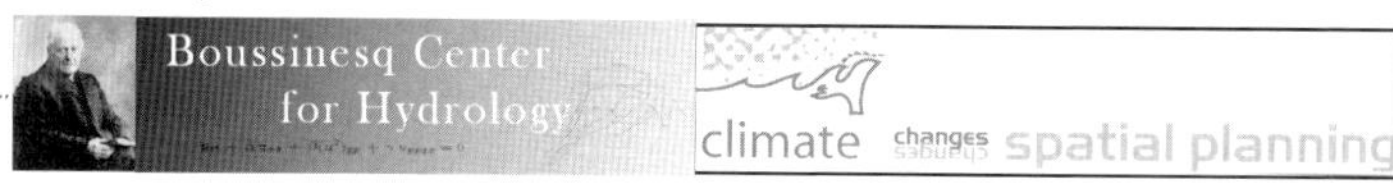

Please send book orders and enquiries to:

Mrs Jill Gash
IAHS Press, Centre for Ecology and Hydrology
Wallingford, Oxfordshire OX10 8BB, UK

jilly@iahs.demon.co.uk
tel.: + 44 1491 692442
fax: + 44 1491 692448/692424

Book prices include postage worldwide. IAHS Members receive discounts on IAHS publications.

See www.IAHS.info for information about IAHS (the International Association of Hydrological Sciences), membership, publications, meetings and other activities.